THREE-DIMENSIONAL CONSTITUTIVE RELATIONS
AND DUCTILE FRACTURE

INTERNATIONAL UNION OF THEORETICAL
AND APPLIED MECHANICS

THREE-DIMENSIONAL CONSTITUTIVE RELATIONS AND DUCTILE FRACTURE

Proceedings of the IUTAM Symposium on Three-Dimensional Constitutive Relations and Ductile Fracture

Dourdan, France, 2–5 June, 1980

Edited by

S. NEMAT-NASSER
*The Technological Institute, Northwestern University,
Evanston, Illinois 60201, U.S.A.*

Sponsored by

International Union of Theoretical and Applied Mechanics (I.U.T.A.M.)
Délégation Générale à la Recherche Scientifique et Technique (D.G.R.S.T.)
Direction des Recherches Etudes et Techniques (D.R.E.T.)
Electricité de France–Septen (E.D.F.)
Association Universitaire de Mécanique (A.U.M.)

NORTH-HOLLAND PUBLISHING COMPANY
AMSTERDAM · NEW YORK · OXFORD

ISBN: 0 444 86108 4

Published by:

North-Holland Publishing Company
Amsterdam · New York · Oxford

Sole distributors for the U.S.A. and Canada:

Elsevier North-Holland, Inc.
52 Vanderbilt Avenue
New York, NY 10017

Library of Congress Cataloging in Publication Data

IUTAM Symposium on Three-Dimensional Constitutive
    Relations and Ductile Fracture, Dourdan, France,
    1980.
    Three-dimensional constitutive relations and
ductile fracture.

    1. Fracture mechanics—Congresses.   2. Metals—
Ductility—Congresses.   I. Nemat-Nasser, S.
II. International Union of Theoretical and
Applied Mechanics.   III. Title.
TA409.I18   1980   620.1'126   80-29065
ISBN 0-444-86108-4 (Elsevier)

Printed in The Netherlands

PREFACE

The development of realistic three-dimensional constitutive relations for the description of flow and ductile fracture of metals is of considerable technological importance and current scientific interest. A fundamental approach for an in-depth development of this area requires careful examination of the involved microscopic processes, as well as of macroscopic material behavior. In an effort to bring together various significant ingredients relevant to flow and ductile fracture of materials and to focus attention on important unresolved research problems in the area, an international symposium on the subject was held at Dourdan, France, during June 2–5, 1980, under the sponsorship of the International Union of Theoretical and Applied Mechanics with direct financial support of: Délégation Générale à la Recherche Scientifique et Technique (D.G.R.S.T.); Direction des Recherches Etudes et Techniques (D.R.E.T.); Electricité de France–Septen (E.D.F.); Association Universitaire de Mécanique (A.U.M.).

The objectives of the Symposium on Three-Dimensional Constitutive Relations and Ductile Fracture were to address with equal emphasis the following three basic areas:

(a) Mechanisms for the formation and propagation of fissures in ductile materials;

(b) Constitutive equations for the plastic behavior of materials in regions close to the ends of fissures;

(c) Global macroscopic consequences of (a) and (b).

These areas were extensively examined, covering theoretical and experimental aspects of various mechanisms involved in the formation, growth, and coalescence or interconnection of voids, the stress and strain fields near the tip of a stationary or mobile crack, and finally, the possibility of formulating global criteria for the initiation and growth of fissures.

The Symposium consisted of seven half-day sessions, each including three general lectures and several contributed papers, and an eighth session totally devoted to contributed papers. The Symposium culminated in a round-table discussion, where various problems covered at the Symposium were examined and directions for future research outlined. Each session was coordinated and guided by a chairman, and a foreman (animateur) presented a prepared critique of the invited lectures, accentuating significant points and providing complementary results. The invited lectures and the comments of the foremen are included in these Proceedings. In addition, a list of contributors and the titles of their contributions are given. Also, a list of all participants is provided.

The Scientific Committee, appointed by the Bureau of International Union of Theoretical and Applied Mechanics, consisted of:

J. Mandel, Ecole Polytechnique (France), Chairman;
J. H. Gittus, U.K. Atomic Energy Authority (U.K.);
F. A. McClintock, Massachusetts Institute of Technology (U.S.A.);
S. Nemat-Nasser, Northwestern University (U.S.A.);
D. Radenkovic, Ecole Polytechnique (France);
B. Storåkers, Royal Institute of Technology (Sweden);
J. Zarka, Ecole Polytechnique (France).

The local arrangements were made through the efforts of Professor Joseph Zarka and Professor P. Habib, together with Mrs. J. Hongre, Mrs. D. Chassaing, Miss B. Meunier, Mr. M. Amestoy, and Mr. A. Ehrlacher of Ecole Polytechnique, and the Symposium was held within the beautiful surroundings of Dourdan, at Le Normont. It is with pleasure that their superb work is gratefully acknowledged. Also, thanks are due to Éva Nemat-Nasser who was in charge of the women's and children's program. The editor is grateful to the U.S. National Science Foundation for a grant awarded to Northwestern University that facilitated his participation in the organization of this Symposium and provided partial support for Mrs. Erika Ivansons who assisted in various editorial tasks. For her ceaseless effort and careful work, the editor is grateful.

Palaiseau, June 1980 S. NEMAT-NASSER

WELCOMING ADDRESS AND INTRODUCTORY REMARKS

Dear Friends,

Welcome to Dourdan. It is a great honor and pleasure for me to receive here so many famous scientists from all over the world. I hope you will enjoy your stay in this village, have a fine weather, and also do some useful work.

The title of our Symposium is Three-Dimensional Constitutive Relations and Ductile Fracture. The main topic in fact is Ductile Fracture. What is ductile fracture? Twenty years ago the standard reply was: "plastic flow precedes breaking". But we know nowadays, that at the microscopic level things are not so simple and that both flow and cleavage processes seem to occur.

About two years ago, when Dr. Zarka asked me to act as chairman of this meeting, I refused at first because I knew nothing about this difficult problem. Well, I still know very little. But what I *do* know is that a close collaboration between metallurgists, rheologists, and mechanicians, and even mathematicians is strongly needed. That is really, I think, the principal objective of our meeting in Dourdan.

Let me, please, return to the title of the Symposium and the words Constitutive Relations. As you know, ductile fracture is associated with nucleation and growth of voids. For this reason the usual continuum equations do not remain valid in the very neighborhood of the crack tips, in regions which we may call "process zones ". Strictly speaking we can not describe regions of this kind in the framework of continuum mechanics. Still, we may perhaps in this framework give a global description. But then we have to introduce some new models which must include such parameters as inclusion spacings, and damage, volumetric expansion, pressure sensitivity, and other effects. That is the reason why the words Constitutive Relations appear in the title of the Symposium.

Dear Friends, I welcome discussions which serve to increase our knowledge of this difficult problem of ductile fracture. Looking at the booklet I have at hand, I am convinced that this will indeed be the case.

I hope you will enjoy your four-day stay in this peaceful place, and now I have great pleasure in handing over the chairmanship for the first session to our well-known colleague Professor Ekkehart Kröner.

J. MANDEL
Ecole Polytechnique
Palaiseau, France

LIST OF INVITED LECTURES

Argon, A. S., I.-W. Chen and C. W. Lau (U.S.A.)
Mechanics and mechanisms of intergranular cavitation in creeping alloys
Beremin, F. M.[1] (France)
Experimental and numerical study of the different stages in ductile rupture: application to crack initiation and stable crack growth
Burdekin, F. M. (U.K.)
Stability of tearing fracture in structural steels
Christoffersen, J. (Denmark)
Microlocalizations
Dragon, A. (Poland)
Limitation to ductility set by inclusion-induced disturbance of yielding
Gittus, J. H. (U.K.)
Constitutive equation for plastic behavior of materials: origins of ductility
Kfouri, A. P. and K. J. Miller (U.K.)
Crack separation energy rates for inclined cracks in an elastic-plastic material
Labbens, R. (France)
Aspects macroscopiques de la rupture ductile
Lehmann, Th. (Fed. Rep. Germany)
On constitutive relations in thermoplasticity
Lippmann, H. (Fed. Rep. Germany)
Ductility caused by progressive formation of shear cracks
Marandet, B. and G. Sanz (France)
A quantitative description of fracture toughness under plane stress conditions by the R-curve method
McClintock, F. A. and J. L. Bassani (U.S.A.)
Problems in environmentally-affected creep crack growth
Nemat-Nasser, S., M. M. Mehrabadi and T. Iwakuma (U.S.A.)
On certain macroscopic and microscopic aspects of plastic flow of ductile materials
Nguyen, Q. S. (France)
A thermodynamic description of the running crack problem

---

[1] F. M. Beremin is a research group including: Y. d'Escatha, P. Ledermann, J. C. Devaux, F. Mudry, A. Pineau and J. C. Lautridou.

Rice, J. R. (U.S.A.)
  Creep cavitation of grain interfaces
Rousselier, G. (France)
  Finite deformation constitutive relations including ductile fracture damage
Stüwe, H. P. (Austria)
  The plastic work spent in ductile fracture
Weertman, J. (U.S.A.)
  Fatigue crack growth theory for ductile material
Yokobori, T., A. T. Yokobori, Jr., H. Sakata and I. Maekawa (Japan)
  Constitutive equations and global criteria for ductile fracture

LIST OF CONTRIBUTIONS*

*Text of contributions not included in this book.
[1] Text included in this book.

CONTENTS

SESSION 8. Chairman: D. Radenkovic

SESSION 1

Chairman: E. Kröner
             Max-Planck-Institut, Universität Stuttgart, Stuttgart,
             Fed. Rep. Germany

Authors: R. Labbens
             A. S. Argon, I.-W. Chen, C. W. Lau
             J. H. Gittus

Discussion: B. A. Bilby

S. Nemat-Nasser, Editor
THREE-DIMENSIONAL CONSTITUTIVE RELATIONS AND DUCTILE FRACTURE
North-Holland Publishing Company (1981) 3–22

ASPECTS MACROSCOPIQUES DE LA RUPTURE DUCTILE

R. LABBENS

Creusot-Loire, Paris, France

Two series of bursting tests of pressurised cylinders, uncracked and cracked, are described, which evidence the different aspects of the initiation and extension of the rupture.

A large safety margin exists when ductile and tough materials are used.

The possibility of describing the phenomena by one crack parameter resulting from continuous solids mechanics is investigated. Criteria for stable or unstable growth should result from the comparison of this parameter with a critical value, property of the material.

0. Introduction

Il paraît naturel qu'une structure soumise à un chargement suffisamment élevé subisse de grandes déformations et finisse par se rompre; car on sait que les règles de dimensionnement ne prévoient pas de telles surcharges. Longtemps les ruptures survenant en service normal ont été déconcertantes, et les ingénieurs ont souvent cherché quelles surcharges soudaines avaient pu les provoquer. On sait maintenant qu'elles résultent de l'instabilité des fissures dont les règles usuelles de dimensionnement ne prévoient pas l'existence.

Avec un matériau ductile des structures fissurées peuvent supporter des surcharges importantes et des déformations plastiques étendues; aussi les ruptures ductiles sont-elles rares en service, et leur étude nécessite souvent des expériences.

Il semble que, les canons exceptés, les ruptures graves soient apparues aux XVIIIème et XIXème siècles avec le développement des chaudières et des appareils à pression. Il est aussi assez simple de faire croître une pression. Aussi beaucoup d'expériences ont-elles porté sur des appareils à pression.

Ce mémoire décrit deux séries d'expériences d'éclatement qui montrent les différents comportements possibles, propagation de la rupture, fragmentation, et la marge de sécurité résultant de la ductilité et de la ténacité des matériaux.

3

Après ces descriptions, on recherche comment la mécanique des solides continus permet de calculer des paramètres caractéristiques des fissures qui doivent être comparés à des valeurs critiques, propriétés des matériaux.

On n'examine que les chargements monotoniquement croissants à l'exclusion des problèmes cycliques et de fatigue.

1. Eclatement des appareils à pression

Les règles de calcul des appareils à pression ont depuis longtemps été énoncées intuitivement (Annales des Mines, 1828), ou expérimentalement (Cook et Robertson, 1911); leur fondement mécanique n'a été bien connu que lorsque la plasticité s'est développée et a été enseignée, en particulier par Hill (1950) et Mandel (1966). La déformation excessive est liée au chargement limite, et l'éclatement à l'instabilité plastique qui n'est pas la rupture, mais en est rapidement suivie si la pression est maintenue. Les règles usuelles de calcul résultent d'un développement limité à deux termes des pressions de ruine et sont valables jusqu'à un rapport des rayons voisin de 1.4 (Langer, 1964).

Une expression de la pression d'éclatement résultant d'une centaine d'expériences fut proposée par Faupel (1956). La pression d'instabilité

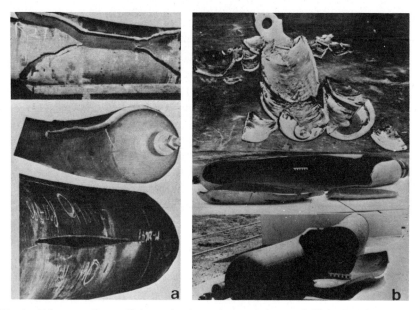

Fig. 1. Eclatement d'appareils à pression à température croissante (Pellini et Puzak, 1963); (a) pression hydraulique, (b) pression de gaz.

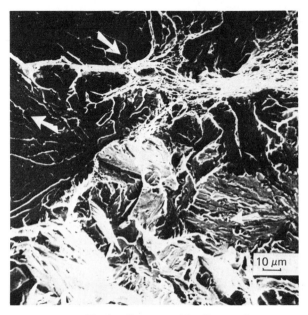

Fig. 2. Cassure semi fragile avec clivages et cupules.

plastique fut calculée indépendamment par Svensson (1958) et Marin et Rimrott (1958). Les expressions sont assez compliquées, mais une formule approchée valable jusqu'à un rapport de rayons 2.5 a été donnée par Svensson. Le développement limité à deux termes conduit pour les aciers ferritiques usuels qui ont un exposant d'écrouissage entre 0.15 et 0.25, comme celui de la formule de Faupel, avec une erreur inférieure à 5%, à

$$p_R\left(\frac{a}{e} + \frac{1}{2}\right) = R_m,$$

p_R = pression d'instabilité plastique;

R_m = résistance à la traction;

a, b = rayons intérieur et extérieur initiaux $(b/a < 1.4)$;

e = épaisseur.

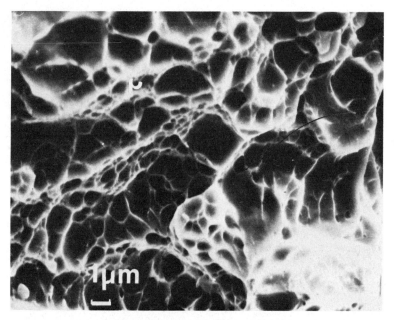

Fig. 3.   Rupture ductile—cupules.

Le premier membre est la valeur moyenne de la contrainte équivalente de Tresca.

Ces résultats ne permettent pas de prévoir comment la rupture se propagera: déchirure ductile limitée, rupture plate avec des lèvres de glissement à 45° plus ou moins développées, bifurcation et projection d'éclats. Depuis 1940, des expériences ont eu pour objet de rechercher les éléments qui déterminent ces diverses possibilités.

En 1953, de Leiris et Bastien étudièrent les conditions de déchirure ductile des bouteilles à oxygène.

Au Laboratoire de Recherches de la Marine Américaine, Pellini et ses collaborateurs (1963) montrèrent l'existence d'une température repère, dite de transition de ductilité nulle, qui permit de définir des domaines de température dans lesquels la rupture est ductile ou fragile; ces limites dépendent du matériau mais aussi de la quantité d'énergie emmagasinée (Fig. 1). Des règles de sécurité purent être énoncées, la rupture étant fort peu probable aux températures du domaine ductile.

Ce mémoire décrit deux séries d'expériences d'un intérêt particulier:

— des essais d'éclatements exécutés en 1942 pour la Marine Nationale sur des bouteilles d'air destinées à des sous-marins;

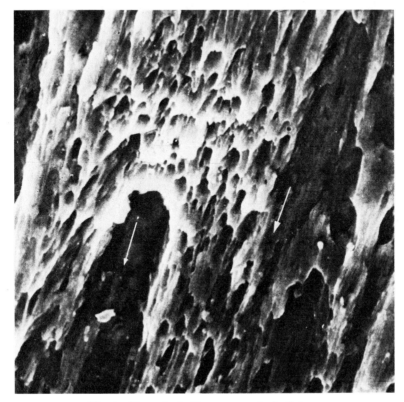

Fig. 4.   Lèvres de cisaillement—cupules inclinées.

— les expériences récentes du Laboratoire d'Oak Ridge de l'Energie Atomique des Etats-Unis (ORNL) sur des cylindres sous pression fissurés.

Avant de décrire ces expériences, il faut noter qu'on rencontrera les divers aspects qui ont été décrits par de Leiris (1944), Sullivan et Kies (1950), et Kies, Sullivan et Irwin (1950) pour les ruptures ductiles ou fragiles:

— cassures plates fragiles avec des chevrons tournés vers l'origine, composées de clivages seuls et plus fréquemment de clivages séparés par des isthmes ductiles à cupules (Fig. 2);

— zones ductiles à cupules résultant de séparations normales avec croissance des vides décrite par McClintock (1971) (Fig. 3);

— lèvres de cisaillement à 45° avec des cupules étirées montrant la déformation des vides par glissement avant la séparation (Fig. 4).

2. Expériences sur des cylindres non fissurés

2.1. L'objet de ces expériences exécutées à l'usine du Creusot, était d'étudier l'influence des propriétés métallurgiques sur la propagation de la rupture, avec ou sans fragmentation ou projection d'éclats. Un compte-rendu complet et un résumé en ont été rédigés par de Leiris (1944, 1970).

Quatre récipients cylindriques, forgés sans soudure, d'environ 3 m de long avec un rayon 270 mm, furent calculés pour une pression de service d'environ 35 MPa et une pression d'éclatement de 100 MPa. Les propriétés étaient les suivantes; en 1942, seule la résilience pouvait donner une indication sur la ténacité et l'éprouvette Charpy U (section 1×0.5 cm) était la seule normalisée (Tableau 1).

Les ruptures se produisirent toutes en domaine plastique, pour de fortes surpressions et se propagèrent comme indiqué sur les développements des cylindres de la Fig. 5.

2.2. Le récipient A avait la résilience la plus élevée. Seule la partie cylindrique s'ouvrit, avec des lèvres à 45° sur toute l'épaisseur, excepté 1 mm au centre. Au milieu du cylindre, l'épaisseur fut réduite de 38.5 à 30 mm.

2.3. Le récipient B avait la même résilience avec une autre composition. Il s'ouvrit dans la partie centrale après gonflement et réduction de l'épaisseur, à partir d'une origine proche de la paroi interne. La rupture présentait d'abord des lèvres à 45° sur toute l'épaisseur; plus loin elle devint plate avec

Tableau 1
Caractéristiques des cylindres non fissurés

Récipient	Composition	Traitement thermique T,R=trempé et revenu R=recuit	Résilience Charpy U 16°C (da J/cm²)	Epaisseur (mm)	Pression d'éclatement Prévue (MPa)	Réalisée (MPa)
A	1.5% NiCr	T, R	L 11 LT 8–9	38.5	101	127
B	CrMo	T, R	L 10, 5–13 T 8–11	38.5	101	122
C	1.5% CrMo	R	L 6–7 T 5–5.5	45–48		120
D	5% Cr	R	L 4–6 T 2–4	44.5		123

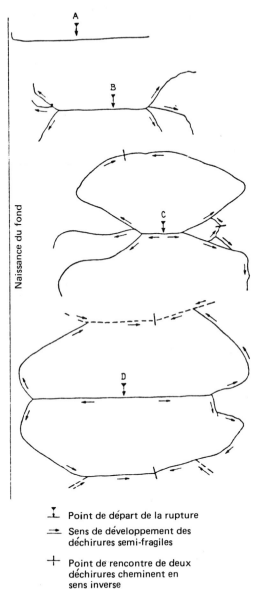

Fig. 5. Déchirure et fragmentation d'appareils à pression (de Leiris, 1970).

des lèvres réduites, se divisa et s'arrêta dans les fonds. Il n'y eut pas de projection.

2.4. Le récipient C était seulement recuit. Il s'ouvrit aussi dans la partie centrale à partir d'une origine plate vers la surface interne, sans réduction d'épaisseur et avec des lèvres de 5 mm seulement. Plus loin la rupture fut complètement plate à chevrons; elle se divisa et se réfléchit sur les fonds; trois fragments furent projetés.

Il faut noter qu'en domain plastique, la contrainte tangentielle maximale est près de la surface interne, mais pas sur celle-ci.

2.5. Le récipient D, recuit, avait une résilience très faible. Il éclata comme le précédent à partir d'une amorce près de l'intérieur, mais sans réduction d'épaisseur, avec des surfaces plates à chevrons, sans lèvres à 45°. La rupture se divisa et se réfléchit et deux gros fragments furent projetés.

2.6. La rupture du récipient A avec la séparation au centre et les lèvres à 45° se conforme presque parfaitement au modèle de glissement symétrique décrit par McClintock (1971).

Le récipient B périt aussi par instabilité plastique, mais l'amorce vers l'intérieur et l'extension partiellement fragile de la cassure montrent que la séparation est plus facile que pour A.

C et D éclatent à peu près à la même pression que A et B mais résistent mal à la séparation par clivage là où la contrainte tangentielle est maximale.

Des initiations analogues furent plus tard observées dans des ruptures par déformation plastique résultant d'une surcharge; de telles observations étaient assez nouvelles en 1942.

L'analyse des phénomènes après la formation du défaut relève de la mécanique de la rupture.

3. Expériences sur des appareils fissurés

3.1.

Dix récipients fissurés furent soumis à une pression hydraulique jusqu'à l'éclatement par le Laboratoire National d'Oak Ridge (ORNL) au titre du Programme Technologie des Aciers en Fortes Epaisseurs (HSST) financé par la Commission de l'Energie Atomique des Etats-Unis. Le compte-rendu en a été publié par Bryan *et al.* (1975).

Tableau 2
Caractéristiques mécaniques des récipients de l'ORNL

Caractéristiques mécaniques à 20°C	Minimales du code ASME (MPa)	Réelles (MPa)
Limite d'élasticité R_e	428	510
Résistance à la traction R_m	570	670
Contrainte admissible $S_m = R_m/3$	190	223
Pression de service $p = S_m \left/ \left(\dfrac{a}{e} + \dfrac{1}{2} \right) \right.$	69	80
Pression d'éclatement	$P_R/p = 3.07$	
K_{IC}	Voir Fig. 6	

Le matériau était l'acier des cuves nucléaires, ASTM A 508 gr. B pour les pièces forgées, A 533 gr. B pour les tôles.

Les dimensions étaient les suivantes: rayon intérieur $a = 343$ mm; epaisseur $e = 151$ mm; $e/a = 0.44$; longueur du cylindre 1370 mm $= 4a$; et les propriétés mécaniques du matériau sont données par le Tableau 2.

Les estimations de K_{IC} de ORNL montrent que la ténacité des cylindres forgés de 150 mm est sensiblement supérieure à celle des tôles A 533 gr. B de 250 mm.

Quatre essais sont particulièrement intéressants. Trois récipients comportant des fissures semi elliptiques de 40 à 50% de l'épaisseur furent rompus sous pression hydraulique à des températures croissantes. Le quatrième avait une fissure très profonde et fut soumis à une pression d'air. Les données sont résumées par le Tableau 3.

3.2. Essai A, à 0°C

Le récipient se rompit à 199 MPa, ou 2.5 fois la pression de service, en état complètement plastique avec des déformations 0, 4% à 180° de la fissure.

La Fig. 7 montre le récipient éclaté. La cassure s'est propagée rapidement dans le plan méridien, sans croissance stable appréciable, ni lèvres de glissement, et s'est divisée en deux branches qui se sont coupées en détachant un éclat. Sur toute la surface des chevrons sont dirigés vers la fissure, et la rupture est complètement semi fragile. La température était au milieu de la transition du K_{IC}.

Il faut noter que la contrainte de membrane était élevée 548 MPa; dans un matériau tenace des contraintes élevées peuvent se développer autour de la fissure et compenser la coupure de la fissure.

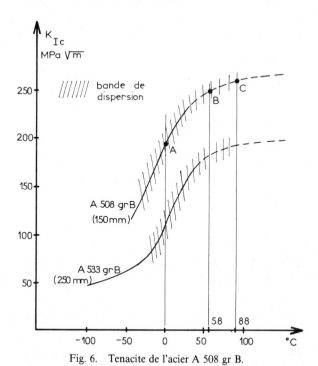

Fig. 6.   Tenacite de l'acier A 508 gr B.

Tableau 3
Caractéristiques des récipients ORNL aux températures des essais

	A	B	C	D
Pression	Hydraulique	Hydraulique	Hydraulique	Air
Température	0°C	58°C	88°C	90°C
Forme de la fissure	Elliptique	Elliptique	Elliptique	Trapèze
Profondeur a' (mm)	64	67	75	147
Longueur $2b$ (mm)	210	203	209	460
a'/e	0.42	0.44	0.49	0.97
Propriétés mécaniques				
R_e (MPa)	514	514	510	480
R_m (MPa)	670	670	605	580
K_{IC} (MPa $\sqrt{m}$)	190	250	275	275
Pression de rupture (MPa)				
calculée sans fissure	245	245	245	230
expérimentale	199	202	227	142

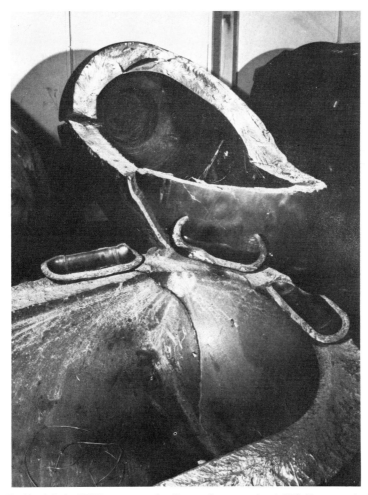

Fig. 7.   Essai A de ORNL—rupture ductile avec fragmentation à 0°C (Bryan *et al.*, 1975).

3.3.   *Essai B, à 58°C*

La température était vers la haut de la transition.

La rupture se produisit à 205 MPa, très près de la rupture de A, mais avec des déformations 0.9% à 180° de la fissure, et après un gonflement.

Sur la Fig. 8, la rupture est très différente de celle de l'essai A. La cassure s'étend dans un plan méridien, mais ne se ramifie pas et s'arrête dans la bride et le fond.

Devant la fissure originelle (Fig. 9), on voit une zone sombre, puis des lèvres de glissement de 30 à 35 mm; une zone plus claire s'étend dans le méridien entre des lèvres qui décroissent et disparaissent.

Le microscope électronique montre une forte densité de cupules devant la fissure et surtout des clivages dans la zone plus claire.

Le caractère ductile de la rupture apparaît à cette température. Devant la fissure originelle, la zone sombre correspond à une croissance stable qui vraisemblablement s'arrêta lorsqu'elle fut assez proche dela surface. La phase suivante est moins sûre. En domaine élastique, le point dangereux aurait été en A, pas en B. Si on admet que ceci subsiste en domaine plastique, ce qui n'est pas sûr, la fissure se serait ouverte en A, avec des glissements à 45° jusqu'à la séparation. La fissure aurait alors été traversante et probablement instable; sa propagation aurait cependant été freinée par des lèvres à 45°.

Il est certain que la croissance stable par séparation des cupules puis les glissements à 45° ont absorbé beaucoup d'énergie. Celà explique que, contrairement à l'essai A, la cassure ne s'est pas divisée et s'est arrêtée. Cependant, la rupture s'est propagée loin, et une pression d'air ajoutant une

Fig. 8.   Essai B de ORNL—rupture ductile puis fragile sans fragmentation à 58°C (Bryan *et al.*, 1979).

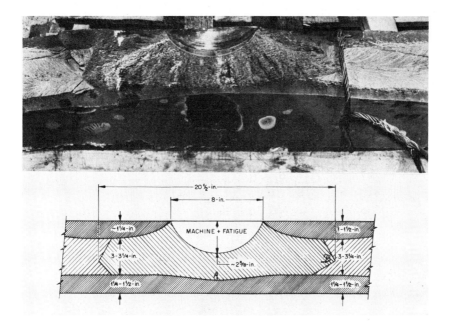

Fig. 9.   Essai B (58°C) de ORNL—détail (Bryan *et al.*, 1975).

énergie extérieure à celle accumulée dans la paroi aurait provoqué la fragmentation suivie de la projection des éclats.

3.4.   Essai C, à 88°C

Ce récipient était forgé avec un joint longitudinal soudé pour essai; la fissure était dans la soudure.

A 88°C, le plateau supérieur du K_{IC} est atteint. La rupture eut lieu à 227 MPa, très près de la pression d'instabilité plastique.

Le compte-rendu ne donne que peu d'enregistrements. A 180° de la fissure, la déformation était 2% assez loin de l'instabilité plastique. Devant la fissure, à l'extérieur 5.2% a été relevé.

Sur la Fig. 10, on voit que la rupture s'est arrêtée rapidement. Il semble qu'une croissance stable se soit produite et que la surface d'ouverture tourna rapidement vers des lèvres à 45° sur toute l'épaisseur.

3.5.   Essai D, à 90°C, sous pression d'air

La fissure était un long trapèze, avec des bases de 460 et 200 mm et une hauteur 147 mm, ne laissant qu'un ligament de 5 mm.

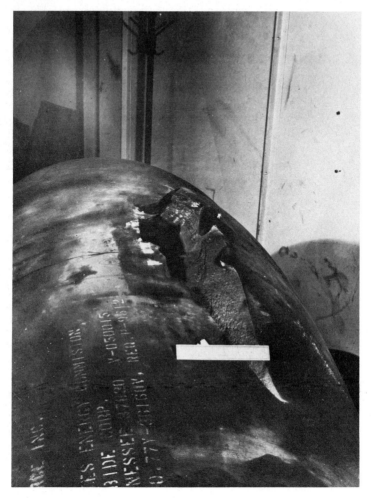

Fig. 10.   Essai C de ORNL—rupture entièrement ductile à 88°C (Bryan *et al*., 1975).

La rupture commença à 142 MPa et fut observée d'abord par télévision; le ligament s'ouvrit lentement en 10 secondes.

La rupture s'étendit fort peu sur les côtés du trapèze, et fut ductile et limitée au ligament. Il n'y eût pas de fragmentation, mais si la fissure n'avait pas été aussi grande, une rupture à 230 MPa aurait probablement été accompagnée de projection d'éclats.

4. Essai d'interpretation des expériences de l'ORNL

Les deux exemples ci-dessus montrent que, même avec des fissures profondes, les matériaux tenaces peuvent supporter des surcharges importantes. On recherche si des méthodes d'analyse permettent de rendre compte des ruptures A, B, C de l'ORNL. Ces trois ruptures sont survenues lorsque le récipient entier était largement plastique, mais avec des zones de grandes déformations près de la fissure plus ou moins étendues.

L'analyse des expériences de 1942 serait analogue à partir de défauts résultant de la déformation.

4.1. Sous certaines conditions qu'on ne répète pas ici, l'intégrale J définit la singularité en domaine plastique et permet de prolonger l'analyse linéaire. Paris *et al.* (1979) et Hutchinson et Paris (1979) ont formulé un critère de stabilité fondé sur la comparaison de J calculée à une courbe de résistance du matériau en déformation plane dite J_R.

Sur la Fig. 11, J_R/R_e et Δa accroissement de la fissure sont des longueurs; les pentes sont sans dimension. Comme Rice (1976) l'avait montré, la partie OAB correspond à l'émoussement avec une pente voisine de 2; au-delà, la partie BC dont la pente est plus faible est la limite inférieure d'un grand nombre d'expériences en déformation plane; elle correspond à une croissance stable et s'arrête en un point où la rupture brutale par clivage se produit. Les pentes de cette courbe et de la courbe J/R_e dépendant du chargement donnent un critère de stabilité. Ce sujet sera étudié par Burdekin au cours de ce Symposium.

J suppose la déformation plane, et n'est donc pas définie dans le plan normal à une fissure quelconque; en termes stricts, on ne peut pas l'utiliser dans un problème à trois dimensions. Il semble cependant que beaucoup d'auteurs admettent que dans certaines limites du plan normal J est suffisamment indépendante du contour pour qu'on puisse la définir. On fait cette hypothèse.

4.2. Pour l'essai A, on calcule au fond de la fissure un facteur d'intensité de contrainte approximatif sans tenir compte de la courbure qui, pour cette épaisseur, aurait un effet minorant.

A la rupture, le K_I est 10 à 15% supérieur au K_{IC} estimé par ORNL, 190 MPa $\sqrt{m}$. A première vue, on peut penser que la théorie linéaire s'applique à peu près jusqu'à la rupture. Mais la théorie linéaire corrigée exige que la zone plastique soit contenue par une zone singulière élastique qui cesse d'exister avant le chargement limite. C'est pourquoi on limite en général la

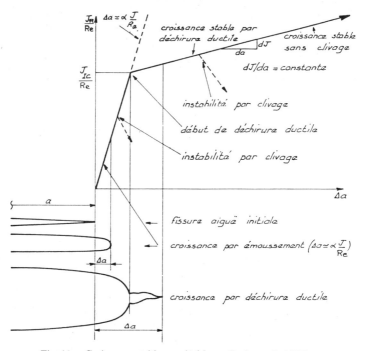

Fig. 11.   Croissance stable par déchirure (Paris *et al.*, 1979).

théorie linéaire corrigée aux deux tiers de ce chargement, soit pour cet essai

$$\tfrac{2}{3}R_e \log_n(b/a) = 125 \text{ MPa.}$$

A cette limite

$$\frac{J}{R_e} = (1 - \nu^2)\frac{K^2}{ER_e} = 0.91\frac{\overline{115}^2}{2 \cdot 10^5 \cdot 514} = 0.12 \cdot 10^{-3} \text{ m.}$$

Au delà, il existe une zone plus ou moins étendue de grandes déformations plastiques.

La condition d'existence d'un seul paramètre, donc de la signification de *J*, énoncée approximativement par McMeeking et Parks (1979) pour une contrainte de traction dominante

$$\text{ligament} = L > 100(J/R_e) = 12 \text{ mm}$$

est largement vérifiée; le rayon dû à l'émoussement est très petit

$$\Delta a = J/2R_e = 0.06 \text{ mm.}$$

On peut donc penser que le récipient A s'est rompu avec une zone de grandes déformations peu étendue sur la branche ascendante OA, limitée assez bas à 0°C par l'apparition du clivage. L'analyse par J n'est peut-être pas ici indispensable mais peut s'imposer dans d'autres problèmes, et rien ne dit que la valeur critique de J en clivage soit $(1 - \nu^2)K_{IC}^2 / E$.

4.3. L'interprétation de l'essai B est plus difficile. On a constaté une croissance stable. Si à 58°C on admet $J_{IC} = 220$ MPa mm, qui est probablement un peu faible mais possible, la condition

$$L > 100\,\frac{J}{R_e} = 100\,\frac{220}{510} = 43 \text{ mm}$$

est encore satisfaite.

Mais la fissure a crû de 67 à environ 120 mm avant la séparation par glissement à 45°; la condition d'existence d'un paramètre unique a cessé d'être vérifiée avant que la fissure n'atteigne 100 mm, sans doute en même temps que la zone de hautes déformations s'étendait.

Il est probable que la proximité de la limite a modifié à la fois les contraintes et la résistance du métal à la séparation normale en avant de la fissure. Il est plausible que la fissure se soit arrêtée en continuant à s'ouvrir jusqu'à ce qu'une ouverture ou un glissement critique déclenche la séparation à 45°. L'auteur ne voit pas clairement si une analyse analogue à celle de McClintock (1971) pour une plaque entaillée sous tension peut être faite, et quel critère de séparation devrait être utilisé.

Après cette séparation, l'instabilité longitudinale ne saurait relever du domaine linéaire, et elle est freinée par les glissements à 45° qui n'en sont pas encore rendus à la séparation. L'analyse qui peut rendre compte de la fin de l'essai B ne semble pas simple.

4.4. Quant à l'essai C, il semble bien qu'il entre dès le début de la séparation dans le domaine des glissements à 45° dont l'analyse précise paraît difficile.

4.5. Les relations entre J et l'ouverture en fond de fissure (C.O.D.) ont été étudiées depuis longtemps. Récemment Turner (1979) a exposé les difficultés soulevées par l'une et l'autre et a énoncé les problèmes particuliers aux solides partout en état plastique, avec des zones de hautes déformations plus ou moins étendues. Mais, à la connaissance de l'auteur, il ne semble pas qu'une méthode d'analyse applicable aux cas où J n'a plus de sens physique, ait été dégagée.

5. Autres méthodes d'analyse

5.1. D'autres modèles de croissance ont été étudiés par éléments finis pour estimer la stabilité d'une croissance; l'une d'elle a été publiée par d'Escatha et Devaux (1979) et le modèle physique de déchirure correspondant est établi au cours de ce Symposium par Beremin. Il ne semble pas que ces méthodes aient été appliquées à des problèmes tridimensionnels complexes tels que les précédents.

5.2. Dans les coques, le modèle du ressort linéaire d'abord étudié par Rice (1972) est maintenant repris par Parks (1980) pour une étude approchée des fissures non traversantes en domaine plastique.

Pour les fissures traversantes, une généralisation du modèle de Dugdale a été étudiée par Erdogan (1979); jusqu'à présent, elle n'a été utilisée que pour le calcul d'ouvertures en fond de fissure (C.O.D.) avec des valeurs critiques qui semblent purement empiriques.

6. Conclusion

Les deux séries d'essais décrites montrent qu'un métal tenace et ductile supporte des surcharges importantes sans rupture, même si une fissure existe. Le comportement après le déclenchement de la rupture et en particulier la projection d'éclats, dépend essentiellement de la position de la température par rapport à la transition de la séparation par clivage à la séparation ductile.

La marge de sécurité mise en évidence dans le domaine ductile peut faire penser qu'une analyse mécanique précise dans ce domaine est un souci purement académique. Celà ne semble pas exact, car on peut rencontrer des chargements moins simples que la pression, de forts gradients thermiques par exemple qui peuvent provoquer des déformations localement très élevées. Il est certain que de telles déformations, tendant à satisfaire localement les dilatations thermiques, doivent diminuer les contraintes et aller dans le sens de la sécurité. En domaine linéaire, on a pu montrer par le calcul et l'expérience que des instabilités provoquées par un gradient thermique peuvent rester locales et s'arrêter (Blauel *et al.*, 1975; Cheverton, 1976). Il est intuitif que cette propriété doit s'étendre au domaine ductile, mais il serait utile de pouvoir le démontrer. Il faut pour celà disposer d'un outil d'analyse et de critères qu'on ne possède encore que partiellement.

Il est vraisemblable que cette analyse de problèmes pratiques, tridimensionnels, sera longtemps fort coûteuse, et qu'on devra industriellement se contenter d'estimations plus approchées. Il serait cependant utile de pouvoir traiter complètement quelques problèmes typiques.

L'étude de tels problèmes, comme l'interprêtation des expériences décrites, doit se faire en trois phases qui coïncident avec les objectifs de ce Symposium:

(a) mécanismes de formation et de croissance des fissures, propriétés correspondantes des matériaux;

(b) calcul de paramètres caractéristiques de la fissure par l'analyse élasto-plastique;

(c) formulation de critères résultant de (a) et (b).

Remerciements

L'auteur présente ses remerciements à l'Ingénieur Général de Leiris et au Professeur Mandel pour l'aide qu'ils lui ont apportée dans les réflexions préalables à la rédaction de ce mémoire. Il y associe le Professeur G. R. Irwin auprès duquel il a acquis l'essentiel de ses connaissances en mécanique de la rupture. Il remercie l'Ingénieur Général Darmon (Marine Nationale) et le Dr. G. D. Whitman (ORNL) pour leur aimable communication de documents.

Bibliographie

Annales des Mines (1828-T. III), "Circulaire Relative aux Ordonnances Royales Concernant les Machines à Vapeur à Haute Pression."

Blauel, J. G., Kalthoff, J. K. and Stahn, D. (1975), "Model Experiments for Thermal Shock Fracture Behavior," *J. Eng. Mater. Technol.*, paper 74-Mat 11.

Bryan, R. H. et al. (1975), "Test of 6 in Thick Pressure Vessels—Series 1, 2, 3, USAEC Report ORNL-5059" (Oak Ridge National Lab., Oak Ridge, Tennessee).

Cheverton, R. D. (1976), "Pressure Vessel Fracture Studies Pertaining to a PWR LOCA-ECC Thermal: Experiments TSE 1 and TSE 2," ORNL Report NUREG/TM 31.

Cook, G. and Robertson, A. (1911), "The Strength of Thick Hollow Cylinders under Internal Pressure," *Engineering*, **15**, 786.

Erdogan, F. (1979), "A Review of Fracture Initiation, Propagation and Stability in Thin Walled Pressure Vessels and Piping," *Structural Mechanics in Reactors Technology*, Berlin, 15.

d'Escatha, Y. and Devaux, J. C. (1979), "Numerical Study of Initiation, Stable Growth and Maximum Load with a Ductile Fracture Criterion Based on the Growth of Holes," *Elastic Plastic Fracture*, ASTM-STP 668, 229.

Faupel, J. H. (1956), "Yield and Bursting Characteristics of Heavy Wall Cylinders," *J. Appl. Mech.*, **78**, 1031.

Hill, R. (1950), *The mathematical theory of plasticity* (Clarendon Press).

Hutchinson, J. W. and Paris, P. C. (1979), "The Theory of Stability Analysis of *J*-controlled Crack Growth," *Elastic Plastic Fracture*, ASTM-STP 668, 37.

Kies, J. A., Sullivan, A. M. and Irwin, G. R. (1950), Interpretation of Fracture Markings, *J. Appl. Phys.*, **21**, No. 7.

Langer, B. F. (1964), *Pressure Vessel Research Committee, Interpretive report of pressure vessel research*, Section 1: Design Considerations, Welding Research Council Bulletin No. 95, 1.

de Leiris, H. (1944), Examen de Quatre Bouteilles Éclatées au Cours d'Essais sous Pression Hydraulique, *Procès Verbal No. 269 LA du 1er Juin 1944* (Service Technique des Constructions et Armes Navales, Paris).

de Leiris, H. (1945), "L'Analyse Morphologique des Cassures," *Association Technique Maritime et Aéronautique*, **44**, 95.

de Leiris, H. et Bastien, P. (1953), "Détermination de la Pression d'Éclatement d'une Capacité à Partir des Caractéristiques du Métal à la Traction," *Rev. Mét.*, **50**, 683.

de Leiris, H. (1970), "Les Ruptures par Décalottage dans les Récipients sous Pression à Corps Cylindriques," *Sciences et Techniques de l'Armement*, 167.

Mandel, J. (1966), *Mécanique des Milieux Continus*, Cours de l'Ecole Polytechnique, T. II, Gauthier Villars.

Marin, J. and Rimrott, F. P. (1958), "Design of Thick Walled Pressure Vessels Based upon the Plastic Range," *Welding Research Council Bulletin* No. 41.

McClintock, F. A. (1971), "Plasticity Aspects of Fracture," *Fracture* III, 2 (Academic Press, New York).

McMeeking, R. M. and Parks, D. M. (1979), "On Criteria for J-Dominance of Crack Tip Fields in Large Scale Yielding," *Elastic Plastic Fracture*, ASTM-STP 668, 175.

Paris, P. C., Tada, H., Zahoor, A. and Ernst, H. (1979), The Theory of the Instability of the Tearing Mode of Elastic Plastic Crack Growth," *Elastic Plastic Fracture*, ASTM-STP 668, 5.

Parks, D. M. (1980), "The Inelastic Line Spring: Estimates of Elastic Plastic Fracture Mechanics for Surface Cracked Plates and Shells," *ASME Pressure Vessel and Piping Conference*, San Francisco.

Pellini, W. S. and Puzak, P. P. (1963), "Fracture Analysis Diagram Procedures for the Fracture-safe Engineering Design of Steel Structures," *U.S. Naval Research Laboratory*, Report 5920.

Rice, J. R. (1972), "The Line Spring Model for Surface Flaws," *The Surface Crack Physical Problems and Computational Solutions*, American Society of Mechanical Engineers.

Rice, J. R. (1976), "Elastic Plastic Fracture Mechanics," *The Mechanics of Fracture*, American Society of Mechanical Engineers.

Sullivan, A. M. and Kies, J. A. (1950), "Fracture Appearance in Various Materials," *Naval Research Laboratory*, Washington, DC.

Svensson, N. L. (1958), "The Bursting Pressure of Cylindrical and Spherical Vessels," *J. Appl. Mech.*, **80**, 89.

Turner, C. E. (1979), "Methods for Post Yield Fracture Safety Assessment," *Post Yield Fracture Mechanics*, ch. 2 (Applied Science Publishers, Amsterdam).

S. Nemat-Nasser, Editor
THREE-DIMENSIONAL CONSTITUTIVE RELATIONS AND DUCTILE FRACTURE
North-Holland Publishing Company (1981) 23–49

MECHANICS AND MECHANISMS OF INTERGRANULAR CAVITATION IN CREEPING ALLOYS

A. S. ARGON, I.-W. CHEN and C. W. LAU

Massachusetts Institute of Technology, Cambridge, MA, U.S.A.

The formation of grain-boundary cavities by vacancy aggregation in creeping alloys at elevated temperatures requires interface tensile stresses in excess of $5 \times 10^{-3} E$. Such high stresses can occur across the faces of grain-boundary particles during early phases of grain-boundary sliding with local power-law creep accommodation, provided that the particles are larger than a critical size and the applied stress is in excess of a threshold value. Thus, cavitation begins as soon as grain-boundary sliding begins. Continued cavity nucleation during creep is a result of the stochastic nature of grain-boundary sliding. Wedge cracks at triple grain junctions normally do not nucleate as a separate entity during creep, but instead result from accelerated cavity growth in response to concentrated tensile stresses that build up there as the shear traction on sliding boundaries is relaxed. Experimental detection of critical size cavities of c.a. 10^{-9} m on grain-boundary particles at or near the time of their nucleation is beyond the resolution of any present experimental technique. Therefore, cavity nucleation measurements having a size resolution at 10^{-7} m in crept 304 stainless steel reflect the kinetics of cavity growth rather than cavity nucleation.

1. Introduction

At low temperatures, where grain-boundary sliding is absent and diffusive matter transport is negligible, ductile fracture involves nucleation of cavities from second phase particles when a critical interface stress is reached (Argon *et al.*, 1975). In the absence of additional embrittling effects, eventual fracture is transgranular and occurs by the growth and linking of such cavities by plastic flow (McClintock and Argon, 1966). At elevated temperatures, under high stresses, and high rates of plastic deformation, where grain-boundary sliding produces a negligible redistribution of stress between grains (Crossman and Ashby, 1975, Chen and Argon, 1979), this low temperature form of fracture persists (Argon and Im, 1975). Under lower applied stresses where significant redistribution of stress between grains due to grain-boundary sliding occurs, the fracture becomes predominantly intergranular and the overall ductility decreases markedly due to the accompanying decrease of matrix plastic deformation. This well known behavior has recently been graphically catalogued in many structural alloys by Ashby *et al.* (1979) and Gandhi and Ashby (1979) by means of fracture mechanism maps. In their pioneering experiments on aluminum, Servi and Grant (1951), and Grant and Mullendore (1965) have demonstrated that such intergranular fracture usually requires the presence of second phase particles on sliding grain-boundaries. This early work and much following work has distinguished between two forms of intergranular creep damage: wedge cracks at triple grain junctions, and isolated round cavities on other sliding boundaries. The experimental observations have suggested that wedge cracks are preferred at high stress while cavities are dominant at lower stresses (Grant, 1971). These observations and much additional detail have been reviewed recently by Perry (1974) together with many of the proposed mechanisms.

As in the case of ductile fracture, in high temperature intergranular fracture too, the major emphasis has gone toward understanding the processes of cavity growth. The most refined current view stated by Needleman and Rice (1980) is that cavity growth along grain boundaries may exhibit three different regimes. At low stress levels producing negligibly low levels of power-law creep in the grain matrix, very small cavities grow in a fully equilibrated "spherical-caps" shape by diffusional flow along grain-boundaries as has been described first by Hull and Rimmer (1959). When cavities become larger and when surface diffusion is slower than boundary diffusion, cavities grow in "crack-like" shapes (Chuang and Rice, 1973). Finally, in the limit when cavities become large, and when the applied stress is high enough to produce significant power-law creep in the grain matrix, the cavities grow primarily by power-law creep. In most other instances, cavity growth occurs by a combination of power-law creep and diffusional flow as has been described by Beere and Speight (1978), Edward and Ashby (1979), and Needleman and Rice (1980).

The process of cavity nucleation in comparison to cavity growth has received comparatively little attention. It has been experimentally studied recently with considerable precision in alpha iron by Cane and Greenwood (1975) and in copper by Fleck *et al.* (1975). Although the data is not plentiful and there is considerable scatter in the measured rates, these investigators have observed cavitation at moderate tensile stress levels of the order of $3 \times 10^{-4} - 3 \times 10^{-3}$ times the shear modulus and found some evidence for a threshold stress for cavitation. While the former investigation could not detect a significant incubation time for cavity nucleation, the latter, based on high voltage electron microscopy, has reported a short incubation time of the order of an hour at temperatures of 0.7–0.75 T_m. Apart from occasional and generally unsatisfactory attempts in seeking the explanation for cavity nucleation in dislocation pile-ups against boundaries (Smith and Barnby, 1967), in sliding boundaries treated as dislocation pile-ups (Brunner and Grant, 1956), or in invoking intersections between sliding boundaries and slip bands (Gifkins, 1959), the cavitation process has been considered in some detail by Raj and Ashby (1975) and by Raj (1978) as a classical nucleation process of a "vacant phase" by the clustering of vacancies on interfaces of grain-boundary particles and on grain-boundary junctions. Raj (1978) in particular in a very detailed treatment has emphasized the possibility of cavitation on nearly completely relaxed grain boundaries under moderate interfacial stresses, for reasons of non-wetting or reasons of geometry, particularly at triple junctions. On the other hand, McLean (1958) had concluded much earlier and on much the same basis, but on less detailed arguments, that intergranular cavitation at elevated temperatures is not possible at the observed rates unless the local interfacial stresses are very nearly equal to the interface strength—even for the mode of cavitation by vacancy clustering under stress. Most engineering alloys that undergo intergranular fracture during creep at high temperature demonstrate very good ductility when tested at high strain rate at either room temperature or elevated temperature, requiring substantial prestrains for cavity formation at particle interfaces. It is now well established that such interface decohesion occurs only when the interface strength is reached by local accentuated strain-hardening that results during the course of the prestrain (Argon and Im, 1975). Hence, we must conclude, together with McLean, that, barring the special case of non-wetting particles, even at elevated temperatures high interface stresses are necessary for intergranular cavitation to begin. The source of these high stresses is the subject of our paper.

2. Nucleation of cavities and microcracks

We consider a stressed, non-coherent interface—typically a grain-boundary—as shown in Fig. 1. We assume that shear stresses transmitted across the boundary have been relieved by boundary sliding, and a local stress σ_n is acting across the boundary. We expect that this local normal stress will be concentrated and quite large, but will always be related to the externally applied system of stresses σ by a boundary value problem involving the grain size, the grain geometry, the areal concentration of non-deformable particles on the boundary, the constitutive creep law of the grain matrix, the diffusional conductance of matter along the boundary, etc. We are here interested in the rate of nucleation of cavities along such a stressed boundary by aggregation of vacancies. This problem has been considered in detail recently by Raj (1978) in the context of classical nucleation theory and gives the following results.

As is the usual result, the rate of nucleation $\dot{\rho}$ is given by an Arrhenius expression

$$\dot{\rho} = \dot{\rho}_0 \exp -(\Delta G^*/kT), \tag{1}$$

where ΔG^* is the free energy of the critical size cavity of spherical-caps shape, having a radius r^* in the plane of the boundary (see Fig. 1) and a characteristic apex half angle α governed by the surface free energy χ_s and boundary free energy χ_B. They have the specific forms given below:

$$\Delta G^* = \Delta G_0/(\sigma_n/E)^2 \quad \text{therefore} \quad \Delta G_0 = E\pi r_0^3/6, \tag{2a,b}$$

$$r^* = r_0/(\sigma_n/E) \quad \text{therefore} \quad r_0 = 2\alpha^2 \chi_s/E, \tag{3a,b}$$

$$\alpha = \cos^{-1}(\chi_B/2\chi_s), \tag{4}$$

$$\dot{\rho}_0 = 2\pi r^* \frac{\delta D_B}{\Omega^{4/3}} \exp\left(\frac{\sigma_n \Omega}{kT}\right). \tag{5}$$

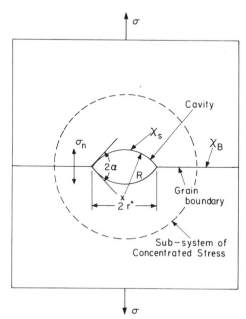

Fig. 1.   Nucleation of a cavity on a stressed interface.

The pre-exponential factor $\dot{\rho}_0$ of Eq. (1), given in explicit form by Eq. (5) is determined, as usual, by the flux of vacancies into the critical size cavity from the surrounding boundary across the periphery of the cavity. In Eq. (5) δ is the effective boundary thickness, D_B the boundary diffusion constant and Ω the atomic volume. Substitution of all of these terms into Eq. (1) results in an explicit form for the cavity nucleation rate given below.

$$\dot{\rho} = \frac{2\pi r_0 \delta D_{B0} E}{\sigma_n \Omega^{4/3}} \exp\left[\frac{\sigma_n \Omega}{kT} - \frac{Q_B}{kT} - \frac{\Delta G_0}{kT}\left(\frac{E}{\sigma_n}\right)^2\right], \qquad (6)$$

where D_{B0} and Q_B are the pre-exponential factor and the activation free energy for boundary diffusion respectively. Evaluation of the surface energy, taken as one half of the ideal interatomic separation work, χ_s by the use of a Lennard-Jones potential gives

$$r_0 = 0.0187\alpha^2 b \qquad (7)$$

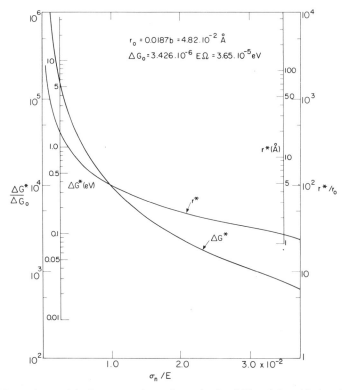

Fig. 2. Dependence of the free-energy for cavity nucleation ΔG^* and the critical cavity radius r^* on the local normal stress σ_n, for γ-iron.

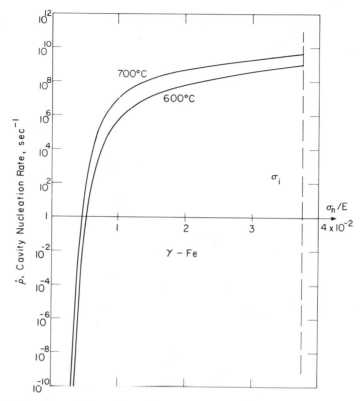

Fig. 3. Dependence of the cavity nucleation rate $\dot{\rho}$ on the local normal stress, at 600 and 700° C, in γ-iron.

and

$$\Delta G_0 = 3.42 \times 10^{-6} E\Omega, \tag{8}$$

where b is the interatomic distance.

It is of interest now to evaluate the expressions for the critical size r^* of the cavity, its free energy ΔG^* and the reciprocal time for cavity nucleation $\dot{\rho}$ by the vacancy clustering process given by Eqs. (3a, b), (2a, b) and (6), as a function of the local stress σ_n acting across the boundary for a typical case such as stainless steel, (or gamma iron) at a temperature range of 600–700°C, where intergranular cavitation is typically observed in service. For this evaluation we use the following magnitudes for the terms under consideration for gamma iron, obtained from values tabulated by Needleman and Rice (1980), $E = 1.40 \times 10^2$ GPa; $\Omega = 1.21 \times 10^{-29}$ m³; $\delta D_{B0} = 7.5 \times 10^{-14}$ m³/sec; $Q_B = 159$ kJ/mole, together with an apex half angle α of one radian. The computed dependences of r^*, ΔG^*, and $\dot{\rho}$ on the normalized stress σ_n/E are shown in Figs. 2 and 3 respectively. Examination of both of these figures shows that for σ_n/E less than $4-5\times10^{-3}$ the rates of cavity nucleation become negligibly small, since for stresses below that value the activation free energy for the critical size cavity becomes impossibly large for thermal fluctuations to provide. At this threshold, stress of about 4×10^{-3} E, giving an activation free energy of about 2 eV, the critical size cavity contains less than 150 vacancies. Under higher stresses, the critical size cavity becomes progressively smaller and approaches atomic dimensions as the stress approaches the interface cohesive strength σ_i.

We conclude from the above discussion that the nucleation of cavities on interfaces or incoherent boundaries requires normal stresses in excess of 4×10^{-3} E. This threshold stress is only a factor of 2.5 under the athermal cavitation stress at room temperature determined by Argon and Im (1975). In long term service applied stresses are usually only about 10% of this level. This establishes that stress concentrations of the order of about 10 must occur across incoherent boundaries in service, and that without them, cavities can not form. That such stress concentrations are produced by sliding boundaries has been known for a long time. Early models by Brunner and Grant (1956), and Smith and Barnby (1967) of these stress concentrations by dislocation pile-ups in elastic media are clearly inadequate as both power-law creep and diffusional flow along grain-boundaries which are always present effectively relax short range stress concentrations. Below we discuss first the transient and steady state conditions of boundary sliding and then the production of high stresses of the required magnitudes even in the presence of power-law creep in the grain matrix, and diffusional flow along the boundaries.

3. Characteristic times for stress redistribution

The concentration of interface stresses by sliding of particle containing grain-boundaries, in the presence of diffusional matter transport along boundaries, elastic unloading of regions near the boundaries, and power-law creep in the surrounding grain matrix, etc., is a very complex problem that has not yet received a general solution. Here, we will consider these processes only approximately to obtain estimates of characteristic times over which they act.

Consider a set of relatively equiaxed grains containing grain-boundary precipitates as sketched in Fig. 4. We assume that all boundaries and interfaces are incoherent and can slide at the temperature under consideration, and that grains have isotropic elastic properties and plastic resistances.

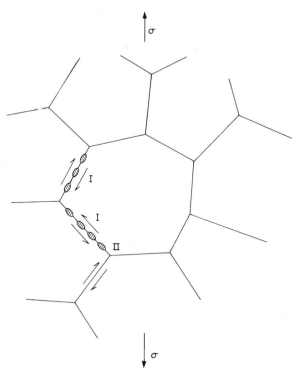

Fig. 4. Intergranular stress concentration problems I and II: at grain-boundary particles, and at triple grain junctions, respectively.

When a tensile stress σ is applied at time $t=0$, a combination of homogeneous elastic, plastic, and creep deformation will be initiated in all grains. For a brief period, tractions of all types are transmitted across all grain-boundaries and interfaces and the stress distribution is uniform everywhere. Within a period τ_{BR} after the application of stress, however, the viscous sliding of all free segments of grain-boundaries, not pinned by particles, relaxes the shear stresses acting across them. This produces a concentration of stress on all particles on sliding boundaries which initiate accelerated power-law creep in the grain matrix around such particles and diffusional flow along the particle boundaries tending to reduce the stress concentrations there. Thus, after a period $t=\tau_{BR}$, the stress distribution in the polycrystalline sample is still largely uniform away from the grain-boundaries. In the boundary, however, the initially uniformly distributed shear tractions are taken up by concentrated stresses at particles. Within an additional increment of time $\Delta\tau_{PL}$ or $\Delta\tau_{DF}$, however, the accelerated power-law creep or diffusional flow around the particles begins to transfer matter around these boundary particles and sets up an initial steady state of stress concentration around them as overall shear displacements across the boundary begin to occur. Such shear displacements along the boundaries, over the particles, over a period of $\Delta\tau'_{PL}$ or $\Delta\tau'_{DF}$ gradually reduce the shear support offered by the boundary particles, relaxing all shear traction along the entire length of the sliding boundaries, and build up stress concentrations at triple grain junctions having ranges equal to the grain size. Thus, after initiation of creep, and as time goes on, stress redistribution occurs over a gradually increasing scale of characteristic dimensions until a final steady state distribution of stress gets established with a wave length on the scale of grains.

The characteristic time τ_{BR} for relaxation of shear stress σ_s on boundary segments of length L between particles of size p (see Fig. 5) can be calculated by elementary methods (Argon *et al.*, 1980) and is

$$\tau_{BR} = \frac{\eta}{\delta}\left(\frac{L}{\pi\mu}\beta(p/L)\right) \simeq \frac{1}{\pi\mu}\left(\frac{kT}{\delta D_B}\right)\left(\frac{L}{b}\right)\beta(p/L), \tag{9}$$

where δ is the thickness of the grain-boundary and η its shear viscosity for which in the last expression we have substituted $\eta \simeq kT/\delta D_B$. In Eq. (9) the function $\beta(p/L)$ represents the interaction of adjacent free boundary segments, and is given by Tada *et al.*, (1973) as

$$\beta(p/L) = \cosh^{-1}\left(\frac{1}{\cos\frac{1}{2}\pi(1-(p/L))}\right). \tag{10}$$

In a three-dimensional case of a flat boundary containing equiaxed randomly placed particles, p/L must be replaced by the areal concentration c of particles on the boundary.

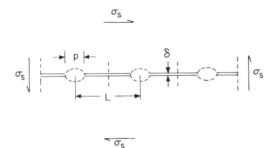

Fig. 5. Relaxation of shear stress along a sliding grain-boundary between boundary particles by viscous sliding of the boundary.

Within approximately another time increment of $\Delta\tau_{\text{DF}}$ or $\Delta\tau_{\text{PL}}$, an initial steady state of stress distribution gets established around the grain-boundary particles as matter begins to be transported around them between oppositely stressed regions of the particle, either by diffusional flow along the particle interfaces or by accelerated power-law creep in the surrounding matrix. The former mechanism dominates for small particles with a size less than a characteristic scaling dimension Λ, given by

$$\Lambda = (\delta D_{\text{B}} \Omega \sigma / kT\dot{\varepsilon})^{1/3}, \tag{11}$$

which is obtained by equating the rate of power-law creep $\dot{\varepsilon}$ to the rate of diffusional boundary flow around a balanced singularity where σ is to be interpreted as the local accentuated tensile stress. Thus, for small particles ($p \ll \Lambda$), a displacement incompatibility across the particles arising from the relief of shear traction over a length L between particles will set up an initial steady state of flow around the particles within a characteristic time increment (Argon *et al.*, 1980),

$$\Delta\tau_{\text{DF}} = \frac{p^3 kT}{8\pi\Omega\mu\delta D_{\text{B}}} \beta(c). \tag{12}$$

On the other hand, for large boundary particles ($p \gg \Lambda$), the same displacement incompatibility across the particles sets up an initial steady state of stress distribution around the particles within an increment of time $\Delta\tau_{\text{PL}}$ of the order of (Appendix I),

$$\Delta\tau_{\text{PL}} = \frac{1}{\pi\dot{\varepsilon}_0} \left(\frac{c}{\sigma/\mu} \right)^{m-1} \beta(c), \tag{13}$$

as matter is being sheared around particles by power-law creep that obeys a stress dependence of

$$\dot{\varepsilon} = \dot{\varepsilon}_0(T) (\sigma/\mu)^m = A(\mu b/kT) D_{\text{v}}(\sigma/\mu)^m. \tag{14}$$

Once an initial steady state of sliding is achieved along boundaries subjected to an initial shear stress, continued matter transport of concentrated shear around particles produces continued boundary sliding to eventually relax all shear tractions across the boundary. The establishment of this final steady state of stress distribution, involving complete eventual shear relaxation, is accomplished within characteristic time increments $\Delta\tau'_{\text{DF}}$ or $\Delta\tau'_{\text{PL}}$ which are roughly multiples d/L of $\Delta\tau_{\text{DF}}$ or $\Delta\tau_{\text{PL}}$ or more precisely

(Argon *et al.*, 1980),

$$\Delta\tau'_{DF} = \frac{p^3 kT(d/L)\beta(1/4)}{8\pi\mu\Omega\delta D_B(1+dD_v/\delta D_B)}$$

$$= 6.43\times10^{-2}\frac{p^3 kT(d/L)}{\mu\Omega\delta D_B(1+dD_v/\delta D_B)}, \tag{15}$$

$$\Delta\tau'_{PL} = \frac{1}{\pi\dot\varepsilon_0}\left(\frac{d}{L}\right)\left(\frac{c}{\sigma/\mu}\right)^{m-1}\beta(1/4). \tag{16}$$

We note that since the developments that have led to τ_{BR}, $\Delta\tau_{DF}$, and $\Delta\tau'_{DF}$ are all linear in stress, these characteristic times are independent of stress level but depend only on microstructural parameters. In comparison, the times that involve power-law creep scale with the characteristic length Λ and are therefore strong functions of the applied stress level. As a representative case we consider gamma iron at two temperatures of 600 and 700° C, and with the following typical magnitudes for the microstructural parameters: $b=2.58\times10^{-10}$ m; $p=4\times10^{-7}$ m; $L=4\times10^{-6}$ m; $d=5\times10^{-5}$ m; $\Omega=1.21\times10^{-29}$ m^3 (∗); $\delta D_{B0}=7.5\times10^{-14}$ m^3/sec (∗); $D_{v0}=1.8\times10^{-5}$ m^2/sec (∗); $Q_B=159$ kJ/mole (∗); $Q_v=270$ kJ/mole (∗); $A=4.3\times10^5$ (∗); $\mu=54$ GPa; $m=3$, and 5; $\beta(p/L)=2.542$ (for $L=4\times10^{-6}$ m) (the quantities designated with an asterisk are due to Needleman and Rice, 1980).

Evaluation of Eq. (9) shows that for 0.4 μm particles with 4 μm spacing where $L/b=1.55\times10^4$, the boundary relaxation times are $\tau_{BR}=119$ and 14 μ sec at 600° C and 700° C respectively. Fig. 6 shows the calculated values of $\Delta\tau_{DF}$ for 0.4 μm size boundary particles for 4 μm spacing ($c=0.1$) at the two different temperatures, and also the stress dependence of $\Delta\tau_{PL}$ for a power-law creep stress exponent of $m=5$. Clearly, in the high stress range, where $\Delta\tau_{PL}<\Delta\tau_{DF}$ for a given particle size and temperature, the stress distribution that will get established on particle interfaces is that governed by power-law creep. Between these times, $\Delta\tau_{PL}$ and the longer times $\Delta\tau'_{PL}(\approx(d/L)\Delta\tau_{PL})$ higher stress concentrations can exist on the interfaces of boundary particles than would be produced in an initial steady state involving matter transport by diffusional flow. Thus, provided that the sizes of the boundary particles and the magnitudes of the applied stresses are not too small, there will exist a critical time window of the order of $\Delta\tau'_{PL}$ during which cavity formation on interfaces of grain-boundary particles is specially favored.

In the next section we discuss two key problems of intergranular stress concentration that control the nucleation of cavities on particle interfaces and their early accelerated growth at triple grain junctions.

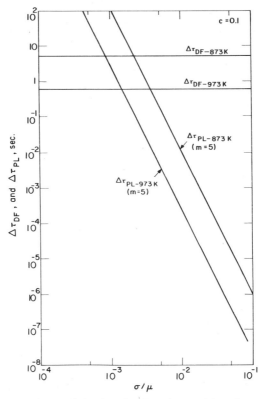

Fig. 6. Dependence of characteristic relaxation times $\Delta\tau_{PL}$ and $\Delta\tau_{DF}$ for power-law creep and diffusional flow respectively in γ-iron for parameters given in the text.

4. Intergranular stress concentration

4.1. Stress concentration at boundary particles

Consider a plane strain geometry of a sliding grain-boundary containing non-deformable hexagon shaped particles of size p, apex half angle ω, and spacing L subjected to a shear stress σ_s as shown in Fig. 7, as an idealization of the actual situation which is necessarily less regular. We assume that the size of these particles and other conditions of temperature, applied stress level, etc., are such that $p \gg \Lambda$, so that $\Delta\tau_{PL} < \Delta\tau_{DF}$, and that therefore, for a period of $\Delta\tau'_{PL}$, an initial steady state of stress distribution is maintained around the particles primarily by power-law creep. Lau and Argon (1977) have shown that in the absence of all diffusional flow along the sliding interfaces, a singularity in normal stress $\sigma_{\theta\theta}$ develops at the particle apex

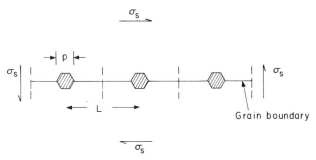

Fig. 7. Particles on a grain-boundary arresting sliding and thereby producing high local stress concentrations.

which is of the form

$$\sigma_{\theta\theta} = \frac{K_\mathrm{I}}{r^{\lambda_\mathrm{I}}} \tilde{\sigma}_{\theta\theta}(\theta = \gamma),\qquad(17)$$

with

$$\gamma = \pi - \omega,\qquad(18)$$

where K_I is a generalized stress intensity factor, r the distance along the inclined particle interface measured from the particle apex, and $\tilde{\sigma}_{\theta\theta}(\theta=\gamma)$ a non-dimensional θ dependent function evaluated at the apex half angle $\theta = \gamma$. The specific values of $\tilde{\sigma}_{\theta\theta}(\theta=\gamma)$, the range exponent λ_I, and the generalized stress intensity factor K_I are all dependent on the creep exponent m and apex half angle ω as has been discussed by Lau and Argon (1977). The dependence of the range exponent λ_I on the creep exponent m for particles with an apex half angle $\omega = 60°$ is given in Table 1. Since the sliding boundary transmits normal tractions, λ_I must be determined from the solution of a background boundary value problem. How this is done will be discussed by us in detail in a separate communication. We can obtain an adequate estimate of $K_\mathrm{I}\tilde{\sigma}_{\theta\theta}(\theta=\gamma)$ from overall tangential equilibrium of average tractions parallel to the boundary on the assumption that properly truncated portions of the distribution given in Eq. (17) apply over the entire inclined interfaces and that there are no other tangential tractions along the

Table 1
Range exponents λ_I for stress concentration on apexes of boundary particles ($\omega = 60°$) (Lau and Argon, 1977)

m	1	3	5	7
λ_I	0.384	0.192	0.132	0.101

boundary between the particles. This gives immediately for the stress distribution $\sigma_{\theta\theta}(\theta=\gamma)$ along the inclined surface of the particle

$$\sigma_{\theta\theta}(\theta=\gamma) = \left[2(1-\lambda_{\mathrm{I}}) \left(\frac{L}{p} \sigma_{\mathrm{s}} \right) \frac{(p/4\cos\omega)^{\lambda_{\mathrm{I}}}}{(\tan\omega)} \right] \frac{1}{r^{\lambda_{\mathrm{I}}}}, \tag{19}$$

where the term in the brackets is the generalized stress intensity factor times $\tilde{\sigma}_{\theta\theta}(\theta=\gamma)$ which shows that the shear stress σ_{s} undergoes an intermediate amplification by a factor L/p, or the reciprocal of the area fraction c of the boundary covered by particles.

In the presence of diffusional flow along the boundaries, however, the singularity at the particle apex is removed by short range boundary diffusion over the apex between oppositely stressed portions of the inclined surfaces, and the stress $\sigma_{\theta\theta}$ is radically modified over a distance Λ from the apex. We assume here that within this region the normal stress $\sigma_{\theta\theta}$ is zero and obtain a modified distribution of $\sigma_{\theta\theta}$ on the inclined faces of the particle interface

$$\sigma_{\theta\theta}(\theta=\gamma) \simeq 0 \quad \text{for} \quad (0 \leqslant r \leqslant \Lambda), \tag{20a}$$

$$\sigma_{\theta\theta}(\theta=\gamma) = \left(\frac{2(L/p)\sigma_{\mathrm{s}}}{\tan\omega} \right) \left(\frac{(1-\lambda_{\mathrm{I}})(p/4\cos\omega)^{\lambda_{\mathrm{I}}}}{1-(4\Lambda\cos\omega/p)^{1-\lambda_{\mathrm{I}}}} \right) \frac{1}{r^{\lambda_{\mathrm{I}}}} \tag{20b}$$

$$\text{for} \quad (\Lambda \leqslant r \leqslant p/4\cos\omega).$$

In Eq. (20b), the first term in brackets gives the nominal intermediate concentration of stress on the inclined faces of the particle due to the absence of shear traction on the portion $(L-p)$ of the boundary between particles, and the second term in brackets together with the range dependent term $r^{-\lambda_{\mathrm{I}}}$ gives the local distribution of normal stress on the inclined faces of the particles themselves.

The maximum value of $\sigma_{\theta\theta}$ of interest for purposes of cavity nucleation is at $r=\Lambda$, and gives the required stress concentration q_{I} (for $\sigma_{\mathrm{s}} \simeq \sigma/2$ on a boundary at 45° to the axis of tensile stress σ)

$$q_{\mathrm{I}} \simeq \left(\frac{1}{c\tan\omega} \right) \times \left(\frac{(1-\lambda_{\mathrm{I}})}{(4\Lambda\cos\omega/p)^{\lambda_{\mathrm{I}}}(1-(4\Lambda\cos\omega/p)^{1-\lambda_{\mathrm{I}}})} \right), \tag{21}$$

for equiaxed particles on a planar boundary where they provide an areal coverage of c. In Eq. (21) above the second term in brackets gives the additional concentration q_{IB} of stress on the inclined faces of the particles.

It is of interest to evaluate these stress concentration factors for gamma iron for the set of parameters given in the preceding section, for 600 and

Table 2
Stress concentration factors q_I, q_{IB}, and threshold stresses
for no-cavitation along boundaries ($\omega = 60°$)

p, (m)	L(m)	c	Λ, (m)	q_I	q_{IB}	σ_{TH}/μ
		$T = 600°$ C, $\sigma/\mu = 3 \times 10^{-3}$, $m = 5$				
4×10^{-7}	8×10^{-6}	0.05	2.11×10^{-8}	15.9	1.38	6.62×10^{-4}
4×10^{-7}	4×10^{-6}	0.1	5.32×10^{-8}	8.80	1.52	1.32×10^{-3}
		$T = 700°$ C, $\sigma/\mu = 1.5 \times 10^{-3}$, $m = 5$				
4×10^{-7}	8×10^{-6}	0.05	3.15×10^{-8}	16.2	1.40	4.47×10^{-4}
4×10^{-7}	4×10^{-6}	0.1	7.94×10^{-8}	10.3	1.79	8.94×10^{-4}

$700°$ C at a typical ratio of tensile stress to shear modulus of $\sigma/\mu = 3 \times 10^{-3}$, and $\sigma/\mu = 1.5 \times 10^{-3}$ respectively, for the two temperatures, and a power-law stress exponent of $m = 5$ appropriate for gamma iron, and for a particle apex half angle $\omega = 60°$. These results are given in Table 2 and indicate that the actual stress concentration factors q_{IB} on the inclined faces of particles range typically between 1.38–1.79, while the main concentration of stress is a result of the geometrical effect of nominal concentration of average tractions given by the reciprocal of the areal concentration c of particles on the sliding boundaries. The two effects together give overall stress concentrations between 8.8 and 16.2 for the cases considered.

When both the applied stress level and the particle size are very small so that Λ is much smaller than the particle size, and $\Delta\tau_{DF} < \Delta\tau_{PL}$, diffusional stress smoothing around particles should be quite effective. Under such conditions, alloys should be resistant against intergranular cavitation and a threshold stress σ_{TH} (in tension) should exist that is obtainable roughly by equating the diffusional smoothing distance Λ in Eq. (11) to the particle size p, giving

$$\frac{\sigma_{TH}}{\mu} < (c) \left(\frac{1}{A} \frac{\Omega}{p^3} \frac{\delta D_B}{b D_v} \right)^{1/m-1}, \tag{22}$$

where c again is the particle area fraction on the boundary. Evaluation of Eq. (22) for the cases considered above gives the values shown in Table 2. They indicate that in alloys with $m \geqslant 5$ the threshold stresses in tension for cavitation-free creep are in the range of $0.5–1.4 \times 10^{-3}$ of the shear modulus, and decrease both with decreasing particle area fraction c, and with increasing temperature.

4.2. Stress concentration at triple grain junctions

As discussed above, after a certain period of time $\Delta\tau'_{PL}$ sliding boundaries of all types cease to support shear tractions requiring that their function of

load support be made up by an intensification of normal stress on
boundaries, with significant long range concentration of such stresses occurring near triple grain junctions. Such normal stress concentration at
triple grain junctions has also been considered by Lau and Argon (1977).
They have shown that for a symmetrically oriented ideal triple junction of
the type shown in Fig. 8, in the absence of all diffusional flow, a singularity
develops in the distribution of normal stress $\sigma_{\theta\theta}$ along the horizontal
boundary that is of the form

$$\sigma_{\theta\theta} = \frac{K_{II}}{r^{\lambda_{II}}} \tilde{\sigma}_{\theta\theta}(\theta=0), \qquad (23)$$

where K_{II} is a generalized stress intensity factor, and $\tilde{\sigma}_{\theta\theta}(\theta=0)$ is again a
non-dimensional angular distribution function evaluated at $\theta=0$. Here too,
K_{II} and λ_{II} are dependent on the creep exponent m. The dependence of the
range exponent λ_{II} on m for an ideal triple junction of 120° angles,
calculated by Lau and Argon (1977) is shown in Table 3 for a number of
creep exponents of general interest. As in the previous case, an adequate
estimate of $K_{II}\tilde{\sigma}_{\theta\theta}(\theta=0)$ can be obtained by assuming that the distribution
of $\sigma_{\theta\theta}$ along the entire horizontal grain boundary can be made up of
segments of the singular distribution given in Eq. (23). This gives a distribution of $\sigma_{\theta\theta}$ on the horizontal boundary in the vicinity of the triple junction
as

$$\sigma_{\theta\theta} = \left[\frac{3}{2}\sigma(1-\lambda_{II})\left(\frac{d}{2\sqrt{3}}\right)^{\lambda_{II}}\right]\frac{1}{r^{\lambda_{II}}}. \qquad (24)$$

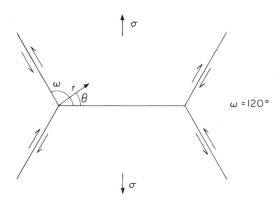

Fig. 8.   Concentrations of stress at triple grain junctions.

Table 3
Range exponents λ_{II} for stress concentration at triple
grain-junctions ($\omega = 120°$) (Lau and Argon, 1977)

m	1	3	5	7
λ_{II}	0.449	0.231	0.150	0.112

In the presence of diffusional flow along the boundaries we expect again to have the singularity removed at $r=0$ and the normal stresses largely relieved over a characteristic length Λ_ℓ near the triple point. This characteristic length is obtained, as in Eq. (11), by equating the local diffusional flow rate at $r = \Lambda_\ell$ to the concentrated power law creep rate governed by the stress distribution given above—making use of the creep law of Eq. (14). This gives

$$\Lambda_\ell = \frac{d}{2\sqrt{3}} \left[\frac{1}{A} \left(\frac{\delta D_B}{b D_v} \right) \frac{\Omega}{\left(d/2\sqrt{3} \right)^3} \right.$$

$$\left. \times \left(\frac{2}{3(1-\lambda_{II})} \frac{1}{(\sigma/\mu)} \right)^{m-1} \right]^{1/[3-\lambda_{II}(m-1)]} . \tag{25}$$

Considering that $\sigma_{\theta\theta} \approx 0$ within this distance of the triple point, we obtain readily a modified normal stress distribution along the horizontal boundary incorporating the effect of diffusional stress smoothing that is

$$\sigma_{\theta\theta} \approx 0 \qquad (0 \leqslant r \leqslant \Lambda_\ell), \tag{26a}$$

$$\sigma_{\theta\theta} = \left[\frac{3}{2}(1-\lambda_{II})\sigma \frac{\left(d/2\sqrt{3} \right)^{\lambda_{II}}}{\left[1-\left(2\sqrt{3}\Lambda_\ell/d \right)^{1-\lambda_{II}} \right]} \right] \frac{1}{r^{\lambda_{II}}}, \tag{26b}$$

$$(\Lambda_\ell \leqslant r \leqslant d/2\sqrt{3}).$$

The maximum stress concentration q_{II} on this distribution that is of interest for purposes of cavity nucleation occurs at $r = \Lambda_\ell$, and is

$$q_{II} = \frac{\frac{3}{2}(1-\lambda_{II})}{\left(2\sqrt{3}\,\Lambda_\ell/d \right)^{\lambda_{II}} \left[1-\left(2\sqrt{3}\,\Lambda_\ell/d \right)^{1-\lambda_{II}} \right]}. \tag{27}$$

The evaluation of this stress concentration for the typical service parameters

Table 4
Stress concentration factors q_{II} near a grain-boundary
triple junction ($\omega = 120°$), $m = 5$

$\Lambda_f(m)$	q_{II}
$T = 600°$ C, $\sigma/\mu = 3 \times 10^{-3}$, $d = 5 \times 10^{-5}$ m	
1.29×10^{-6}	2.10
$T = 700°$ C, $\sigma/\mu = 1.5 \times 10^{-3}$, $d = 5 \times 10^{-5}$ m	
2.14×10^{-6}	2.12

for gamma iron chosen earlier are given in Table 4. Clearly, since the fraction of boundary length of the total that undergoes relaxation of shear stress in this case is of order 0.3–0.5, there is only a moderate concentration of nominal stress near the triple junctions. In our judgement, these stress concentrations are less than adequate to nucleate cavities on the horizontal boundaries or on particle interfaces on these boundaries, if no further concentration of stress is present due to sliding. They are, however, of importance in accelerated cavity growth in these regions once cavities have nucleated as a result of sliding.

5. Experimental results on cavitation

5.1. Experimental procedure

To confirm the theoretical predictions on cavity nucleation, experiments were carried out on 304 stainless steel specimens in tension. This type of stainless steel exhibits very good fracture ductility at room temperature indicating that the carbides it contains, both in the grain matrix as well as on grain-boundaries, are well adhered to the matrix, and should not undergo premature cavitation at elevated temperature due to "non-wetting" of particles—a possibility that has been emphasized by Raj (1978).

The experimental techniques used in specimen preparation in control of grain-boundary carbide size by aging, for reliable measurement of cavity densities on grain-boundaries, and cavity sizes on intergranularly separated specimens after the experiment was over have been described in detail by Chen (1980) and by Argon et al. (1980), and can not be repeated here. It must suffice merely to state that the grain size of the aged material was $d = 40$–$50\ \mu$m, and that the grain boundary carbides at an average spacing $L = 1.4\ \mu$m had an average size of $0.4\ \mu$m.

Hour-glass shaped specimens were crept in a specially constructed creep testing machine in a vacuum of better than 5×10^{-6} Torr pressure at 600

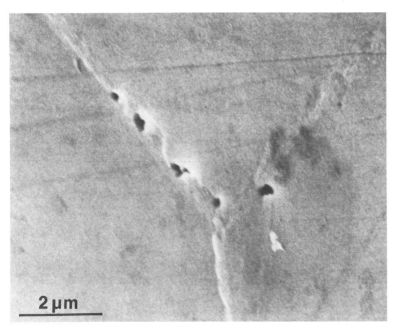

Fig. 9.   Cavities on sliding grain-boundaries in a creeping 304 stainless steel alloy revealed by
SEM in a specimen subjected to a two-stage creep history.

and 700° C. A typical example of cavities in the size range of 0.1 μm that
could be detected by SEM on specimens subjected to a special two stage
creep history is shown in Fig. 9. Fig. 10, on the other hand, shows cavities
on grain boundaries revealed by a post-experiment intergranular separation
technique pioneered by Cane and Greenwood (1975). It clearly shows the
carbides that gave rise to the cavities.

5.2.   Experimental results

The increase in the measured linear concentration of cavities, averaged
over all grain boundary orientations, is shown in Fig. 11 for tests at 600 and
700° C and at a variety of nominal stress levels along the length of the
contoured specimens. The same information replotted as a function of
nominal stress level after different test times is shown in Fig. 12. For
comparison it is interesting to note that the time for the principal transient
in the creep strain-time relationship is of the order of 50 hours, and has no
relevance in the cavitation process.

A point of special interest is the dependence of the cavity concentration
on the apparent inclination of the boundary to the tensile direction. When

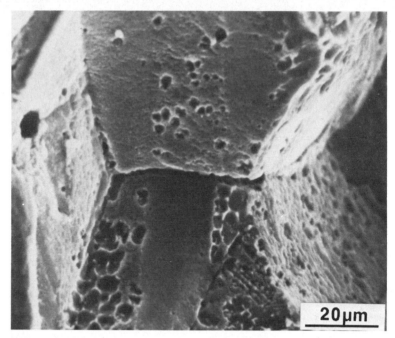

Fig. 10. Creep cavities in crept (and sensitized) 304 stainless steel, revealed by SEM on intergranular fracture surfaces produced at cryogenic temperatures at high strain rate.

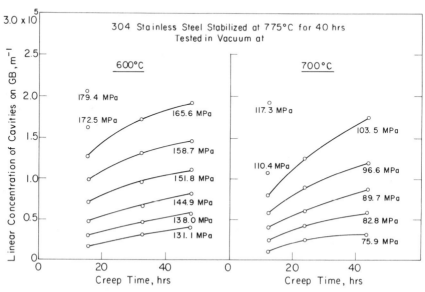

Fig. 11. Increase in linear concentration of cavities on grain-boundaries with creep time, at several stress levels, at 600 and 700° C.

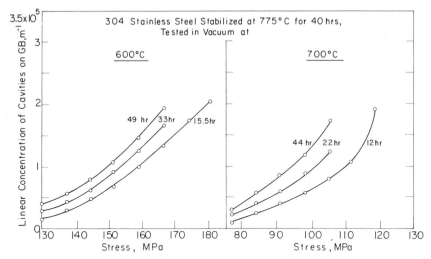

Fig. 12. Increase in linear concentration of cavities on grain-boundaries with tensile stress in 304 stainless steel, at several stages in the creep life.

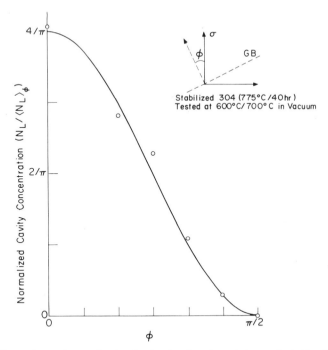

Fig. 13. Dependence of normalized cavity density along boundaries on the inclination ϕ of the boundary with the tensile stress.

the normalized linear cavity concentration at both 600 and 700° C is plotted against the apparent angle of inclination ϕ of the boundary, all data at all nominal stress levels can be superimposed on a basic distribution curve shown in Fig. 13.

6. Discussion

We have demonstrated in Section 2 above that at elevated temperatures, where vacancy nucleation and mobility is high, the high level of interface stress that is normally required for decohesion at particles at low temperatures is reduced to somewhat lower levels as it becomes possible to nucleate cavities by clustering of vacancies at such interfaces. When this problem of cavity formation is treated as a classical nucleation problem, we find that the decohesion stress must be above a threshold of 4–5×10^{-3} E at customary creep temperatures of 0.45–0.55 T_m. We find that, at around this threshold range, the critical size cavity as a lenticular entity has a major axis dimension of less than 20 Å and is made up of a cluster of less than 150 vacancies. These results lead us to conclude that at the time of nucleation the measureable dilatation Δ in the creeping alloys is of order

$$\Delta = n\Omega c / \pi d (r^*)^2, \tag{28}$$

where n is the number of vacancies in the critical size cluster, Ω the atomic volume, c the areal concentration of critical size cavities on the boundary (roughly equal to the ratio of the critical cavity diameter $2r^*$ to the spacing L of critical size cavities along a sampling line in the boundary or as revealed on a surface section), and d the average grain size. Using for $r^* = 10^{-9}$ m, $L = 0.5 \times 10^{-5}$ m, with a grain size of $d = 5 \times 10^{-5}$ m, we calculate a density change of only 5×10^{-9} which is about four orders of magnitude smaller than what is detectable by the most refined dilatometric methods. The critical size cavities are also two orders of magnitude smaller than the limit of resolution of our special technique of two stage creep with an intervening repolishing stage. The cavities reported by Fleck *et al.* (1975) using high voltage microscopy were also in the same size range as those detectable by our two stage creep method. Thus, we conclude that performing meaningful measurements on nucleation rates of critical size cavities is well beyond what is possible by any currently available technique of sampling.

Raj and Ashby in several publications on this subject (Raj and Ashby, 1971; Raj, 1975; Raj and Ashby, 1975; Raj, 1978) have sketched a picture of an almost instantaneous build-up of elastic stress distributions, with large stress concentrations being dismantled by boundary diffusional flow. We maintain that since plastic deformation and creep can respond to local loading situations quite rapidly, and since even diffusional flow is exceedingly effective in stress leveling over very short distances, elastic stress distributions as initial conditions are never relevant. On the other hand, as we have demonstrated above, for relatively large particle sizes, power-law creep can establish an initial steady state of stress distribution across the particle interfaces along sliding boundaries over time periods much shorter than that given by Eq. (12). Thus, this power-law stress distribution, modified only minimally by the very short wave length Fourier components at the singularities, will be the relevant one for cavity nucleation. It is this stress distribution that we have calculated for the large particles on grain boundaries (Eq. (20b)). The effective stress concentrations q_I that result from this mechanism around large boundary particles are larger than those in the corresponding case of small particles and diffusional flow, and are of the right magnitudes as can be seen from Table 2. They are responsible for the nucleation of cavities. Until the shear traction along the entire sliding boundary is relaxed during a time period $\Delta\tau'_{PL}$, these high stresses should remain. This time increment, which is roughly of the order of a multiple d/L of the time for the establishment of the stress distribution, is comparatively very short and only of the order of seconds. This indicates that actual interface stresses that are necessary must be high enough to be compatible with high local cavitation rates of the order of $1-100 \text{ sec}^{-1}$, which, as Fig. 3 indicates, establishes them to be in the range of $5 \times 10^{-3} E$ or larger. It is important to note that our estimate for these times is based on elastic unloading of the regions near the grain boundaries of an otherwise inelastically deforming set of grains. Other contributions to the sliding, such as those that always accompany transcrystalline diffusional currents (Lifschitz sliding) are, however, likely to be small over such short periods of time. Thus, it is clear that cavitation is accomplished during relatively small boundary sliding displacements that can occur even in response to the relief of residual stresses locked in a polycrystalline sample deformed at low temperature when subsequently annealed—as has been observed by Dyson *et al.* (1976) in a Nimonic 80 A alloy.

We see that when our experimental results on cavity densities are extrapolated to short times, they appear to originate at zero time, indicating that some cavitation starts almost immediately upon the onset of deformation at elevated temperature, or more correctly, the onset of boundary sliding, in keeping with our expectations. The continued nucleation of cavities with time is at first unsettling and might be considered evidence against the necessity of high interface stresses. The clue to this riddle comes from the pioneering experiments of Chang and Grant (1953) on boundary sliding. They showed that even though the external creep curve is smooth, the elementary processes of sliding and creep inside individual grains are of a stochastic nature. Individual grains are continually undergoing large transients and quiescent periods as, no doubt, the intercrystalline tractions change by local recovery effects, kinking at triple junctions, migration of parts of boundaries to more favorable positions, etc. We note that at both test temperatures after longer times the cavitation rate shows a decrease, probably due to exhaustion of available sites. Furthermore, the exhaustion is present at all stress levels, consistent with our expectation that of the broad size distribution of particles increasingly larger fractions become available under higher stresses as the critical particle size escaping complete diffusional stress smoothing becomes smaller for larger applied stresses. From Fig. 11 we note further that when the cavity density is extrapolated to smaller stresses, they appear to go to zero around 60 MPa at 700° C and, less clearly, around 100 MPa at 600° C. These threshold stresses are $1.85 \times 10^{-3} \mu$ at 600° C and $1.10 \times 10^{-3} \mu$ at 700° C respectively, which are somewhat higher than, but within a factor of 2 of our estimates given in Table 2 for the appropriate power exponent of 5.

Apart from the agreement between the experimental results and our expectations we note that the stress dependence of the cavitation rate shown in Fig. 12 and the dependence indicated through the change with inclination of boundary angle are quite weak and very far from the strong exponential dependence required by the nucleation theory that should have been reflected even through the apparent "filtering process" of boundary sliding. We believe that this strong stress dependence is only present if critical size cavities in the range of c.a. 2×10^{-9} m on boundaries could be sampled. As we have discussed above, this is a nearly impossible task. The cavities that our experimental technique can unambiguously distinguish are nearly two orders of magnitude larger than this critical size. Therefore, we are of the opinion that the stress dependence which we have reported reflects more the rate of growth of cavities as they become observable rather than the rate of nucleation. Since cavities in the sub-micron size range grow primarily by diffusional flow by the Hull and Rimmer (1959) mechanism that is linear in the normal traction acting across the boundary, the $\cos^2 \phi$ dependence of the cavity density as a function of boundary inclination angle ϕ becomes clear. A corrolary to this observation is the development of wedge cracks. As Fig. 14 shows, wedge cracks, contrary to general knowledge, do not nucleate as a separate entity but begin as isolated cavities in the triple point region by the same mechanism of high stresses arising from boundary sliding. They subsequently grow very rapidly and link together into a wedge crack due to the concentrations of stress that we have discussed earlier and given as Eq. (26b).

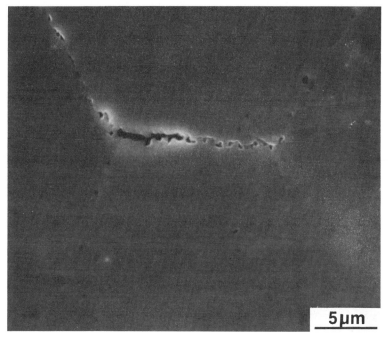

Fig. 14. Early stage in the development of a wedge crack at a triple grain junction, where cavities along the triple point region nucleated earlier are shown to rapidly link together by accelerated growth.

7. Conclusions

(1) Nucleation of cavities during creep on grain boundaries by aggregation of vacancies requires local interface normal stresses in excess of $5 \times 10^{-3} E$, for which stress concentrations between 10–20 are needed.

(2) Such stress concentrations can be produced in early stages of grain boundary sliding when relatively coarse particles impede the smooth sliding and cause concentrated power-law creep around the particles setting up high interface normal stresses on particle interfaces.

(3) Cavitation is accomplished during very short periods of time at the beginning of boundary sliding, and in later transients arising from the stochastic nature of grain-boundary sliding.

(4) Wedge cracks are a result of accelerated linking of growing cavities in the triple point region of stress concentration, and are not a separate phenomenon.

Acknowledgment

This research has been supported by the U.S. Department of Energy under Contract EG-77-S-02-4461.

References

Argon, A. S., Im, J. and Safoğlu, R. (1975), "Cavity Formation from Inclusions in Ductile Fracture," *Met. Trans.*, **6A**, 825.

Argon, A. S. and Im, J. (1975), "Separation of Second Phase Particles in Spheroidized 1045 Steel, Cu–0.6% Cr Alloy, and Maraging Steel in Plastic Straining," *Met. Trans.*, **6A**, 839.

Argon, A. S., Chen, I.-W. and Lau, C. W. (1980), "Intergranular Cavitation in Creep: Theory and Experiments," in: *Creep-Fatigue-Environment Interactions*, R. M. N. Pelloux and N. Stoloff, Eds. (AIME, New York) 46.

Ashby, M. F., Gandhi, C. and Taplin, D. M. R. (1979), "Fracture Mechanism Maps and Their Construction for fcc Metals and Alloys," *Acta Met.*, **27**, 699.

Beere, W. and Speight, M. V. (1978), "Creep Cavitation by Vacancy Diffusion in Plastically Deforming Solids," *Met. Sci.*, **12**, 172.

Brunner, H. and Grant, N. J. (1956), "Calculation of the Contribution made by Grain-Boundary Sliding to Total Tensile Elongation," *J. Inst. Met.*, **85**, 77.

Cane, B. J. and Greenwood, G. W. (1975), "The Nucleation and Growth of Cavities in Iron during Deformation at Elevated Temperatures," *Met. Sci.*, **9**, 55.

Chang, H. C. and Grant, N. J. (1953), "Inhomogeneity in Creep Deformation of Course Grained High Purity Aluminum," *Trans. AIME J. Metals*, **5**, 1175.

Chen, I.-W. (1980), "Creep Cavitation in 304 Stainless Steel," Ph.D. Thesis, Dept. Mater. Sci. Eng., M.I.T., Cambridge, MA.

Chen, I.-W. and Argon, A. S. (1979), "Grain-Boundary and Interphase Boundary Sliding in Power-Law Creep," *Acta Met.*, **27**, 749.

Chuang, T.-J. and Rice, J. R. (1973), "The Shape of Intergranular Cracks Growing by Surface Diffusion," *Acta Met.*, **21**, 1625.

Crossman, F. W. and Ashby, M. F. (1975), "The Non-Uniform Flow of Polycrystals by Grain-Boundary Sliding Accommodated by Power-Law Creep," *Acta Met.*, **23**, 425.

Dyson, B. F., Loveday, M. S. and Rodgers, M. J. (1976), "Grain-Boundary Cavitation under Various States of Applied Stress," *Proc. Roy. Soc.*, **A349**, 245.

Edward, G. H. and Ashby, M. F. (1979), "Intergranular Fracture During Power-Law Creep," *Acta Met.*, **27**, 1505.

Fleck, R. G., Taplin, D. M. R. and Beevers, C. J. (1975), "An Investigation of the Nucleation of Creep Cavities by 1 MV Electron Microscopy," *Acta Met.*, **23**, 415.

Gandhi, C. and Ashby, M. F. (1979), "Fracture Mechanism Maps for Materials which Cleave: fcc, bcc, and hcp Metals and Ceramics," *Acta Met.*, **27**, 1565.

Gifkins, R. C. (1959), "Mechanisms of Intergranular Fracture at Elevated Temperatures," in: *Fracture*, B. L. Averbach *et al.*, Eds. (M.I.T. Press, Cambridge, MA) 579.

Grant, N. J. (1971), "Fracture under Conditions of Hot Creep Rupture," in: *Fracture*, vol. 3, H. Liebowitz, Ed. (Academic Press, New York) 484.

Grant, N. J. and Mullendore, A. W. (1965), *Deformation and Fracture at Elevated Temperatures* (M.I.T. Press, Cambridge, MA).

Hull, D. and Rimmer, D. E. (1959), "The Growth of Grain-Boundary Voids under Stress," *Phil. Mag.*, **4**, 673.

Lau, C. W. and Argon, A. S. (1977), "Stress Concentrations Caused by Grain-Boundary Sliding in Metals Undergoing Power-Law Creep," in: *Fracture 1977*, vol. 2, D. M. R. Taplin Ed., (University of Waterloo Press, Waterloo, Canada) 595.

McClintock, F. A. and Argon, A. S. (1966), *Mechanical Behavior of Materials*, (Addison-Wesley, Reading, MA) 625.

McLean, D. (1958), "Point Defects and The Mechanical Properties of Metals and Alloys at High Temperatures," in: *Vacancies and Other Point Defects in Metals and Alloys* (The Institute of Metals, London) 187.

Needleman, A. and Rice, J. R. (1980), "Plastic Creep Flow Effects in Diffusive Cavitation of Grain-Boundaries," *Acta Met.*, **28**, 1315.

Perry, A. J. (1974), "Cavitation in Creep," *J. Mater. Sci.*, **9**, 1016.

Raj, R. (1975), "Transition Behavior of Diffusion-Induced Creep and Creep Rupture," *Met. Trans.*, **6A**, 1499.

Raj, R. (1978), "Nucleation of Cavities at Second Phase Particles in Grain-Boundaries," *Acta Met.*, **26**, 995.

Raj, R. and Ashby, M. F. (1971), "On Grain-Boundary Sliding and Diffusional Creep," *Met. Trans.*, **2**, 1113.

Raj, R. and Ashby, M. F. (1975), "Intergranular Fracture at Elevated Temperature," *Acta Met.*, **23**, 653.

Servi, I. and Grant, N. J. (1951), "Creep and Stress Rupture Behavior of Aluminum as a Function of Purity," *Trans. AIME*, **191**, 909.

Smith, E. and Barnby, J. T. (1967), "Nucleation of Grain-Boundary Cavities during High-Temperature Creep," *Metal Sci.*, **1**, 1.

Tada, H., Paris, P. C., and Irwin, G. R. (1973), *The Stress Analysis of Cracks Handbook*, 7.1.

S. Nemat-Nasser, Editor
THREE-DIMENSIONAL CONSTITUTIVE RELATIONS AND DUCTILE FRACTURE
North-Holland Publishing Company (1981) 51–73

CONSTITUTIVE EQUATION FOR PLASTIC BEHAVIOR OF MATERIALS: ORIGINS OF DUCTILITY

J. H. GITTUS

United Kingdom Atomic Energy Authority, Springfields Nuclear Power Development Laboratories, Salwick, Preston, U.K.

At high temperature and at the high strain rates that occur near fissures, metals commonly deform by dislocation creep. The constitutive equation for the process is developed and it is shown that a minimum total free energy is associated with the formation of a dislocation cell structure. The cell or sub-grain size and dislocation density are calculated, as well as the creep rate, as functions of temperature and stress. The predicted sub-grain size agrees well with experimental measurements.

1. Introduction

Fracture mechanics has developed out of an analysis of the strength of flawed, brittle materials. The original guiding principle was an analysis of the energy of the system. In the extension of fracture mechanics to ductile materials this approach has been preserved. Here we shall first determine the theoretical equations for plastic deformation by dislocation creep. These are then used to arrive at the theoretical magnitude of the free energy of a creeping specimen which, it is shown, has three components. One of these, the energy of the dislocation network, depends on the dislocation cell diameter, itself a function of the applied stress (and a weak function of temperature). We are then able to examine the hypothesis that the origin of ductility is the fact that in creep a steady state is attained characterized by a minimum free energy.

2. Theoretical equation for dislocation creep in the Frank network

Materials undergoing creep develop a three dimensional dislocation network which does not change significantly during the steady state phase of flow. In this section we develop the theory of plastic deformation, by dislocation creep, in a material where the dislocations form a Frank network.

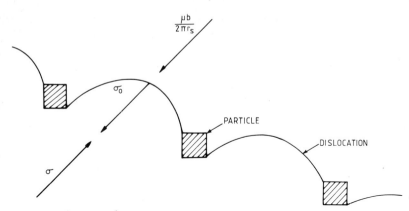

Fig. 1. Dislocation held up by dispersed particles and subject to the external stress (σ), a friction stress due to the particles (σ_0) and a stress due to neighboring dislocations ($\mu b/2\pi r_s$).

Dislocations, as they glide, are impeded by a variety of *friction* stresses. An example is stress due to dispersed particles (Fig. 1). Here the dislocations have to increase in length in order to ride over or circumvent the obstacles and there is a threshold stress determined by the change in the total energy of the configuration which this increase in dislocation length involves.

The stress due to dispersed particles then is one of a class of friction stresses: others are the Peierls stress, the stress due to solute drag, and jog drag.

. It has been proposed by a number of workers that the nonconservative motion of vacancy adsorbing and vacancy emitting jogs in the screw components of glide dislocations can exert a retarding effect on glide. The chemical force f_e per unit length acting on a vacancy emitting jog is given by,

$$f_e = \frac{kT}{b^2} \ln\left(\frac{C_e}{C_0}\right).$$ (1)

Here C_e is the local and C_0 the equilibrium concentration of vacancies. The corresponding equation for a vacancy adsorbing jog is,

$$f_a = \frac{kT}{b^2} \ln\left(\frac{C_0}{C_a}\right).$$ (2)

Consider the case in which C_e and C_a are predominantly controlled by the rates at which vacancies can diffuse to or from the jogs. This is the problem of a moving source (or sink) of vacancies. In general the velocity of the vacancy emitting jogs will be greater than that of the vacancy adsorbing jogs for a given force causing jog climb. However, that force results from the stress acting upon the arcs of dislocation (length l) separating adjacent jogs and upon the curvature of those arcs. This curvature will tend to adjust itself locally so that, at equilibrium, jogs are subjected to forces which give the same velocity to the vacancy emitters as to the vacancy adsorbers. The theoretical dislocation velocity (at stress$=F$) then becomes,

$$v = b\sinh(\alpha F)/\alpha A, \tag{3}$$

which at low stresses reduces to

$$\alpha F \to 0: \ V = Fb/A. \tag{4}$$

Eq. (4) is algebraically identical with the equation for an analogous case of solute drag already examined (Eq. (2) of Gittus, 1974) and so we can immediately write down the creep equation for the low stress case since it is the same for jog drag as it was for solute drag (i.e. Eq. (12) of Gittus, 1974)

$$\sigma \to 0: \ \dot{\varepsilon}_v = \left(8\pi^3 c_j\right)\frac{D_y \mu b}{kT}\left(\frac{\sigma}{\mu}\right)^3\left[1 - \frac{1}{\left(kT/A2\pi D_v bc_j\right)+1}\right]. \tag{5}$$

In these equations, following Barrett and Nix (1965),

$$\alpha = b^2 l/kT, \tag{6}$$

$$A = b^2/20\alpha D_v, \tag{7}$$

where b is Burgers vector, k is Boltzmann's constant, T is the absolute temperature, D_v is the volume self-diffusion coefficient. In addition, in Eq. (5), μ is the elastic modulus and c_j is the thermal jog concentration given by the following equations, due to Friedel (1964):

$$c_j = \exp\left(-\chi\mu b^3/kT\right),$$

$$0.2 > \chi \geqslant 1/8\pi. \tag{8}$$

In the case where αF is large, Eq. (4) becomes a poor approximation and we obtain, instead of Eq. (5), the following:

$$\sigma = \left(\frac{\dot{\varepsilon}_s}{B}\right)^{1/3} + \frac{1}{\alpha}\sinh^{-1}\left[\left(\frac{\dot{\varepsilon}_s}{B}\right)^{1/3}\frac{\mu^2\alpha AB}{4\pi^2}\right]. \tag{9}$$

At low stresses Eq. (9) reduces to Eq. (5), as it should. The notation,

$$B = 8\pi^3 c_j\left(D_v b/\mu^2 kT\right) \tag{10}$$

has been used in writing down Eq. (9).

Eq. (9) then is the required expression for the steady-state creep rate of material containing a three dimensional network of dislocations whose movements are influenced not only by the internal stresses which the network imposes on each of its member links, but also by the nonconservative drag of screw jogs.

2.1. Activation energy for creep

The activation energy for steady-state creep, according to Eq. (9), is largely determined by that of volume self-diffusion. This is because the main source of temperature dependence in the equation is contained in the factor B. The temperature dependence of the factor $\mu^2 \alpha AB/(4\pi^2)$ is only that of the jog concentration and this is small, since the activation energy predicted by Eq. (8) is typically only a few kilojoules per mole. Taking the factor $\mu^2 \alpha AB/(4\pi^2)$ to be largely independent of temperature, we see then that at constant external stress the quotient $\dot{\varepsilon}/B$ must be constant if Eq. (9) is to be satisfied, i.e.

$$\sigma = \text{constant}: \dot{\varepsilon}_s \propto B, \tag{11}$$

and so from Eq. (10) (to the first approximation) we expect the activation energy for the steady-state creep process to be equal to that of volume self-diffusion. This conclusion is independent of temperature and stress. It is the same as the conclusion reached in the case of solute drag.

2.2. Exponent of stress

Over limited ranges of stress and temperature it is often observed that the following equation, attributed to Nutting and Scott-Blair, provides a reasonable approximation to dislocation creep data:

$$\dot{\varepsilon}_s = A'\sigma^n \exp(-Q_{app}/RT). \tag{12}$$

Here A' is a constant for the material, as is n, whilst Q_{app} is the apparent activation energy for the process. Differentiating Eq. (12) we get, for the effect of changing the stress at constant temperature,

$$n = \frac{\sigma}{\dot{\varepsilon}_s} \frac{d\dot{\varepsilon}_s}{d\sigma}. \tag{13}$$

Making the convenient substitutions,

$$y = \left(\frac{\dot{\varepsilon}_s}{B}\right)^{1/3} \frac{\mu^2 \alpha AB}{4\pi^2} \tag{14}$$

and

$$C = 4\pi^2/\mu^2 \alpha AB, \tag{15}$$

we can differentiate Eq. (9) and use the result to define n in Eq. (13),

$$n = \frac{3\left[\left(\sinh^{-1} y/y\right) + \alpha C\right]}{\left(1+y^2\right)^{-1/2} + \alpha C}. \tag{16}$$

Evidently the condition for n to exceed 3 is that,

$$\sinh^{-1} y/y > \left(1+y^2\right)^{-1/2}, \tag{17}$$

and this condition is met for all finite values of y. It is only as y approaches zero or infinity that n approaches 3: for intermediate values of y (i.e. for finite values of the externally applied stress) n is greater than 3. As αC is non-negative, it follows that for a given value of y the maximum value of n will always be obtained with $\alpha C = 0$ and the larger αC the more closely will n approximate 3. At the limit where αC is negligible (corresponding to very high stresses) we get,

$$\sigma \to \infty \ (y \to \infty \text{ and } \alpha C = 0) : n \to 3\ln(2y). \tag{18}$$

In Fig. 2 the function $\sinh^{-1} y$ is plotted against the value of n for various values of αC. Also plotted as a broken line is the locus of the maximum values of n. The latter is obtained by differentiating Eq. (16) and setting $dn/dy = 0$, then,

$$n_{max} = 3\left(\frac{\left(1+y^2\right)^{3/2}\left(\sinh^{-1} y/y\right)\left(1+y^2\right)}{y^2}\right), \tag{19}$$

which is the equation of the broken line. The limits of n_{max} can be seen,

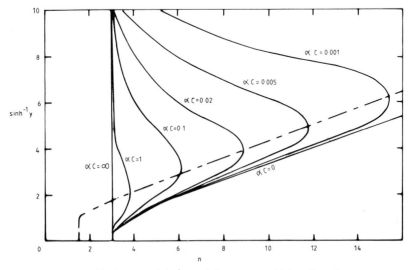

Fig. 2. Relationship between $\sinh^{-1} y$ and the stress sensitivity of steady state creep, n.

from Eq. (19), to be,

$$y \to 0: n_{max} \to 1.5,$$

$$y \to \infty: n_{max} \to 3(\sinh^{-1} y - 1) \to 3 \ln(2y) - 3. \tag{20}$$

The requirement that αC be finite and non-negative imposes the further limitation that $n_{max} \nleq 3$.

2.3. Numerical examples

To illustrate the results likely to be obtained by evaluating Eq. (9) consider first the case of gold at half its absolute temperature of melting, Table 1. In the third column of the table the theoretical value of stress has been shown as the sum of two terms,

$$\sigma = \sigma_\perp + \sigma_j, \tag{21}$$

where

$$\sigma_\perp = (\dot{\varepsilon}_s / B)^{1/3}, \tag{22}$$

and

$$\sigma_j = \frac{kT}{b^2 \ell} \sinh^{-1} \left[\left(\frac{\dot{\varepsilon}_s}{B} \right)^{1/3} \frac{b^3 \pi c_j}{10 kT} \right], \tag{23}$$

so that $\sigma_\perp$ denotes that part of the applied stress which would be required to produce a creep rate of $\dot{\varepsilon}_s$ in the absence of jog drag whilst σ_j denotes the contribution of jog drag to the external stress which has to be applied.

The effect of the magnitude of ℓ, the spacing between screw jogs, on the predicted external stress is now clear: the larger the value of ℓ, the smaller the predicted stress will be. In Table 1 it was assumed that $\ell = b/c_j$. Suppose now that $(\dot{\varepsilon}_s / B)^{1/3} = 10^{-2} \mu$. This would imply a very high dislocation density (around 5×10^{16} m^{-2}) and may be regarded as an extreme case. Let $b^3 = 30 \times 10^{-30}$ m^3 and let $\mu = 10^{11}$ N/m^2 at 500 K and with $c_j = 1$ and $\ell = b$. These values have been set at levels designed to make the effect of jog drag (σ_j) large compared with that of the dislocation self stresses ($\sigma_\perp$). Then

Table 1
Comparison of the actual and theoretical values of stress which produces a given creep rate in gold at $T_m/2$

Creep rate (s^{-1})	Actual experimentally determined stress	Theoretical stress, given by Eq. (21)	
		($\sigma_\perp$)	(σ_j)
3.147×10^{-6}	$10^{-3}\mu$	$0.882 \times 10^{-3}\mu + 0.009 \times 10^{-3}\mu$	
9.95×10^{-12}	$10^{-4}\mu$	$1.304 \times 10^{-5}\mu + 0.004 \times 10^{-5}\mu$	

from Eq. (23),

$$\sigma_j/\sigma = 0.24, \tag{24}$$

whilst from Eq. (16),

$$n = 3.17, \tag{25}$$

which implies that it is unlikely that there will be cases in which the contribution of jog drag is more than one quarter of the external stress needed to produce a given creep rate. In the examples of the table, $\sigma_j/\sigma < 0.01$ and $n \sim 3.0$ and it transpires that for all practical purposes the linear approximation (Eq. (5)) originally derived for the case of solute drag is adequate for jog drag too.

3. Modification to the creep mechanism due to dislocation cell (sub-grain) formation

In the previous section we limited attention to a material in which the dislocations had formed into the three-dimensional Frank network. We shall now broaden our analysis to include the case in which the dislocations form into a cell structure (Gittus, 1976).

Experiments show that characteristic networks of dislocations tend to form during creep. The networks outline cells within which there are relatively few free dislocations. The cells are sub-grains.

As the networks do not exert significant long-range stresses, most of the free dislocations within the cells are unaffected by the stresses produced by dislocations in the cell walls (Friedel, 1964). They are still, however, subject to the internal stress produced by line tension. This internal stress is of magnitude $\mu b/L$ where μ is the shear modulus, b is the magnitude of the Burgers vector and L is the mean distance between neighboring free dislocations. In addition, if they are moving, the dislocations experience the friction stress (F) due to jog drag. These two stresses act in conjunction to oppose the glide of the dislocations and at equilibrium they are just counterbalanced by σ, the externally applied stress,

$$\sigma - \mu b/L = F. \tag{26}$$

Holt (1970) has reviewed data on the cell structure and has advanced a theory relating the cell diameter to the mean distance between dislocations. Both the data and his theory indicate that L is of the order of the cell diameter and that the latter is proportional to the mean spacing ($\rho^{-1/2}$) of the dislocations, imagining them to be uniformly spread out throughout the volume of the material, i.e.

$$L = K\rho^{-1/2}. \tag{27}$$

Here ρ is the dislocation density and K a constant. For copper, deformed at temperatures between 15 and 700°C under shear stresses of from 10.69 MN/m^{-2} (at the highest temperature) to 78.97 MN/m^2 (at the lowest temperature), Staker and Holt (1972) later found that K had a mean value of 16. Other workers have reported values of this order of magnitude for other materials, including for example, iron and aluminum.

Turning to the value of the friction stress, F: here we continue to use Eqs. (3), (6) and (7). Then we can show that the shear stress σ needed to produce a strain rate $\dot{\varepsilon}$ is given by

$$\sigma = \frac{1}{K^2}\left(\frac{\dot{\varepsilon}}{B}\right)^{1/3} + \frac{1}{\alpha}\sinh^{-1}\left[\left(\frac{\dot{\varepsilon}}{B}\right)^{1/3}\mu^2\alpha ABK^4\right], \tag{28}$$

where

$$B = c_j\left(D_v b/\mu^2 kT\right). \tag{29}$$

3.1.   Creep of gold

As an example of the comparison between actual and theoretical creep rates we shall reconsider first the case of gold. Table 2 shows the results obtained. An empirical equation was fitted to the dislocation creep data. This was solved at the two stress levels of column (a) to produce estimated creep rates. The latter are therefore estimates based on experimental data and were used in Eq. (28) to produce the purely theoretical values of column (b). In arriving at the latter, the value of K was set equal to 13. This is the order of magnitude indicated by the summary of the literature by Staker and Holt (1972) on a number of pure metals (they find $K = 16$ for Cu and

Table 2

Comparison of the actual and theoretical values of stress which produce a given creep rate in gold at $T_m/2$; predicted and actual values of n

Actual, experimentally determined stress, σ (a)	Theoretical stress (Eq. (28), $K=13$, $\ell = b/c_j$) (b)	$\dfrac{d(\ln \dot{\varepsilon})}{d(\ln \sigma)} = n$	
		(c) Theoretical	(d) Experimental
$10^{-3}\mu$	$1.0 \times 10^{-3}\mu$	7.2	5.5
$10^{-4}\mu$	$0.9 \times 10^{-4}\mu$	3.0	5.5

pure Fe, $K=13$ for 0.007% C–Fe and $K=8$ for 3% Si–Fe). The predictions are not particularly sensitive to K in this range. For example if $K=12$, 13 or 14 then $\sigma/\mu=0.99\times10^{-3}$, 1.0×10^{-3} or 1.03×10^{-3}, respectively, in the case where the expectation is $\sigma/\mu=1.0\times10^{-3}$. The tabulated value of the parameter n has been derived by the procedure of Sections 2.2 and 3.2. The stresses in the table show excellent accord between theoretical expectation and the results inferred from experiment: this type of agreement is found for a number of other pure metals too. Moreover the more general characteristics of Eq. (28), including the type of stress sensitivity and temperature dependence which it predicts, are in accord with experience, as the next three sections reveal.

3.2. Exponent of stress, n, in the Nutting–Scott-Blair equation

Making the substitutions,

$$y=(\dot{\varepsilon}/B)^{1/3}\mu^2\alpha ABK^4 \tag{30}$$

and

$$C=1/\mu^2\alpha ABK^6, \tag{31}$$

we can differentiate Eq. (28) and use the result to define n in the first differential of Eq. (12). In this way it is found (c.f. Eq. (16)):

$$n=\frac{3\left[(\sinh^{-1}y/y)+\alpha C\right]}{(1+y^2)^{-1/2}+\alpha C}. \tag{32}$$

Eq. (32) was used to produce the values of n for gold in Table 2. The value of 5.5 given in that table for the results of experiment is a mean value since it is commonly observed (as the present theory predicts) that n is stress dependent.

The graph of Eq. (32) is identical to Fig. 2 and indeed Eq. (32) reduces, as it should, to the corresponding equations for a three-dimensional Frank network if we set K equal to unity in Eqs. (30) and (31). In that case, however, the value of n for all reasonable values of the other parameters does not differ significantly from 3 and we find that we are always in the vicinity of the n-axis in Fig. 2. It is the formation of the cellular dislocation network that gives rise to values of n which (like that for gold in Table 2) can theoretically rise above 3 at the higher stresses.

3.3. The exponent, m, in Garofalo's equation

In the previous section we calculated the stress dependence of creep rate at a specific stress. The result was the value of n which is the slope of the

Table 3
Values of the exponent m in Eq. (33); data from (Garofalo, 1965)

			Copper			
°K	673	723	773	823	903	973
m^a	3.57	3.36	3.39	3.38	2.66	2.26

	Aluminum			Al–3.1% Mg	Austenitic stainless steel	
°K	447	533	920	531	977	1089
m^a	5	4.55	1.24	2.26	3.64	3.50

[a] Mean of all these m values $= 3.23$. Theoretical value $= 3.00$.

tangent to the curve of log stress versus log strain rate at a specific stress. The slope was predicted to change with stress. The form of this change in stress sensitivity is such that, as the value of K is increased, the curve moves towards an asymptotic or limiting shape. From Eq. (28) the equation of this limiting curve is,

$$\dot{\varepsilon} = B\left[CK^2 \sinh(\alpha\sigma)\right]^m, \tag{33}$$

$$m = 3, \tag{34}$$

where C is defined by Eq. (31) and B by Eq. (29). Equation (33) is precisely the form of equation which Garofalo (1965) found fitted the creep of several metals. His results are summarised in Table 3 which reveals that the value of m which best fitted the data that he analysed is, as predicted, of order 3. The average of the tabulated values is in fact 3.23.

Garofalo also found that the activation energy for creep was close to that for volume self-diffusion and this too is predicted by Eq. (33) when expanded into the form,

$$\dot{\varepsilon} = \frac{c_j D_v b}{\mu^2 kT}\left[\frac{20kT}{b^3 c_j K^4} \sinh(\alpha\sigma)\right]^3. \tag{35}$$

As the activation energy for jog formation (Eq. (8)) is typically an order of magnitude lower than that for self diffusion (contained in D_v) it is the latter which, according to Eq. (35), should largely determine the temperature dependence of the creep rate.

3.4. The activation energy for creep

Consider a case in which the effect of the first term of Eq. (28) upon the stress is negligible. Suppose that the temperature is suddenly altered and that we wish to keep the stress constant. Then we shall have to adjust the strain rate so as to keep the factor $(\dot{\varepsilon}c_j^{\,3}/B)^{1/3}$ constant. The strain rate must in fact change in proportion to the product $Bc_j^{\,-3}$ and so the activation energy for steady-state creep, as determined in a temperature change experiment, will be equal, at this extreme, to $Q_{vol}-2Q_j$. Here Q_{vol} is the activation energy for bulk diffusion and Q_j the (much smaller) activation energy for thermal jog formation. A similar argument for the case in which the applied stress is entirely dictated by the first term of Eq. (28), leads to the conclusion that, in that case, the activation energy for steady-state creep will be $Q_{vol}+Q_j$. From this it is concluded that the activation energy for creep will generally lie between $(Q_{vol}-2Q_j)$ and $(Q_{vol}+Q_j)$. Its average value will be close to that for volume self diffusion, a conclusion which receives support from the literature on experiments that have been conducted with many pure metals. For gold at $T_m/2$ and for $\sigma/\mu=10^{-3}$, for example, the theoretical value of the activation energy for creep, given by Eq. (28) is 170.5 kJ/mole whereas that for volume diffusion is 174.3 kJ/mole.

3.5. Theoretical relationship between cell diameter and strength at low strain rates

As the strain rate goes towards zero, Eq. (28) gives the approximation

$$\frac{\sigma}{F}=1+\frac{1}{\mu^2 ABK^6},\tag{36}$$

which together with Eq. (26) yields the expression:

$$\sigma=\frac{\mu b}{L}\left[1+\frac{c_j^{\,2}K^6}{20}\right].\tag{37}$$

Now a review of the creep properties of a range of materials (Staker and Holt, 1972) indicates that,

$$\sigma\sim 10(\mu b/L).\tag{38}$$

From this equation and Eq. (37) we then obtain,

$$K\sim\left[13.4/c_j\right]^{1/3}.\tag{39}$$

If then, for example, we take the theoretical value $c_j=0.0104$ used above in the case of gold, we find that $K=11$, which is of the order of magnitude found experimentally (Staker and Holt, 1972).

4.   Theoretical value of the cell size for minimum free energy

The machine which applies stress and strain to a cracked or flawed specimen does several kinds of work. Firstly it stretches the specimen elastically and this is recoverable work: a component of the free energy. Then it produces recoverable plastic strain due to reversible movements of dislocations (bowing, for example). Again this is part of the free energy, unlike the nonrecoverable steady-state creep strain. A further component of the free energy of the system is the energy of the dislocation assembly. Finally we have the energy of the new crack surface produced as the machine stretches the specimen. Any generalization of fracture mechanics to the case of ductile, time-dependent fracture must consider all of these components of the energy balance.

4.1.   Thermodynamic considerations

Consider a specimen undergoing dislocation creep at a constant strain rate, $\dot{\varepsilon}$. Characteristically it will, at the onset of the creep process, contain a random three-dimensional dislocation network. With the passage of time the dislocations form themselves into cells. Typically the cell diameter and dislocation density both move towards an equilibrium value during this initial period (McElroy and Szkopiak, 1972). The cell size, dislocation density and K value then remain constant during the ensuing period of steady-state creep, characterized by the fact that (at constant strain rate) the stress needed to cause creep remains constant. The constancy of the cell size and the large extensions that can occur during this steady-state, strongly suggest that it is a condition of thermodynamic equilibrium, characterized by a minimum free energy.

The initial stage of creep is then seen to be one in which the glide and climb of dislocations, produced by stress and temperature, permit them to move into the configuration (characterized by cells of a certain size) which minimizes the free energy. When once that configuration has been attained, the cell size and K value cease to change (for any such change would raise the free energy above its minimum value) and so the stress to cause creep becomes constant.

At this thermodynamic equilibrium,

$$(\delta F)_{T,V} \geqslant 0. \tag{40}$$

Here, the lower case delta (δ) is used to indicate that the free energy is unchanged or increased for any small (finite) displacement of the system.

4.2. Components of the free energy of the specimen

The dislocations which are contained in the creep specimen make an important contribution to its free energy. That contribution depends on the dislocation density and upon the arrangement of the dislocations. The quantity K is a measure of the characteristics of the dislocation arrangement. If $K=1$ then the dislocations are arranged in a uniform three-dimensional network. If $K>1$ then the dislocation arrangement consists of cells having most of the dislocations in their walls or boundaries. The larger K, the bigger the cell diameter (for a given dislocation density).

Another component of the free energy of the specimen is the elastic plus viscoelastic strain energy produced by the externally applied stress. If this was the only component, then the lower the applied stress the lower the free energy. So at a fixed strain rate the dislocations would tend to move into that arrangement which minimizes the stress that the external machine has to exert. The condition of thermodynamic equilibrium could then be the condition of minimum creep strength. We can show that there is a certain value of K (of order $c_j^{-1/3}$) that produces this minimum creep strength. If we now consider both components of the free energy (that due to dislocations and that due to elastic plus viscoelastic strain) then we can calculate a more exact estimate of the value of K that minimizes the free energy.

4.3. Free energy contributed by dislocations

The total energy per unit volume due to the dislocations in the cell boundaries, $\overline{E}$, is given by:

$$\overline{E}=\rho\mu b^2/2. \tag{41}$$

Here ρ is the smeared density of dislocations, dimensions L^{-2}. We can then show (Gittus, 1977),

$$\overline{E}=\left(\frac{\dot{\varepsilon}}{2K_1 b}\right)^{2/3}\frac{\mu b^2}{2K^2}. \tag{42}$$

According to Eq. (40) at thermodynamic equilibrium we shall expect to find that the free energy change produced by any arbitrary change in K is positive or zero. Forming then the differential of $\overline{E}$ with respect to K,

$$\frac{\mathrm{d}\overline{E}}{\mathrm{d}K}=-\left[\left(\frac{\dot{\varepsilon}}{2K_1 b}\right)^{2/3}\frac{\mu b^2}{2}\right]2K^{-3} \tag{43}$$

$$=-\frac{1}{\mu}\left(\frac{\varepsilon}{B}\right)^{2/3}K^{-3}, \tag{44}$$

where B is given by Eq. (29).

4.4.   Free energy due to elastic plus viscoelastic straining

Considering next the component of free energy due to elastic and viscoelastic deformation. In a specimen subjected to a constant strain rate, the machine causing deformation does both recoverable and nonrecoverable work. The nonrecoverable work is done to produce the creep strain. Part of the recoverable work is used to produce elastic and viscoelastic strains. When the external stress is removed, not only does creep cease but the elastic strain is instantaneously recovered. The viscoelastic strain is recovered, too, but this does not occur instantaneously since it involves time-dependent processes. The energy stored in the material as a result of the elastic and viscoelastic strains is seen therefore to be available to do external work: it is part of the free energy of the specimen.

The actual change (dE_{el}) in free energy produced when a specimen under stress σ deforms recoverably is given by,

$$dE_{el} = \sigma \, de. \tag{45}$$

Here de is an infinitesimal increment of recoverable strain. If μ_R is the modulus relating strain e to stress σ (i.e. μ_R is, in the terminology of linear viscoelastic theory, the relaxed modulus), then Eq. (45) becomes,

$$dE_{el} = \sigma \, d(\sigma/\mu_R). \tag{46}$$

To arrive at the K-dependence of the viscoelastic energy, use will be made of the creep equation developed above (Eq. (28)). At the low stress limit this equation becomes,

$$\sigma = \left(\frac{\dot{\varepsilon}}{B}\right)^{1/3}\left(K^{-2} + \frac{c_j{}^2 K^4}{20}\right). \tag{47}$$

Our procedure will be to use Eq. (47) to derive a value for the change in free energy in Eq. (46). A value of the relaxed viscoelastic modulus, μ_R, is therefore needed. Friedel (1964) has indicated how μ_R depends on the geometry of the dislocation assembly. For the case in which the dislocations have formed themselves into cells he shows, theoretically, that the bowing of the dislocations under the influence of the applied stress will supplement the elastic strain, giving a relaxed modulus of the following form:

$$\mu_R = \mu/(1 + C'K^2). \tag{48}$$

From equation (46) with μ_R defined by Eq. (48),

$$dE_{el} = \frac{\sigma}{\mu}\left[d\sigma(1 + C'K^2) + 2K \, dK \, C'\sigma\right]. \tag{49}$$

To obtain a value for $d\sigma$ we differentiate Eq. (47) obtaining,

$$d\sigma = dK \left(\frac{\dot{\varepsilon}}{B} \right)^{1/3} \left[-2K^{-3} + \frac{c_j^2 K^3}{5} \right]. \tag{50}$$

Eq. (50) may now be used to define $d\sigma$ in Eq. (49). Equation (47) is used to define σ in Eq. (49). The value of dE_{el} is now defined in terms of C', c_j and K. Upon adding $d\bar{E}$, a value for the total change in free energy dF is derived,

$$dF = d\bar{E} + dE_{el}, \tag{51}$$

i.e.,

$$\frac{dF}{dK} = \frac{-1}{\mu} \left(\frac{\dot{\varepsilon}}{B} \right)^{2/3} K^{-3}$$

$$+ \frac{1}{\mu} \left(\frac{\dot{\varepsilon}}{B} \right)^{2/3} \left(-2K^{-3} + \frac{c_j^2 K^3}{5} + \frac{3}{10} C' c_j^2 K^5 \right)$$

$$\times \left(K^{-2} + \frac{c_j^2 K^4}{20} \right). \tag{52}$$

The condition set by Eq. (40) is met if $dF/dK = 0$ and d^2F/dK^2 is positive. Eq. (52), with $dF/dK = 0$, gives,

$$c_j^2 = \left[-b + \sqrt{(b^2 - 4ac)} \right] / 2a, \tag{53}$$

where

$$a = 10^{-2} K^7 + \frac{3C'K^9}{200}, \tag{54}$$

$$b = \frac{3C'K^3}{10} + \frac{K}{10}, \tag{55}$$

$$c = -K^{-3} - 2K^{-5}. \tag{56}$$

Differentiating Eq. (53) a second time we find that the condition for d^2F/dK^2 to be positive is met. So Eqs. (53)–(56) correspond to a state of thermodynamic equilibrium. From them the theoretical value of K may be calculated.

4.5.   Theoretical relationship between K, C' and c_j

In Fig. 3, Eq. (53) has been used to provide the theoretical relationship between K and c_j for $C' = 0$, 0.005 and 0.05. $C' = 0.05$ is the maximum theoretical value of this parameter, corresponding to the case in which all of the dislocations in the cell walls bow out reversibly under stress, contributing to the viscoelastic strain (Friedel, 1964). In practice the measured ratios of relaxed to unrelaxed modulus correspond to C' values of order 0.005 (i.e. only about a tenth of the dislocations bow in this manner).

J. H. Gittus

For materials in use at the temperatures where dislocation creep is the dominant deformation mechanism, we generally have $0.1 > c_j > 0.001$ and so theoretically $20 > K > 5$: the values of K that have been measured experimentally do in fact generally lie in this range.

4.6. Approximations to K

For finite values of C' and for $K > 5$, Eq. (53) reduces to the following approximation:

$$K \sim [0.03C']^{-1/12} c_j^{-1/3}. \tag{57}$$

For $C' = 0$ and for $K \geqslant 1$ the following alternative approximation to Eq. (53) is valid:

$$K \sim 1.58 c_j^{-0.4}. \tag{58}$$

To illustrate the range of validity of these approximations, Eqs. (57) and (58) have been used to produce estimates of c_j (c_j (approx.)) for selected values of K and C'. The results are compared in Table 4 with the exact values of c_j, derived from the full solution of Eq. (53). Only in the case of $K = 1$ for $C' = 0.005$ and 0.05 are the approximations poor: for most purposes, the approximations are acceptable and lead to some simplification of the creep equation.

Table 4
Theoretical relationship between the value of K and those of c_j and K' for various values of C' [a]

C'	K	c_j	c_j (approx.)	K'
0	1	3.61	3.61 -	1.65
	5	5.50×10^{-2}	5.63×10^{-2}	3.36
	13	5.11×10^{-3}	5.13×10^{-3}	7.30
	20	1.75×10^{-3}	1.74×10^{-3}	10.80
0.001	10	9.42×10^{-3}	9.04×10^{-3}	5.44
0.005	1	3.60	9.12	1.65
	5	5.20×10^{-2}	7.24×10^{-2}	3.11
	13	3.80×10^{-3}	4.17×10^{-3}	4.48
	20	1.25×10^{-3}	1.15×10^{-3}	6.60
0.05	1	3.49	5.13	1.61
	5	3.87×10^{-2}	4.07×10^{-2}	2.17
	13	2.37×10^{-3}	2.34×10^{-3}	2.36
	20	6.87×10^{-4}	6.46×10^{-4}	2.51

[a] If $C' = 0$ then $K \sim 1.58 c_j^{-0.4}$ and $K' \sim 1 + K/2$.

4.7. *Example: creep of gold at* $T_m/2$

In Table 2 we examined the actual and theoretical creep characteristics of gold at half its absolute melting point ($T_m/2$). A value of $K=13$ was used. The calculations will now be repeated using the theoretical value of K given by Eq. (53). The value of C' does not exert a large effect on K at this level of c_j (see Fig. 3): we use the approximation $C'=0$ corresponding to negligible anelastic strain (relaxed modulus=shear modulus). Then Fig. 3 shows,

$$K=10. \tag{59}$$

Using this value in the creep equation (Eq. (28)) we obtain the results of Table 5. Agreement at the lower stress is not as good as was the case when we used the higher value ($=13$) of K but certainly the predictions are of the correct magnitude and provide support for the theory.

4.8. *Theoretical relationship between cell diameter and strength*

Both K and c_j can now be eliminated from Eq. (37) providing that approximation (57) is valid. Assuming this to be the case then substitution

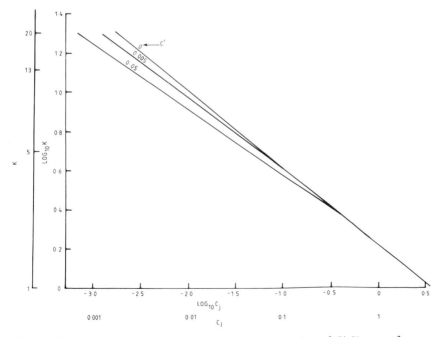

Fig. 3. Theoretical value of K as a function of c_j for various values of C'. Here $C'K^2=(\mu-\mu_R)/\mu_R$ where μ and μ_R are the shear modulus and the relaxed modulus, respectively. Theoretically $0 \leqslant C' \leqslant 0.05$. In practice its value is usually less than 0.005.

Table 5
Comparison of the actual and theoretical values of stress which produce a given creep rate in gold at $T_m/2$; predicted and actual values of n; theoretical value used for K; $\ell = b/c_j$

| Actual, experimentally determined stress, σ | Theoretical stress (Eq. (28)) | $\dfrac{d(\ln \dot{\varepsilon})}{d(\ln \sigma)} = n$ | |
		Theoretical	Experimental
$10^{-3}\mu$	$0.96 \times 10^{-3}\mu$	5.0	5.5
$10^{-4}\mu$	$0.37 \times 10^{-4}\mu$	3.0	5.5

of Eq. (57) into Eq. (37) gives:

$$\sigma = \frac{\mu b}{L}\left[1 + (12C')^{-1/2}\right]. \tag{60}$$

Now, theoretically, $C' \leqslant 0.05$ and so from Eq. (60),

$$\sigma \geqslant 2.3\mu b/L. \tag{61}$$

In fact experiments (Staker and Holt, 1972) indicate,

$$\sigma \sim K'\mu b/L, \tag{62}$$

with

$$K' \sim 10. \tag{63}$$

Equation (61) is an approximation valid for certain finite values of C' and K. An exact theoretical value of K' can be obtained by solving Eq. (37) for K' $(= 1 + c_j{}^2K^6/20)$ using the values of K and c_j from Table 4. This has been done and the theoretical values of K' so obtained are there tabulated. The theoretical K' values are seen to be of the expected magnitude.

5. Relation between free energy and cell size (Gittus, 1979)

We can plot graphs for the theoretical effect of cell diameter of the free energy and its two components: that due to elastic deformation and that due to the dislocations. As an example, such a plot is shown for zirconium in Fig. 4. A temperature of 900 K was used, with $c_j = 0.0017$. The figure shows that the total free energy of the specimen passes through a minimum for a critical value of the cell size: smaller or larger cells lead to higher free energies. The minimum occurs because of the interplay of the two components of the free energy. Thus the elastic strain energy changes continuously as the cell size is increased, while simultaneously the energy due to the

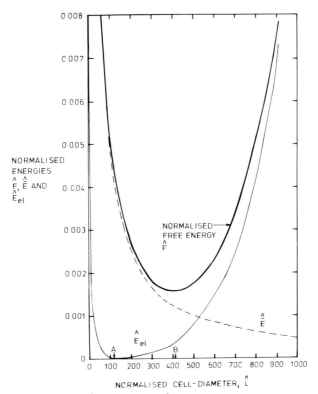

Fig. 4. Effect of cell diameter $\hat{L}$ on free energy $\hat{F}$ for a specimen of zirconium creeping at 900 K and a fixed strain rate. $\bar{E}$ is the contribution of the dislocations to the free energy and $\hat{E}_{el}$ is the contribution of the elastic strain energy. The curve for F is then the sum of $\hat{E}_{el}$ and $\bar{E}$. (See (Gittus, 1979) for a description of the normalization procedures adopted here.) Point A is the minimum in $\hat{E}_{el}$; point B is the minimum in $\hat{F}$. The curve for $\bar{E}$ is a rectangular hyperbola and has no minimum.

dislocations is falling. The minimum value of the free energy occurs at a cell size where the elastic energy has not yet risen to a value high enough to compensate for the fall that has occurred in the energy due to the dislocations.

In Section 4 it was shown that there was a value of K for a given temperature such that the free energy was a minimum. It was shown that this value of K is similar in magnitude to values derived experimentally $(20 > K > 5)$ for a variety of materials. We note that, corresponding to this "optimum" value of K there is always, at a given stress, a unique value of sub-grain size. It was also noted earlier that the theoretical value of K' is of the order of 10 and that this again agrees well with experimental data.

The present example is in good agreement with these earlier generalizations since, at the minimum value of the free energy, we find that:

$$K=20.25 \quad \text{and} \quad K'=11.12.$$

6. Free energy release as a criterion for creep cracking

We can show that the equation for the total free energy of a creeping specimen may be written in the form,

$$F=E\left(1+(K/K')^2\right), \tag{64}$$

where

$$E=\sigma^2/2\mu.$$

That is to say E is the strain energy of the specimen. Now Table 4 shows that K/K' approximately equals 2 for cases where a well defined cell structure forms in the creeping specimen. Therefore,

$$F_{cell} \sim 5E, \tag{65}$$

whilst for materials in which the dislocations do not form cells ($K=1$), Eq. (64) becomes,

$$F_{net} \sim 1.36E. \tag{66}$$

If we assume that this free energy is all available to propagate a crack during creep then Eq. (65) implies that there is critical crack length for crack growth during creep and that this can be as little as one-fifth the time-independent value.

7. Discussion and conclusions

The key conclusion of this work is that steady-state creep is a state of minimum free energy. It follows that any small perturbation that would be expected to precipitate failure will be strongly resisted because it increases the free energy and this is clearly the origin of ductility—the tendency for a material to exhibit large and reproducible uniform elongations (here exhibited in creep). One can show that the free energy in the super-plastic state is even lower than that in steady-state dislocation creep and this undoubtedly explains why superplastic materials are even more ductile.

The second, important outcome of the work is an understanding of dislocation cell structure and sub-grain size. Previous workers had been unable to account, numerically, for the stress dependence of sub-grain size (cell diameter) and we have now remedied this deficiency. The cell diameter is little affected by temperature and the possibility now exists of inferring the stress distribution round a notch or flaw from microscopic measurements of the cell diameters. A kind of "finite element" analysis using the naturally occurring mesh of dislocations.

Finally we have shown how to calculate from theory the free energy of a specimen undergoing creep, including elastic, viscoelastic and dislocation network terms. This opens the door to the extension of fracture mechanics to include the case of a creeping, cracked specimen.

Appendix

Stimulated by discussion with Professor Bilby, I have arrived at the following alternative derivation of the equation for strain rate (see Fig. 5).

In a time interval dt a number dn_m of dislocations are generated. Each travels a distance L (the cell diameter) to the cell wall, where it is adsorbed, adding to the cell wall dislocation density. In the steady state, with which we are concerned, for every one of the dn_m dislocations generated at a source in time dt, a dislocation is adsorbed (having moved across the cell) by the cell wall. So the mobile dislocation density, ρ_m, is increased by the dn_m newly generated dislocations and simultaneously diminished by the dn_a mobile dislocations that have reached the cell wall and there been adsorbed, i.e.

$$d\rho_m = dn_m - dn_a = 0 \quad \text{so} \quad dn_a = dn_m.$$

Consider next the strain contributed by this number, dn_m, of freshly generated mobile dislocations as they traverse the cell diameter. This is given by,

$$d\varepsilon = dn_m bL. \tag{67}$$

Now at equilibrium, for every mobile dislocation that enters the cell boundary, a cell boundary dislocation is eliminated by recovery. In this

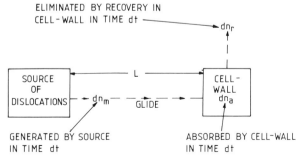

Fig. 5. Generation glide absorption and elimination of dislocations at equilibrium $|dn_m| = |dn_a| = |dn_r|$.

manner the cell boundary dislocation density is maintained constant. Equation (A11) of Gittus (1976) is,

$$(dr_b/dt)_{recovery} = K_1/r_b, \tag{68}$$

whilst from Eqs. (A3) and (A4),

$$1/r_b = K\sqrt{\rho}, \tag{69}$$

and

$$\frac{dr_b}{dt} = -\frac{1}{2K}\frac{1}{\rho^{1.5}}\frac{d\rho}{dt}. \tag{70}$$

Now the loss of dn_r dislocations reduces the cell wall density by $d\rho$, i.e.

$$d\rho = dn_r. \tag{71}$$

Hence, from Eqs. (68, 69, 70 and 71),

$$dn_r = -2K^2K_1\rho^2 dt. \tag{72}$$

Eq. (72) tells us the rate at which recovery eliminates dislocation from the cell walls. From Eq. (67),

$$dn_a = dn_m = d\varepsilon/bL. \tag{73}$$

Eq. (73) gives the rate at which mobile dislocations enter the cell walls.

At equilibrium the dn_a mobile dislocations that are adsorbed in the cell walls are equal to the $-dn_r$ that are eliminated from the cell walls by recovery and so from Eqs. (72) and (73),

$$dn_a + dn_r = 0,$$

i.e. from Eqs. (72) and (73),

$$\dot{\varepsilon} = 2K^2K_1\rho^2 bL, \tag{74}$$

which, with (A3) and (A4) yields,

$$\dot{\varepsilon} = 2K_1b/r_b^3 \tag{75}$$

which is Eq. (A21) of Gittus (1976).

References

Barrett, C. R. and Nix, W. D. (1965), "A Model for Steady-State Creep Based on the Motion of · Jogged Screw Dislocations," *Acta Met.*, **13**, 1247.

Friedel, J. (1964), *Dislocations* (Pergamon Press, Oxford).

Garofalo, F. (1965), *Fundamentals of Creep and Creep Rupture in Metals* (MacMillan, New York).

Gittus, J. H. (1974), "Theoretical Equation for Steady-State Dislocation Creep: Effect of Solute Drag," *Acta Met.*, **22**, 1179.

Gittus, J. H. (1976), "Theoretical Equation for Steady-State Dislocation Creep: Effects of Jog Drag and Cell Formation," *Phil. Mag.*, **34**, 401.

Gittus, J. H. (1977), "Theoretical Value of the Ratio (K) of Cell Diameter to Dislocation Spacing for a Material undergoing Dislocation Creep," *Phil. Mag.*, **35**, 293.

Gittus, J. H. (1979), "Theoretical Relationship between Free Energy and Dislocation Cell Diameter During Creep," *Phil. Mag.*, **39**, 829.

Holt, D. L. (1970), "Dislocation Cell Formation in Metals," *J. Appl. Phys.*, **41**, 3197.

McElroy, R. J. and Szkopiak, Z. C. (1972), *Int. Met. Rev.*, **17**, 175.

Staker, M. R. and Holt, D. L. (1972), "The Dislocation Cell Size and Dislocation Density in Copper Deformed at Temperatures between 25 and 700°C," *Acta Met.*, **20**, 569.

S. Nemat-Nasser, Editor
THREE-DIMENSIONAL CONSTITUTIVE RELATIONS AND DUCTILE FRACTURE
North-Holland Publishing Company (1981) 75–79

DISCUSSION ON SESSION 1

B. A. BILBY

University of Sheffield, Sheffield, U.K.

While the wide range of topics covered by the three papers in this session must surely make for a lively discussion, it is not easy to select common themes on which it might be based. Of the three papers, I shall concentrate mainly on those by Dr. Gittus and by M. Labbens, since that by Professor Argon and his collaborators was unfortunately only available in abstract at the time of writing.

The paper by Dr. Gittus draws together and extends a number of recent published works on dislocation creep by this author. In these, his attempt to estimate the dislocation cell size in steady-state creep by the following argument raises some interesting questions. A crystalline material creeps by the motion of a small density $\rho_m = 1/L^2$ of mobile dislocations subjected to an applied stress σ, a friction stress F due to jog drag and an internal stress $\mu b/L$ due to line tension, where $\sigma - F = \mu b/L$. Most dislocations are in cells of size L and produce no far reaching stress; their mean density is $\rho_b = 1/Lr_b$, where $r_b \ll L$ is the spacing of dislocations in the cell walls. Thus $\rho = \rho_b + \rho_m$ where $\rho_m \ll \rho_b$. Following Holt (1970) it is assumed that $L = k\rho^{-1/2}$. The dislocation spacing $r_b = L/k^2$ in the cell walls increases at a rate $(\partial r_b/\partial t)_{rec.}$ due to climb, while the mobile dislocations causing the creep strain ε reduce r_b by accumulating in the walls.

The author uses the criterion that the total change of r_b should be zero to determine a k-dependent [1] steady-state creep rate (Eq. (28) of the paper) which is the corner-stone of his work.

The writer had some difficulty with the original formulation of this condition (Gittus, 1976) but a new one developed by Dr. Gittus at the meeting (see the Appendix to the present paper of Dr. Gittus) yields a clearer presentation of his original result.

There is however an important point of principle, which arises in connection with this paper. Dr. Gittus refers to the state of steady-state creep as one of "thermodynamic equilibrium"; and he determines the k to be expected by showing that in the class of states of steady creep characterised by k, there is a member which minimizes (at small constant creep strain rate) the sum of the elastic energy due to the applied stress and an estimate

[1] I use a small k to avoid confusion with the stress intensity factor K.

of the elastic self energy of the dislocations in the cell walls. The energy due to the applied stress may include a correction due to the use of a relaxed viscoelastic modulus which introduces a further k dependence (as in the present paper and in several earlier ones) or it may not (Gittus, 1979). The sum of these energies is referred to as the free energy and the k which minimizes this sum is identified with that observed. In this procedure Dr. Gittus sets on one side the steady dissipation and entropy production which is occurring as work done by the external forces is converted into heat. To some extent in materials science Dr. Gittus's procedure is used all the time; he merely poses the matter rather starkly. There is an elaborate theory of elastic forces on dislocations, cracks, inhomogeneities and other defects. This is used to help in the interpretation of situations where, increasingly as the temperature rises, there are a variety of dissipative processes, particularly those associated with the movement of atoms by diffusion (transport processes in general, recovery, recrystallisation, many aspects of chemical reaction and phase transformation). It is all, like the real world, a slightly unwholesome and vastly complicated mixture of mechanics, kinetics and the principles of equilibrium.

Dr. Gittus is in good company with his procedure but he is not quite on the side of the angels in speaking of thermodynamic equilibrium. Rayleigh's "principle of minimum dissipation" is mentioned by Onsager (1931) in his early attempts to justify Kelvin's treatment (1854) of the thermoelectric effects, and there has been and continues to be much discussion of the principles to be used in handling such problems[2]. It would be helpful to have the views of experts on whether irreversible thermodynamics can help in situations of this kind for they are of considerable interest at the present time. Dr. Gittus's k is a parameter controlling an internal structure in a specimen in which dissipation is occurring. Towards the end of his paper he also refers to the propagation of creep cracking and the phenomenon of superplasticity. The length of a creep crack is another example of a structural parameter which is to be determined in the presence of a process of dissipation. Other similar situations arise in the determination of crack paths; of the limit of stable crack growth in plastic-elastic or creeping solids; and of the behaviour of substances showing flow-softening and flow-hardening. It has been noted (Eshelby, 1975) that in the linear viscous case the C^* integral gives minus half the rate of change of the total dissipation rate with respect to a parameter characterizing the position of a defect, while Riedel and Rice (1979) have recently given a careful discussion of stress and strain fields near a crack tip under creep conditions and of the relevance of C^* and other parameters in characterizing them. As Eshelby

[2] For a recent survey see (Lavenda, 1978).

remarks however, the suggestion of Jeffery (1922) that the behavior of defects might be inferred by minimizing this dissipation rate (at constant applied strain rate or, equivalently, maximizing it at constant stress) does not seem to be justified by any respectable physical principle.

In an earlier paper (Gittus, 1977) the author refers to a suggestion of Takeuchi (1969) that the dislocation distribution in a creeping specimen might be determined by minimizing the applied stress at a given strain rate; this is equivalent to minimizing the dissipation rate. In such a purely mechanical approach, there are still no clear principles unless we first establish limit theorems based on classes of creep laws and the principles of mechanics[3]. For given creep laws there are a number of such limit theorems governing the work done by external forces for classes of fields satisfying given boundary conditions (see, for example, Leckie and Martin, 1967; Leckie and Ponter, 1970). It might be possible to make some progress along these lines once the general form of the transient and steady creep laws predicted by the structural theory are established. The steady state creep stress $\sigma(\dot{\varepsilon}, k)$ obtained from the author's theory (or from the modification suggested above) does have a minimum with respect to k; this is at a smaller cell size than that obtained when the self energy of the cell walls is included.

The paper by M. Labbens is salutary. It reminds us first of what we should be about; these are some of the important real problems which require solution and which make our studies not simply a quest for understanding but an urgent need to which society allocates very considerable resources (it is a constant irony that real advances in the solution of such problems often stem from the concern to understand rather than to meet the immediate demand). Secondly, his paper reminds us, if we need reminding, how complicated the real problems are compared with our simple models of them, even when we limit ourselves to a purely mechanical description.

M. Labbens reviews some important tests on pressure vessels and concludes that with tough ductile metals a considerable overpressure is required for failure, even when large defects are present. He believes that there is need for further investigation of more complex loadings, for example, those involving thermal stress. What has always to be considered, however, is whether this tough ductile behaviour will continue always to manifest itself during all conditions throughout the service life, or whether through some microstructural change or interaction with the environment, a dangerous reduction in toughness or ductility may result. Clearly the greatest possible collaboration between the materials scientist and the engineer is required in approaching these problems and it might be a useful thing for there to be

[3] I am indebted to Dr. I. C. Howard for the suggestion that such limit theorems might be relevant.

some discussion of whether we have the right balance of effort towards understanding the mechanics and the microstructure. More specifically one might ask whether sufficient effort on the theoretical side is being directed to models of crack propagation which take full account of its discontinuous nature.

In the work of de Leiris to which M. Labbens refers there is an interesting quotation from Portevin:

"Dans l'étude des causes d'un accident, l'examen de la cassure joue, à notre avis, un rôle essentiel".

A fracture surface shows many interesting features, though the heart of the theoretician may tremble at their complexity. It is just because of this complexity that the practical engineer seeks characterizing parameters like K and J which try to circumvent our areas of ignorance; or turns to simplified semi-empirical approaches based, for example, on the DBCS model. Nevertheless, phenomena in this area of complex microstructure may suddenly tip the balance of toughness and change a slow stable growth to a fast catastrophic one, as indicated by the lines showing the intervention of cleavage on the J integral-R curves in Fig. 11 of M. Labbens paper. I believe it would be valuable for us to discuss the present state of our understanding of this transition from stable growth to fast fracture. Can we succeed in treating it as a problem of mechanical stability using suitable characterizing parameters? And if we can, how can we monitor the situation so that we can ensure that the parameters we have selected maintain their values in service?

In discussing this transition some early work of Irwin (1950) and his collaborators in which similar fracture markings were produced in a wide variety of materials are of interest to those who look for unified explanations of nature. Irwin notes that fracture is essentially discontinuous, starting continually from small initiators ahead of the main crack front, whence fractures propagate forward and backward at different levels, continually tearing together to make a complicated pattern of dimples, clam shells, herringbones, chevrons, river lines and other exotic appearances. The shapes of many of these features (ellipses, parabolas, hyperbolas) can be interpreted by considering the intersection of crack fronts which move at different levels engulfing or escaping from one another.

Irwin in 1950 summarized the transition thus[4]:

"The transition from slow to moderately fast fracture in hardened steel is accompanied by finer and finer scaled elements initiating closer and closer to the leading edge of the main crack as the local velocity of propagation increases. Fewer and fewer main elements persist with individually smooth faces."

[4](Kies, Sullivan and Irwin, 1950).

It might be interesting to have the comments of today's experts on this description and their views on the outstanding problems in this field.

Finally I must apologize to Professor Argon and his co-workers for devoting relatively little time to their important paper of which only the abstract has been available to me. I believe that the detailed discussion of it which will doubtless ensue in this session and in others where related topics arise will only reinforce the main conclusion which I draw from all three of these papers. This is that in the study of fracture the engineering mechanics and the observation of microstructure must go hand in hand if we are to make the fastest progress; and that our enormous and growing computing power must blind us neither to the importance of the general laws we have already established about Nature on the one hand nor to the need for continued careful observation of her behavior on the other.

I am much indebted to my colleagues Professor J. D. Eshelby and Dr. I. C. Howard for valuable discussions.

References

Eshelby, J. D. (1975), *J. Elasticity*, **5**, 321.
Gittus, J. H. (1976), *Phil. Mag.*, **34**, 401.
Gittus, J. H. (1977), *Phil. Mag.*, **35**, 293.
Gittus, J. H. (1979), *Phil. Mag.*, **39**, 829.
Holt, D. L. (1970), *J. Appl. Phys.*, **41**, 3197.
Jeffery, G. B. (1922), *Proc. Roy. Soc.*, **A102**, 161.
Kelvin, (1854) *Proc. Roy. Soc. Edin.*, p. 123.
Kies, J. A., Sullivan, A. M., and Irwin, G. R. (1950), *J. Appl. Phys.*, **21**, 716.
Lavenda, B. H. (1978), *Thermodynamics of Irreversible Processes* (Macmillan, London).
Leckie, F. A. and Martin, J. B. (1967), *J. Appl. Mech.*, **34**, 411.
Leckie, F. A. and Ponter, A. R. S. (1970), *J. Appl. Mech.*, **37**, 426.
Onsager, L. (1931), *Phys. Rev.*, **37**, 405.
Riedel, H. and Rice, J. R. (1979), Report C00-3084/64, Division of Engineering, Brown University.
Takeuchi, S. (1969), *J. Phys. Soc. Jap.*, **27**, 929.

SESSION 2

Chairman: J. de Fouquet
                  E.N.S.M.A., Poitiers, France

Authors: A. P. Kfouri, K. J. Miller
                  J. Weertman
                  F. A. McClintock, J. L. Bassani

Discussion: T. Mura

S. Nemat-Nasser, Editor
THREE-DIMENSIONAL CONSTITUTIVE RELATIONS AND DUCTILE FRACTURE
North-Holland Publishing Company (1981) 83–109

CRACK SEPARATION ENERGY RATES FOR INCLINED CRACKS IN AN ELASTIC-PLASTIC MATERIAL

Alex P. KFOURI and Keith J. MILLER

University of Sheffield, Sheffield, U.K.

Plane strain elastic-plastic finite element analyses have been conducted on a tensile-loaded plate containing an inclined center crack having five different crack inclination angles varying from 0°, i.e. normal to the applied tensile stress, to 45°. In addition a pure shear Mode II load was applied to the plate with the crack at zero inclination. The yield stress for the material which obeyed the Von Mises yield criterion was 310 MN/m² with linear hardening at a tangent modulus of 4830 MN/m². Qualitative and quantitative information is presented on Mode I, Mode II and total crack tip opening displacements, on plastic strains and on maximum stresses near the crack tip. The analyses include calculations of J contour integrals and of crack separation energy rates G^Δ for coplanar crack extensions. Estimates of G^Δ for nonplanar crack growth in cases when the plastic flow is small are also made and a criterion based on maximum G^Δ is proposed.

1. Introduction

Structural failure is frequently caused by the birth and growth of a crack. In fatigue the birth is due to alternating shear stresses whilst the growth can occur by different modes, namely Modes I, II and III. In long life fatigue the direction of growth and the dominance of a particular mode can change with time. For example in uniaxial fatigue cracks grow initially by Stage I involving Mode II with an increasing participation of Mode I. Eventually the crack growth is by Stage II with Mode I dominating. In cyclic torsion loading however, all crack growth is by Stage I with Mode II operative along the surface but radial crack growth into the bulk of the material is by Mode III. Thus along the crack front there is a varying degree of mixed mode growth.

It has frequently been shown that Stage I cracks change direction and become Stage II. However, recently it has also been shown that Stage II cracks can change to Stage I. It is not yet clearly understood why cracks change direction but this must be linked to the state of strain in the crack tip region which is itself dependent on the bulk stress-strain state.

This paper considers mixed mode loading on a crack in an elastic-plastic material. The behavior of inclined cracks under mixed mode loading has aroused considerable interest in recent years. Investigations of inclined cracks in brittle materials under uniform environmental tensile stress date back at least to those by Erdogan and Sih (1963). Several fracture criteria have been proposed for the prediction of the direction of new nonplanar crack growth initiating from the tip of an existing flaw and for assessing the fracture resistance of a particular configuration. The analyses have been based on linear-elastic fracture mechanics (LEFM) and results from theoretical solutions have generally compared favorably with experimental data obtained from fracture tests on brittle materials, typically polymethylmethacrylate (PMMA). However, since yielding almost invariably takes place in the crack tip region of real engineering materials it may be appropriate to extend the concepts of elastic-plastic fracture mechanics (EPFM) to cover the case of combined Mode I and Mode II loadings.

2.   The problem

The situation is. illustrated in Fig. 1a showing a plate under plane strain conditions, loaded by applied stresses σ_P and σ_Q. The central crack of length $2a$ is inclined at an angle θ to the plane normal to σ_P. Using a coordinate system x, y with x parallel and y normal to the crack the situation is presented in Fig. 1b, where σ_N and σ_L are now applied stresses normal and parallel to the crack. In addition there is an applied shear stress τ with the positive sign as indicated in the figure. By simple tensor transformations the applied stresses in the two configurations are related as follows:

$$\sigma_N = \tfrac{1}{2}(\sigma_P + \sigma_Q) + \tfrac{1}{2}(\sigma_P - \sigma_Q)\cos 2\theta, \tag{1a}$$

$$\sigma_L = \tfrac{1}{2}(\sigma_P + \sigma_Q) - \tfrac{1}{2}(\sigma_P - \sigma_Q)\cos 2\theta, \tag{1b}$$

$$\tau = \tfrac{1}{2}(\sigma_P - \sigma_Q)\sin 2\theta. \tag{1c}$$

Figure 1a presents the most general two-dimensional biaxial state of a homogeneous environmental stress referred to axes coincident with the principal applied stresses σ_P and σ_Q. The uniaxial tensile loading case studied by Erdogan and Sih and by most subsequent workers is a practically important special case, suitable for a preliminary investigation. With $\sigma_Q = 0$, Eqs. (1) become:

$$\sigma_N = \sigma_P \cos^2\theta, \qquad \sigma_L = \sigma_P \sin^2\theta, \qquad \tau = \sigma_P \sin\theta \cos\theta. \tag{2}$$

The load on the crack tip region can be expressed in terms of the Mode I and Mode II stress intensity factors which, for the case of a crack in an

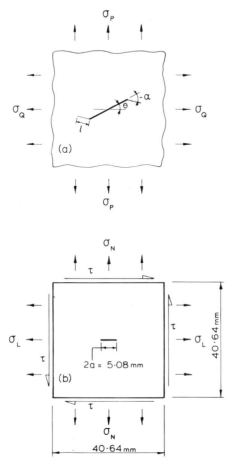

Fig. 1. An inclined crack (a) in a biaxial principal stress field and (b) a center-cracked plate (CCP) loaded by an equivalent stress system.

infinite medium, are given by,

$$K_I = \sigma_P \sqrt{\pi a} \cos^2 \theta, \qquad K_{II} = \sigma_P \sqrt{\pi a} \sin \theta \cos \theta, \qquad K_{III} = 0. \quad (3)$$

Hence the applied shear stress τ which is anti-symmetric relative to the crack plane contributes a Mode II component of the stress intensity factor. The lack of symmetry under mixed loading modes results in a nonplanar crack extension at an angle α from the original crack plane. There are several equivalent expressions in the literature for the stresses near the tip of a crack in a linear elastic material, e.g. Iron and Steel Institute (1968), Rice (1968a), Williams (1957), Swedlow (1976). Using polar coordinates, r, ω with the origin at the tip of the crack, the stresses are obtained by superimposing the contributions due to the Mode I and Mode II loads. Thus,

$$\sigma_r = K_I(2\pi r)^{-1/2}\left[(5/4)\cos(\omega/2)-(1/4)\cos(3\omega/2)\right]$$
$$- K_{II}(2\pi r)^{-1/2}\left[(5/4)\sin(\omega/2)-(3/4)\sin(3\omega/2)\right]$$
$$\sigma_P\cos 2\theta\cos^2\omega + \cdots, \tag{4a}$$

$$\sigma_\omega = K_I(2\pi r)^{-1/2}\left[(3/4)\cos(\omega/2)+(1/4)\cos(3\omega/2)\right]$$
$$- K_{II}(2\pi r)^{-1/2}\left[(3/4)\sin(\omega/2)+(3/4)\sin(3\omega/2)\right]$$
$$- \sigma_P\cos 2\theta\sin^2\omega + \cdots, \tag{4b}$$

$$\tau_{r\omega} = K_I(2\pi r)^{-1/2}\left[(1/4)\sin(\omega/2)+(1/4)\sin(3\omega/2)\right]$$
$$+ K_{II}(2\pi r)^{-1/2}\left[(1/4)\cos(\omega/2)+(3/4)\cos(3\omega/2)\right]$$
$$+ 0.5\sigma_P\cos 2\theta\sin 2\omega + \cdots. \tag{4c}$$

Each of Eqs. (4) includes on the extreme right-hand side (RHS) a nonsingular term which can be significant only in cases where the considered crack tip region is not infinitesimal. This term which depends on the geometry and loading is for the uniaxial case considered here.

For coplanar crack extension, the components G_I, G_{II} and G_{III} of the Griffith energy release rate G are given by,

$$G_I = K_I^2/E', \qquad G_{II} = K_{II}^2/E' \quad \text{where} \quad E' = E/(1-\nu^2), \tag{5}$$

and

$$G = G_I + G_{II} = J, \tag{6}$$

where J is the path independent integral calculated on a contour surrounding the crack tip (Rice, 1968b; Eshelby, 1951). The literature on the inclined crack problem is now considerable and for general reviews the reader is referred to Palaniswamy and Knauss (1974), Swedlow (1976), and Corlett (1977). In addition the branched or angled crack has received much attention; see Bilby, Cardew and Howard (1978) and the list of references therein. More recent additions include Chiang (1978), Howard (1978), Lo (1978). In Fig. 1a, the branch of length l is inclined at an angle α to the main crack. The load is in the form of applied stress intensity factors K_I and K_{II} on the tip of the main crack without the branch. Using coordinates parallel and normal to the branch, the stress intensity factors at the tip of the branch are k_1 and k_2 and the corresponding energy release rates are g_1 and g_2 for crack growth in the direction α. Bilby and Cardew (1975) have expressed k_1 and k_2 in terms of K_I, K_{II} and certain coefficients K_{ij} obtained by means of quadratures (Khrapkov, 1971), for values of α varying from $0°$ to $90°$. Thus,

$$k_1(\alpha) = K_{11}(\alpha)K_I + K_{12}(\alpha)K_{II}, \tag{7a}$$

$$k_2(\alpha) = K_{21}(\alpha)K_I + K_{22}(\alpha)K_{II}. \tag{7b}$$

The branch length l can be infinitesimal or finite. In the first case the nonsingular terms on the RHS of Eqs. (4) are irrelevant in the context of LEFM. In the second case they play a significant role. Bilby and Cardew used the parameter $\lambda = (2l/a)^{1/2}$ to include the influence of the constant terms in the stresses of the main crack near the tip.

3. Fracture criteria for the inclined crack

Brief mention will be made of some of the main fracture criteria which have been proposed for predicting the fracture angle α and the fracture resistance of the configuration.

(a) Perhaps the simplest is the maximum stress fracture criterion originated by Erdogan and Sih (1963), stating that the crack extension starts radially from the tip of the main crack in a direction α_s perpendicular to the direction of greatest tension, i.e. for $\omega = \alpha_s$, σ_ω in Eq. (4b) takes a maximum value $\sigma_{\omega M}$ which in turn is equal to a critical value σ_c where σ_c is a material property.

(b) A modification of the maximum stress criterion was suggested by Williams and Ewing (1972) who stipulated that the maximum stress $\sigma_{\omega M}$ be calculated at a small but *finite* characteristic distance from the tip of the crack. See also Finnie and Saith (1973), Ewing and Williams (1974a, b) and the discussion by Sih and Kipp (1974). In calculating σ_ω, the effects of the nonsingular terms on the extreme RHS of Eqs. (4) are taken into account.

(c) The minimum strain energy criterion proposed by Sih (1974) states that the initial crack growth takes place in the direction along which a strain-energy-density factor defined by Sih possesses a minimum value. Crack initiation occurs when this factor attains a critical value. A modification of this criterion by Sih and Kipp stipulates a small finite core region surrounding the crack tip used in the calculation of the strain energy density factor. This is analogous to Williams and Ewing's characteristic distance in criterion (b) above.

(d) A fracture criterion based on the branched crack analysis proposed by Bilby and Cardew (1975) states that the crack extension emanating from a tip of the main crack will occur at an angle α_0 such that $k_2(\alpha_0) = 0$ and $g_1(\alpha_0) = G_c$ (or $k_1(\alpha_0) = K_{IC}$) where G_c and K_{IC} correspond to the fracture toughness of the material.

(e) Finally the maximum energy release rate criterion of Palaniswamy and Knauss (1974) requires that the angle of fracture α_M will be such that $g(\alpha_M) = g_1(\alpha_M) + g_2(\alpha_M)$ is a maximum and crack extension takes place when $g(\alpha_M) = G_c$.

Again in criteria (d) and (e) the branch length ℓ must be specified and can be either infinitesimal or finite. In a discussion on Chiang's contribution, Tanaka and Ichikawa (1979) suggest that for an infinitesimal value of ℓ the distinction between $g(\alpha)$ calculated from the tip of the branch and $G(\alpha)$ for nonplanar crack extension emanating from the tip of the main crack in the direction α, becomes blurred.

The loads to fracture and the fracture angles calculated by the different criteria are not dramatically different from one another especially in the range $\theta = 0$ to $60°$ (Swedlow, 1976). Within the same range, the departure of the direction of the initial crack extension from the normal to the applied stress is measured by the sum $(\theta + \alpha)$ which varies approximately from $0°$ to $-15°$, with the largest negative values corresponding to the region $\theta = 20°$ to $30°$, e.g. Fig. 9 of the paper by Palaniswamy and Knauss (1974). The correlations of the theoretical predictions with experiments, have been generally encouraging. Fracture tests were carried out on brittle and quasi-brittle materials, namely PMMA (Williams and Ewing, 1972), glass and polyurethane (Palaniswamy and Knauss, 1974) plexiglass and a Zn aluminum alloy (Sih, 1974). In some cases the agreement was enhanced by using an optimal value in the finite distance from the tip criteria, compared with the criteria involving only an infinitesimal crack tip region. For instance Williams and Ewing found that a best fit between their theoretical and experimental results was obtained with a value of $(2r/a)^{1/2} \simeq 0.1$.

All the analyses mentioned so far were on an ideal linear elastic material in spite of the known occurrence of inelastic phenomena in the crack tip region such as the crazing in PMMA reported by Palaniswamy and Knauss. An elastic-plastic analysis providing qualitative and quantitative phenomenological information seems desirable at this stage and provides the motivation for the present exercise.

4. The analyses

Analyses on the plate shown in Fig. 1b were carried out using an elastic-plastic finite element program based on the initial stress approach. The runs were made on the University of Cambridge IBM 370/165 computer. The material properties were: Young's modulus $E = 207$ GN/m², Poisson's ratio $\nu = 0.3$, yield stress $\sigma_y = 310$ MN/m² and linear hardening was applied with a tangent modulus $H' = 4830$ MN/m². The material

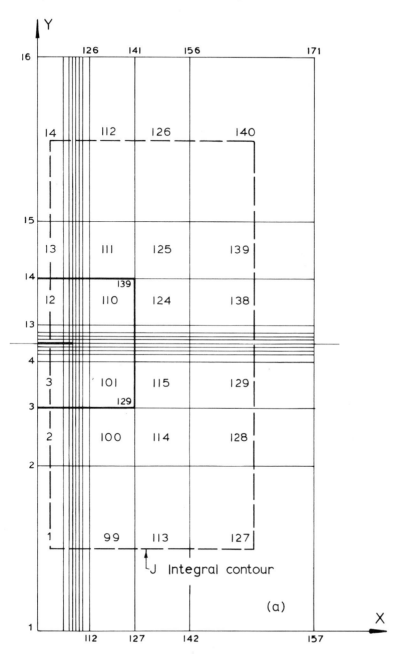

Fig. 2a.   Finite element idealization of one half of the plate; σ_P equals 71.46 MN/m².

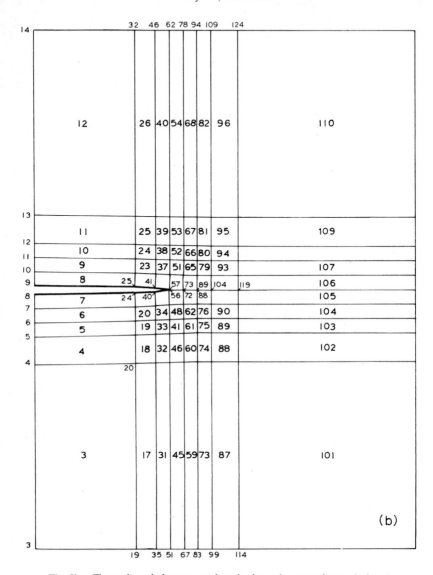

Fig. 2b.　The node and element numbers in the region near the crack tip.

obeyed the Von Mises yield criterion. Plane strain conditions were assumed. Earlier analyses on a center-cracked plate (CCP) of the same material and having the same dimensions as the present one but with the crack normal to σ_p have been described by Kfouri and Miller (1976). As only Mode I loading was involved it was then possible to take advantage of the symmetry of the configuration by confining the analysis to one quadrant of the plate only. This is not the case in the present situation since symmetry does not prevail. However, by using the anti-symmetric property of the configuration which looks the same when viewed upside down, it was possible to restrict the analyses to one half of the plate (Kfouri, 1979b). The finite element idealization of the right half is shown in Fig. 2a and the region near the crack tip is shown to a larger scale in Fig. 2b.

Five different crack inclinations were examined under uniaxial tensile load σ_P, namely $\theta = 0°$, $11.25°$, $22.5°$, $33.75°$ and $45°$. In addition a pure shear load was applied parallel to the crack. In each case the analyses started with an elastic solution which adjusted automatically the load to a value σ_{P0} (or τ_0 in the case of the pure shear load) equal to 95% of the load required to cause incipient yielding at the crack tip. This was followed by the iterative incremental-load elastic-plastic analyses using loading steps of $0.08 \, \sigma_{P0}$. At the end of each loading step information was obtained on the development of the crack tip plastic zones, the distribution of plastic strains within the zones, the nodal displacements, in particular the Mode I and the Mode II crack tip opening displacements (CTOD) and the stress at the centers of the elements. The J integral was calculated around the contour shown in Fig. 2a passing through unyielded elastic material at all loads. In addition, the two components G_I^{Δ} and G_{II}^{Δ} of the crack separation energy rates, G^{Δ} were evaluated using a modification of the crack tip release technique described by Kfouri and Miller (1974 and 1976) and Kfouri (1979a). Three successive growth steps each of size $\Delta a = 0.254$ mm were achieved by detaching gradually in six release steps the coupled pair of nodes in the crack plane (56, 57), (72, 73) and (88, 89) consecutively. The work absorbed during the release was calculated for each mode, namely ΔW_I for Mode I and ΔW_{II} for Mode II, giving $G_I^{\Delta} = \Delta W_I / \Delta a$ and $G_{II}^{\Delta} = \Delta W_{II} / \Delta a$. The crack extensions were made at constant load at one of the values of σ_P given by σ_{P0}, $1.56\sigma_{P0}$, $2.12\sigma_{P0}$, $2.68\sigma_{P0}$ and $3.24\sigma_{P0}$. Note however, that σ_{P0} varies with the crack inclination angle θ.

5. The results

The results of the analyses are summarized in Tables 1 and 2 and are presented pictorially in Figs. 3–8.

In Table 1, the first five columns indicate the crack inclination and the applied load. Columns 6 to 9 give the horizontal displacement u_L and the vertical displacement v_L of node 40 on the lower crack surface next to the crack tip, followed by the horizontal displacement u_U and the vertical displacement v_U of node 41 on the upper crack surface. Nodes 40 and 41 coincide initially, before loading. Columns 10 and 11 give the average plastic equivalent strains $\bar{\varepsilon}_{AV}$ in the lower and upper elements 35 and 36

Table 1
Results of elastic-plastic finite element analyses on a plate with a central inclined crack

θ°	σ_P	σ_N	σ_L	τ	u_L	v_L	u_U	v_U	ε_AV EL35	ε_AV EL36	ε_TIP ND56	ε_TIP ND57	σ_1M	δ_I/Δα	δ_II/Δα	δ/Δα	u_L	v_L	u_U	v_U	ε_AV EL63	ε_AV EL64	ε_TIP ND88	ε_TIP ND89	σ_1M	δ_I/Δα	δ_II/Δα	δ/Δα
	N/mm²				(NDS 40 & 41) ×10³ mm				(×10⁴)				N/mm²	(×10³)			(NDS 72 & 73) ×10³ mm				(×10⁴)				N/mm²	(×10³)		
									BEFORE NODE RELEASES								AFTER TWO NODE RELEASES											
0	71·46	71·46	0	0	-0·65	-0·63	-0·65	-0·63	0	0	0	⊃	224·3	5·0	0	5·0	-1·91	-1·54	-1·91	-1·54	29·2	29·2	49·2	49·2	513·4	12·1	0	12·1
	191·5	191·5	0	0	-1·34	-2·47	-1·34	-2·47	28·6	28·6	95·3	95·3	463·6	19·4	0	19·4	-0·80	-0·53	-0·59	0·69	0	0	0	0	230·2	4·8	0·8	4·9
11·25	66·72	64·18	2·54	12·77	-0·65	-0·51	-0·46	0·62	0	0	0	0	214·9	4·5	0·8	4·5	-1·17	-0·84	-0·87	1·05	5·8	5·8	22·9	13·9	346·5	7·4	1·2	7·5
	104·1	100·1	3·96	19·91	-0·98	-0·89	-0·64	1·04	6·0	4·2	23·9	16·7	330·2	7·6	1·4	7·7	-1·53	-1·13	-1·14	1·32	14·7	17·8	39·3	31·2	451·0	9·6	1·5	9·7
	141·4	136·0	5·38	27·05	-1·38	-1·38	-0·72	1·57	13·5	13·5	54·0	52·9	413·6	11·6	2·3	11·8	-1·91	-1·44	-1·35	1·51	24·2	29·2	49·8	42·1	453·6	11·3	1·9	11·5
	178·8	172·0	6·81	34·21	-1·64	-1·99	-0·72	2·23	26·5	22·9	84·4	90·2	453·6	16·6	3·6	17·0	-2·43	-1·52	-1·83	1·79	41·3	34·1	61·4	51·3	539·7	13·0	2·4	13·2
	216·2	207·9	8·23	41·36	-1·96	-2·80	-0·64	3·06	36·3	43·3	119·7	129·3	489·1	23·1	5·2	23·7												
22·5	63·33	54·06	9·27	22·39	-0·56	-0·37	-0·23	0·58	0	0	0	0	225·6	3·8	1·3	4·0	-0·68	-0·38	-0·32	0·65	0	0	0	0	259·9	4·1	1·4	4·3
	98·80	84·33	14·47	34·93	-0·87	-0·68	-0·30	0·94	6·1	2·0	24·5	8·1	307·2	6·4	2·3	6·8	-0·97	-0·61	-0·45	1·00	6·0	3·6	22·8	7·8	335·2	6·3	2·0	6·6
	134·3	114·6	19·66	47·47	-1·21	-1·05	-0·25	1·38	13·0	11·1	51·5	44·4	384·1	9·6	3·8	10·3	-1·31	-0·83	-0·65	1·29	13·1	15·6	37·8	23·6	418·5	8·3	2·6	8·7
	169·7	144·9	24·86	60·01	-1·57	-1·53	-0·11	1·89	21·6	22·9	78·6	84·0	432·2	13·5	5·7	14·6	-1·64	-1·03	-0·87	1·47	22·4	27·8	46·8	35·4	465·7	9·8	3·0	10·2
	205·2	175·2	30·05	72·55	-1·95	-2·13	-0·12	2·51	32·3	38·6	110·1	127·5	479·8	18·3	8·1	20·0	-2·10	-1·21	-1·12	1·72	32·2	43·3	58·3	50·8	518·6	11·5	3·9	12·1
33·75	62·63	43·30	19·33	28·93	-0·42	-0·25	-0·02	0·52	0·6	0	0	0	234·3	3·0	1·7	3·5	-0·49	-0·24	-0·02	0·58	0·1	0	0·2	0	25·7	3·2	1·9	3·7
	97·70	67·54	30·16	45·13	-0·65	-0·47	0·08	0·83	6·3	2·2	25·0	2·2	331·3	5·1	2·9	5·9	-0·66	-0·39	0·02	0·92	6·3	12·7	22·8	4·0	35·8	5·2	2·7	5·9
	132·8	91·79	40·98	61·33	-0·94	-0·76	0·25	1·18	13·4	8·7	50·3	34·8	398·8	7·6	4·7	8·3	-0·89	-0·54	-0·05	1·19	13·1	25·3	35·5	19·4	420·7	6·8	3·3	7·6
	167·8	116·0	51·81	77·53	-1·25	-1·12	0·49	1·56	21·9	19·2	76·9	75·4	475·4	10·5	6·9	11·2	-1·08	-0·66	-0·04	1·44	23·0	25·3	44·5	31·8	454·0	8·3	3·7	9·1
	202·9	140·3	62·63	97·73	-1·56	-1·58	0·81	1·98	33·3	31·5	111·2	116·8	514·5	14·0	9·3	16·8	-1·40	-0·84	-0·24	1·67	33·8	42·6	56·3	52·5	488·1	9·9	4·6	10·9
45	65·82	32·91	32·91	32·91	-0·21	-0·14	0·28	0·44	0	0	0	0	242·4	2·3	2·0	3·0	-0·23	-0·12	0·30	0·51	0·1	0	0·2	0	259·9	2·5	2·1	3·3
	102·7	51·35	51·35	51·35	-0·35	-0·30	0·48	0·71	6·3	6·3	25·4	0	346·3	4·0	3·3	5·1	-0·25	-0·20	-0·53	0·81	6·6	6·6	22·6	0·4	361·1	4·0	3·1	5·1
	139·6	69·78	69·78	69·78	-0·54	-0·50	0·76	0·98	14·2	6·3	50·3	25·3	423·0	5·8	5·1	7·8	-0·31	-0·29	0·65	1·08	14·4	8·8	34·6	15·0	427·1	5·4	3·8	6·6
	176·4	88·21	88·21	88·21	-0·70	-0·80	1·14	1·26	25·1	14·4	79·7	57·5	480·0	8·1	7·3	10·9	-0·27	-0·33	0·75	1·35	24·2	20·0	42·8	29·9	460·5	6·6	4·0	7·7
	213·3	106·6	106·6	106·6	-0·84	-1·25	1·61	1·59	40·7	23·2	116·5	91·2	504·5	11·2	9·6	14·8	-0·29	-0·41	0·94	1·60	38·0	34·4	60·9	50·8	480·6	7·9	4·8	9·2
PURE SHEAR	0	0	0	43·90	-0·33	0·20	0·33	0·20	6·2	6·2	24·5	24·7	218·7	0	2·6	2·6	-0·35	0·25	0·35	0·26	0·0	0·0	0·1	0·1	255·9	0·1	2·6	2·6
	0	0	0	68·49	-0·60	0·60	0·60	0·60	16·2	16·2	64·3	64·6	305·1	0	4·7	4·7	-0·53	0·53	0·53	0·40	5·5	5·5	21·3	21·4	306·4	0	4·2	4·2
	0	0	0	93·07	-0·96	0·29	0·96	0·29	16·1	16·1	64·3	64·6	351·9	0	7·6	7·6	-0·59	0·56	0·59	0·56	17·7	17·8	36·4	36·7	313·9	0	4·6	4·6
	0	0	0	117·7	-1·35	0·28	1·34	0·27	28·1	28·3	108·7	109·3	379·2	0	10·6	10·6	-0·73	0·70	0·70	0·74	36·9	36·9	63·0	63·7	322·4	0	5·8	5·8
	0	0	0	142·2	-2·03	0·13	2·00	0·12	49·6	50·1	186·8	188·7	402·7	0	15·9	15·9	-0·85	0·84	0·86	0·84	62·8	63·0	86·6	88·6	346·7	0	6·7	6·7

Table 2
Values of J, G_I^Δ and G_{II}^Δ and $G^\Delta(\alpha_M)$ for the extension of variously inclined cracks

$\theta°$	$(\sigma_p/\sigma_o)^2$ $\sigma_o = 7522\,\mathrm{N/mm}^2$	J (N/mm) BEFORE NODE RELEASE'S	J_{AV} MEAN OF 1st & 2nd NODE RELEASE'S	G_I^Δ FIRST NODE RELEASE	G_{II}^Δ FIRST NODE RELEASE	G_I^Δ (N/mm) SECOND NODE RELEASE	G_{II}^Δ SECOND NODE RELEASE	G_I^Δ THIRD NODE RELEASE	G_{II}^Δ THIRD NODE RELEASE	$\dfrac{G^\Delta(o)}{G_o}$	$\dfrac{G}{G_o}$	$\dfrac{J}{G_o}$	$\alpha_M°$	$\dfrac{G^\Delta(\alpha_M)}{G_o}$
0	0·903	0·159	—	0·602	0	0·659	0	—	—	0·903	0·903	0·903	0	0·903
	6·482	1·247	1·464							3·002	6·482	6·967	0	3·003
11·25	0·887	0·133	0·159	0·150	0·006	0·160	0·006	0·164	0·006	0·757	0·757	0·757	-20	0·810
	1·915	0·329	0·391	0·300	0·011	0·333	0·011	0·334	0·010	1·568	1·842	1·865	-20	1·660
	3·536	0·624	0·736	0·414	0·022	0·482	0·015	0·494	0·012	2·270	3·401	3·510	-20	2·404
	5·650	1·033	1·214	0·509	0·037	0·585	0·020	0·610	0·015	2·758	5·435	5·789	-20	2·932
	8·258	1·597	1·904	0·613	0·053	0·684	0·027	0·736	0·023	3·243	7·944	9·077	-20	3·480
22·5	0·709	0·106	0·126	0·107	0·018	0·114	0·019	0·116	0·019	0·606	0·605	0·606	-35	0·753
	1·725	0·262	0·312	0·220	0·034	0·240	0·033	0·241	0·032	1·249	1·473	1·495	-35	1·516
	3·186	0·496	0·588	0·308	0·062	0·368	0·046	0·382	0·040	1·892	2·720	2·819	-35	2·262
	5·092	0·822	0·971	0·378	0·096	0·458	0·055	0·492	0·046	2·348	4·346	4·551	-30	2·796
	7·442	1·256	1·510	0·452	0·137	0·546	0·071	0·586	0·065	2·823	6·352	7·234	-35	3·396
33·75	0·693	0·084	0·099	0·068	0·029	0·073	0·031	0·074	0·032	0·479	0·479	0·479	-45	0·696
	1·687	0·206	0·246	0·148	0·058	0·156	0·058	0·157	0·057	0·984	1·166	1·187	-45	1·398
	3·116	0·392	0·447	0·206	0·098	0·245	0·077	0·256	0·072	1·477	2·154	2·254	-45	2·036
	4·979	0·648	0·773	0·266	0·141	0·321	0·083	0·345	0·076	1·855	3·442	3·732	-40	2·475
	7·277	0·989	1·201	0·334	0·180	0·400	0·101	0·425	0·098	2·296	5·031	5·798	-40	3·052
45	0·766	0·066	0·079	0·040	0·038	0·042	0·040	0·043	0·042	0·383	0·383	0·383	-55	0·627
	1·864	0·164	0·195	0·086	0·078	0·091	0·078	0·093	0·077	0·782	0·932	0·952	-55	1·260
	3·442	0·312	0·373	0·129	0·122	0·148	0·100	0·154	0·097	1·149	1·721	1·818	-50	1·794
	5·500	0·519	0·621	0·183	0·158	0·212	0·105	0·231	0·104	1·464	2·750	3·027	-50	2·180
	8·039	0·798	0·971	0·227	0·187	0·268	0·124	0·288	0·133	1·805	4·020	4·732	-50	2·663
PURE SHEAR $(\sigma_o/\sigma_o)^2$	0·341	0·058	0·068	0	0·068	0	0·072	0	0·075	0·341	0·341	0·341	-75	0·500
	0·829	0·147	0·176	0	0·134	0	0·127	0	0·125	0·603	0·829	0·877	-75	0·885
	1·531	0·292	0·353	0	0·186	0	0·140	0	0·147	0·664	1·531	1·758	-75	0·975
	2·447	0·510	0·645	0	0·266	0	0·187	0	0·202	0·886	2·447	3·217	-75	1·301
	3·576	0·910	1·139	0	0·361	0	0·222	0	0·265	1·051	3·576	5·678	-75	1·542

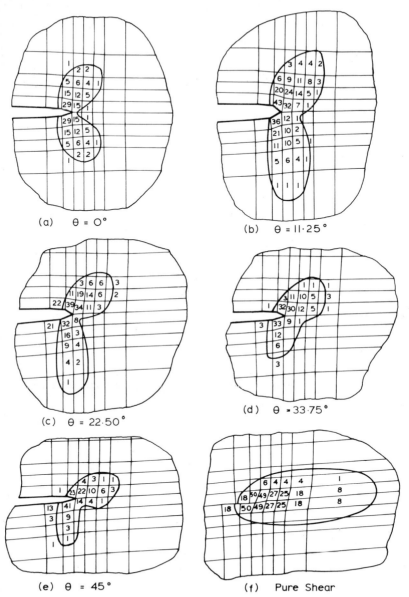

Fig. 3. Plastic zones for different crack inclinations; uniaxial loading (a to e), pure shear loading (f) and after the release of crack tip nodes (g and h).

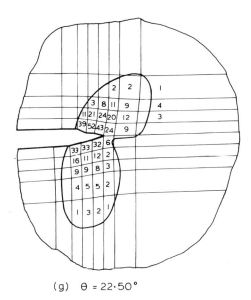

(g)   θ = 22·50°

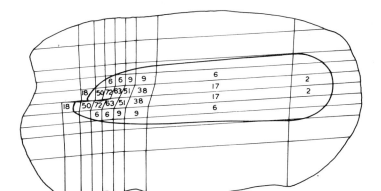

(h)   Pure Shear

Fig. 3.   Continued.

containing the crack tip respectively, while columns 12 and 13 give the
calculated plastic equivalent strains $\bar{\varepsilon}_{TIP}$ at the coincident crack tip nodes 56
and 57, respectively. Column 14 gives the largest attained value σ_{1M} of the
main principal stresses calculated at the centers of the elements. Columns 15
and 16 give the components of the CTOD for each mode, $\delta_I = (v_U - v_L)/\Delta a$
and $\delta_{II} = (u_U - u_L)/\Delta a$. Column 17 gives the total CTOD, i.e. the magni-
tude of the vector sum of δ_I and δ_{II}, $\delta = (\delta_I{}^2 + \delta_{II}{}^2)^{1/2}$. Columns 18 to 29

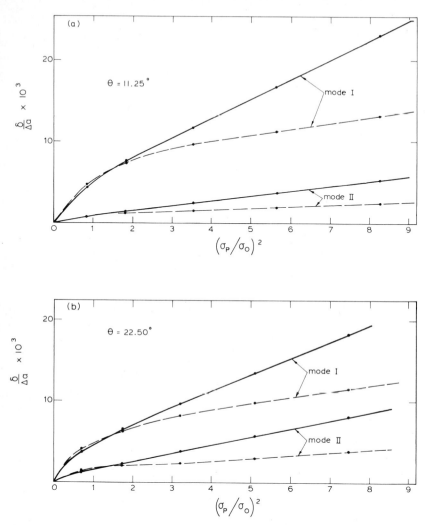

Fig. 4. The effect of applied tensile load on vertical and horizontal distances separating crack tip nodes before (——) and after (---) crack extension.

give similar information to that contained in columns 6 to 17 but the latter columns apply to the situation after two crack growth steps. Hence u_L and v_L in columns 18 and 19 apply to node 72, u_U and v_U in columns 20 and 21 apply to node 73. Columns 22 and 23 giving $\bar{\varepsilon}_{AV}$ apply to elements 63 and 64 respectively and columns 24 and 25 giving $\bar{\varepsilon}_{TIP}$ apply to the coincident crack tip nodes 88 and 89 respectively. The value of σ_P at incipient yielding when $\theta = 0$ is $\sigma_0 = 75.22$ MN/m². At this value of the load the energy release

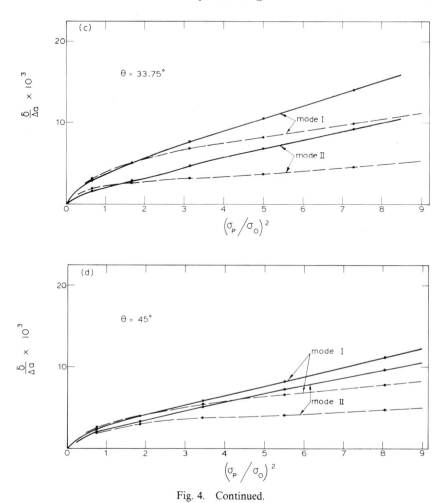

Fig. 4.   Continued.

rate G is equal to G_0 given by Eq. (5a) when K_{I} takes the value $K_{\mathrm{I}0} = \sigma_0 K_{\mathrm{I}}^* a^{1/2}$. The nondimensional stress intensity factor K_{I}^* would be equal to $\sqrt{\pi} = 1.772$ in the case of an infinite plate. Using the previously determined value of 1.785 for K_{I}^* and taking $a = 2.921$ mm as the average half-crack length during the second crack tip node release, a value of G_0 equal to 0.2315 N/mm is obtained. In Table 2, the second column gives the applied tensile load in the nondimensional form $(\sigma_{\mathrm{P}}/\sigma_0)^2$ or $(\tau/\sigma_0)^2$ in the case of the pure shear mode. Column 3 gives the value of J calculated over the contour shown in Fig. 2a before crack tip releases and column 4 gives the average of the values of J over the same contour after one and two crack tip

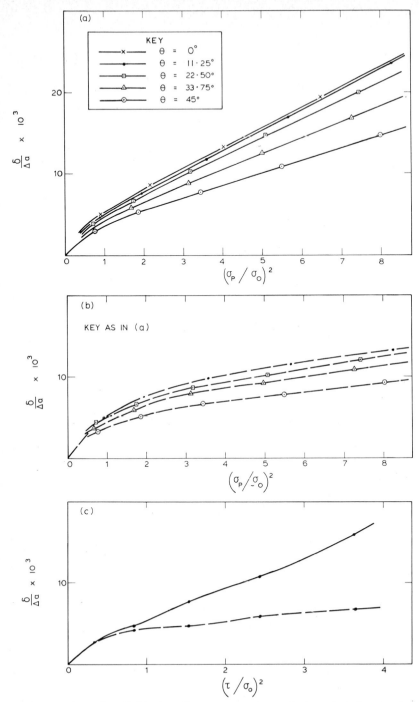

Fig. 5. The effect of applied load on distances separating crack tip nodes before (——) and after (- - -) crack extension.

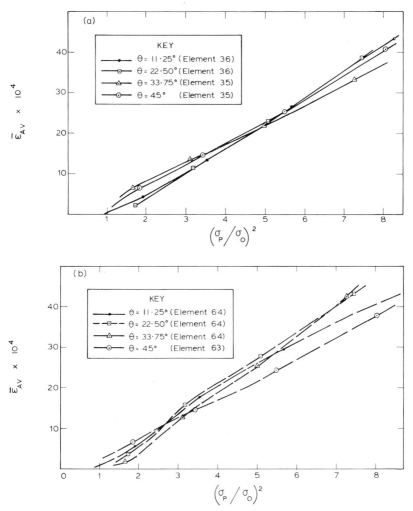

Fig. 6. The effect of applied load on average equivalent plastic strains in crack tip element before (——) and after (---) crack extension.

releases, i.e. corresponding to $a = 2.921$ mm. Columns 5 to 10 give consecutive values of G_I^Δ and G_{II}^Δ calculated during the first, second and third crack tip node releases.

The elastic states with $\theta = 0$ can be looked upon as reference states. Under tensile loading, theory predicts that $G_I = G_I^\Delta = J$ while under pure shear load $G_{II} = G_{II}^\Delta = J$. In the actual analyses the three quantities are only approximately equal because of numerical discrepancies inherent in finite element methods.

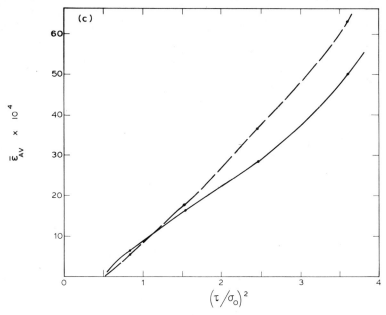

Fig. 6. Continued.

Correction factors based on the elastic solutions were applied, namely 1.054 for G_I^Δ and 1.097 for G_{II}^Δ. The correction factor A, applicable to J in the case of mixed loading modes, was taken as the weighted average of the values, 1.102 and 1.154 for the two separate modes, respectively, Thus,

$$A = 1.102 \cos^2 \theta + 1.154 \sin^2 \theta.$$

Columns 11 and 13 give the corrected values of G^Δ and J, normalized with respect to G_0, respectively. Column 12 gives the normalized theoretical values of the energy release rates $G/G_0 = (G_I + G_{II})/G_0$ for coplanar extension of the main crack, obtained by means of Eqs. (5) using $K_I^* = 1.785$ and $a = 2.921$ mm.

Figures 3a–3e show the crack tip plastic zones for the crack under uniaxial tension at the different inclinations $\theta = 0$, 11.25°, 22.5°, 33.75° and 45° while Fig. 3f shows the crack tip plastic zone for the crack when a pure shear stress τ is applied. The deformations are shown with an exaggerated factor of 50. The load corresponding to each plastic zone is the last (largest) value given in Table 1 for each inclination angle, and for the pure shear mode in the case of Fig. 3f. The numbers in the elements are the average plastic equivalent strains in the elements times 10^4. Figures 3g and 3h show the crack tip plastic zones after two coplanar crack growth steps. Figure 3g corresponds to the same inclination, $\theta = 22.5°$ and applied load as in Fig. 3c,

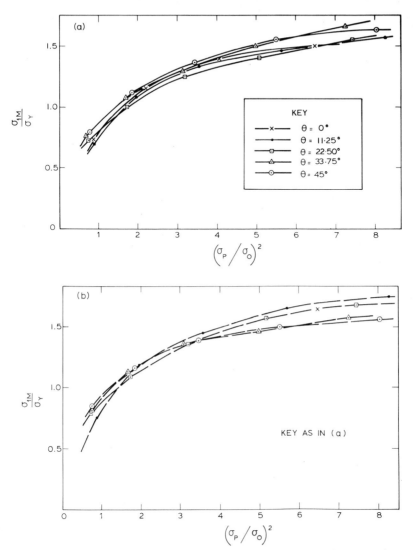

Fig. 7. Largest principal stress at the center of the crack tip elements against applied tensile load at different crack inclination angles (a) before and (b) after crack extension.

while Fig. 3h corresponds to the pure shear mode at the same applied stress as in Fig. 3f. The plastic regions shown in Figs. 3g and 3h are not active everywhere as they include a wake region in which elastic unloading has occurred.

In Figs. 4 to 8 the abscissae represent the normalized load $(\sigma_P/\sigma_0)^2$ with the exception of cases where the load is a shear, given by $(\tau/\sigma_0)^2$. Figures

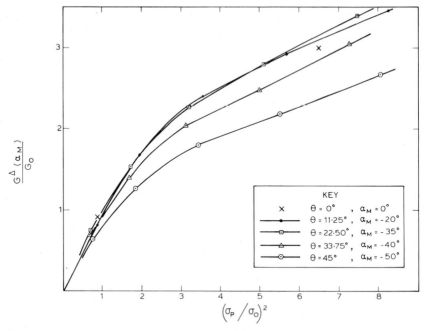

Fig. 8. Estimates of maximum crack separation energy rates for nonplanar crack extension against applied tensile load for different crack inclinations.

4a–4d give the Mode I and Mode II components of the CTOD, $(\delta_I/\Delta a)$ and (δ_{II}/Δ_a) both times 10^3, respectively, for four different crack inclinations. The solid lines refer to the situation before the release of crack tip nodes and the broken lines after two crack growth steps. Values of the CTOD $(\delta/\Delta a)$ times 10^3 are shown in Fig. 5a before crack extension and in Fig. 5b after two crack growth steps. In Fig. 5c, the solid line gives $(\delta_{II}/\Delta a)$ times 10^3 before crack extension and the broken line gives the same quantity after two crack growth steps. Figure 6a gives the average plastic strains times 10^4 in elements 35 or 36 containing the crack tip for the different crack inclinations before crack extension and Fig. 6b in elements 63 or 64 after two crack growth steps. Figure 6c gives the average plastic strains times 10^4 for the pure shear mode, the solid line referring to the state before crack extension and the broken line after two crack growth steps. Figures 7a and 7b give the maximum attained main principal stresses at the centers of the elements for the different inclinations under tensile loading, before crack extension and after two crack growth steps, respectively. Columns 14 and 15 of Table 2 and Fig. 8 refer to nonplanar crack extensions and will be explained later.

6. Discussion

In linear elastic materials the load on the crack tip region of inclined and branched cracks has been characterized by the components of the stress intensity factors K_I and K_{II} acting on the main crack (Bilby and Cardew, 1975). In the case of cracks with branches of finite size the characterization is incomplete without taking into account the additional effect of the nonsingular stress term.

$$\sigma_T = (\sigma_L - \sigma_N) = (\sigma_Q - \sigma_P)\cos 2\theta.$$

This is also the case when the material is elastic-plastic since fracture processes are thought to occur in a crack tip zone of finite dimensions and crack tip plasticity parameters such as the plastic zone size are known to be influenced by the biaxiality of the stress state (Miller and Kfouri, 1974; Kfouri and Miller, 1976 and 1978; Eftis *et al.*, 1977). For mixed loading modes, the usefulness of the characterizing parameters in fracture assessments depends on the adoption of a realistic fracture criterion predicting unequivocally the fracture angle α. Used on its own, J calculated on a contour surrounding the main crack gives insufficient characterization since, as Eq. (6) shows, J is equal to the energy release rate G, i.e. the *sum* of G_I and G_{II}, for coplanar crack extension only. Compare in passing the values of G^Δ/G_0, G/G_0 and J/G_0 in columns 11, 12 and 13 of Table 2. For each crack inclination, at the lowest applied load corresponding to the elastic response, there is almost complete agreement of the three values in all cases. In the single-mode reference states with $\theta=0$, used in establishing the numerical correction factors this is necessarily the case. However the values are the same also at all the other inclinations.

Normalizing G^Δ, J and G with respect to G_0 much reduces the dependence on the finite element mesh used.

Note that at incipient yielding the normal stress at a distance $\Delta a/2$ ahead of the crack tip is of the order of σ_y, i.e.

$$K_{I0}(\pi\Delta a)^{-1/2} = \rho\sigma_y, \tag{8a}$$

where ρ is a number of order unity. Substituting K_{I0} in Eq. (8a) for K_I in Eq. (5a) and replacing E and ν by their values,

$$G_0 = 1.381 \times 10^{-5}\rho^2\sigma_y{}^2\Delta a. \tag{8b}$$

For the existing finite element idealization, substitution of G_0, σ_y and a in (8b) by their actual values results in $\rho \simeq 0.829$. Corresponding values of ρ for meshes with $\Delta a = 0.127$ mm and $\Delta a = 0.0635$ mm are 0.834 and 0.839, respectively (Kfouri and Miller, 1976). When only Mode I loading is applied to the crack with θ and σ_Q equal to zero the responses to the applied loads, as reflected by the finite element analyses, are purely elastic when $G \leqslant G_0$, and elastic-plastic when $G > G_0$.

Figures 3b to 3e show that when the crack is inclined the upper and lower plastic regions emanating from the crack tip have somewhat different shapes and the directions in which they extend outwards from the crack tip are affected by the inclination of the applied stress σ_P relative to the crack plane. For instance when $\theta = 45°$, Fig. 3e shows that the part of the plastic zone above the crack spreads in a direction approximately parallel to the crack while the region below the crack extends in a downward direction, normal to the crack. Under pure shear loading the crack extends parallel to the crack as shown in Figs. 3f and 3h. Certain features of the CTOD and of the plastic strains in the crack tip region which have already been reported in the analyses involving Mode I uniaxial and biaxial loading (Miller and Kfouri, 1977) are repeated in the case of the inclined crack. Figures 4 and 5 show that after an initial transition stage the CTOD increases linearly with $(\sigma_P / \sigma_0)^2$. The crack profile after crack extension (Fig. 3g) shows the effect of the permanent distortion caused by crack tip blunting processes during the initiation stage before crack extension (Rice and Johnson, 1970). After crack extension the CTOD is smaller and the crack profile near the crack tip is sharper than before crack extension has taken place (Rice, 1975). Figures 5a and 5b reflect these trends and show in addition that under constant uniaxial tension the CTOD decreases with increasing values of the inclination angle θ. For the pure shear loading mode Fig. 5c again indicates smaller values of the CTOD after crack extension than the corresponding values before crack extension and this is also true of the plastic equivalent strain $\bar{\varepsilon}_{TIP}$ at the crack tip. Here it must be emphasized that values of $\bar{\varepsilon}_{TIP}$ which emerge from the calculations are only of qualitative interest enabling comparison to be made in different situations but the precise conditions at the crack tip are not known. Table 1 and Fig. 6 show that the values of $\bar{\varepsilon}_{AV}$ and $\bar{\varepsilon}_{TIP}$ induced by pure Mode II loading are much greater than the corresponding values for pure Mode I loading, given equal values of τ and σ_P. Alternatively, for equal values of the pure Mode I and Mode II strains, σ_P must be equal to 1.85 τ approximately. As the crack inclination increases σ_N decreases but τ and σ_L increase. The increase in the biaxiality parameter σ_T results in higher stresses in the crack tip region but also in reduced plastic flow and smaller crack tip plastic zone sizes. A consequence of these conflicting effects is that the value of θ corresponding to the largest plastic strain in the crack tip region is not clearly defined in Figs. 6a and 6b. Figures 7a and 7b show that the maximum attainable stress occurs at positive values of the inclination angle θ.

7. Nonplanar crack extension

As in the case of LEFM, the application of EPFM to the angled crack problem requires the adoption of a suitable fracture criterion. It is now proposed to extend the maximum energy release rate criterion to the requirements of EPFM. This is done by replacing the energy release rate G concept in criterion (e) above by the crack separation energy rate G^Δ. Since G^Δ needs to be calculated for nonplanar crack extension, the dependence on the direction α is indicated by the notation $G^\Delta(\alpha)$. The quantities in columns 5 to 12 of Table 2 now correspond to $G_I^\Delta(0)$, $G_{II}^\Delta(0)$ or $G^\Delta(0)$ as the case may be, according to the new notation. The modified criterion will be referred to as the maximum crack separation energy rate criterion for crack propagation. It will be recalled that the quantity G^Δ is based on the stipulation of a finite growth step Δa (Kfouri and Miller, 1976; Kfouri and Rice, 1978) but that Δa is generally considered small compared to a. When the material is linear elastic G^Δ is equal to G and the maximum crack separation energy rate criterion is equivalent to the maximum energy release rate criterion.

The evaluation of $G^\Delta(\alpha)$ for different values of α and θ by finite elements methods presents a considerable effort probably requiring different finite element idealizations to cover each case. However, it may be possible to obtain estimates of $G^\Delta(\alpha)$ for nonzero values of α if $G_I^\Delta(0)$ and $G_{II}^\Delta(0)$ are known, by extrapolating our existing knowledge based on elastic conditions into states involving moderate crack tip plasticity. Thus, noting that in the case of a linear elastic material $K=\{EG^\Delta/(1-\nu^2)\}^{1/2}$, Eqs. (7a) and (7b) can be rewritten,

$$\left[G_I^\Delta(\alpha)\right]^{1/2}=K_{11}(\alpha)\left[G_I^\Delta(0)\right]^{1/2}+K_{12}(\alpha)\left[G_{II}^\Delta(0)\right]^{1/2}, \qquad (9a)$$

$$\left[G_{II}^\Delta(\alpha)\right]^{1/2}=K_{21}(\alpha)\left[G_I^\Delta(0)\right]^{1/2}+K_{22}(\alpha)\left[G_{II}^\Delta(0)\right]^{1/2}. \qquad (9b)$$

The coefficients $K_{ij}(\alpha)$ have been calculated for a linear elastic material and Eqs. (9) therefore give correct answers in the state of incipient yielding. Now it can be argued that for moderate plastic flow the form of Eqs. (9) will remain virtually unchanged but the coefficients $K_{ij}(\alpha)$ will vary continuously as the plastic flow occurs. Therefore using Eqs. (9) with the initial coefficients K_{ij} will provide a first approximation for cases involving only a small amount of plastic flow.

A short computer program based on Eqs. (9) was written to calculate $G_I^\Delta(\alpha)/G_0$ and $G_{II}^\Delta(\alpha)/G_0$ for the values of α varying from 0 to 90° at 5° intervals. This was done for all the inclination angles, using the corrected values of $G_I^\Delta(0)$ and $G_{II}^\Delta(0)$ from columns 7 and 8 of Table 2, respectively. The coefficients $K_{ij}(\alpha)$ were provided by Howard (1978). The results are shown in Fig. 8 which gives the maximum value of the crack separation energy rate $G^{\Delta}(\alpha_M)$ for the different inclination angles θ. The approximate values of α_M are also shown inset under the key. As in the case of the elastic analyses negative values of α_M are obtained, with values of $(\theta + \alpha_M)$ varying from $-12.5°$ to $-5°$. A little surprisingly the most critical crack inclination is $\theta = 22.5°$ and not $\theta = 0°$ as intuition might have suggested. Note however, that the linear elastic analyses also show this. For instance, using the criterion (d) or (e) the situation presented by an inclination of $\theta = 22.5°$ with α_0 or α_M approximately equal to $-37.5°$ is more dangerous than the case $\theta = \alpha = 0$. The predictions of the fracture angle α_M for different crack inclinations θ using the EPFM maximum G^{Δ} criterion for moderate plastic flow, are not dramatically different from the LEFM predictions based on one of the criteria (a), (b), (d) or (e). This fact perhaps helps to explain the success of the predictions obtained from linear-elastic analyses in spite of the known occurrence of irreversible processes in the crack tip regions of the quasi-brittle materials used in the experiments.

8. Implications for fatigue crack propagation

The types of fracture discussed so far occurred in materials subjected to monotonically increasing loads. In ductile materials the large crack tip plastic strains associated with Mode II loading compared with Mode I loading suggest that crack tip phenomena sensitive to the intensity of crack tip plasticity would be increased by the inclusion of a large Mode II component to the cyclic loading. The ratio σ_P/τ of approximately 1.85 required to produce equal equivalent plastic strains in the crack tip region under pure Mode I and pure Mode II loading, respectively, may have some bearing on the observed transition from Stage I crack propagation on a plane of maximum shear, to Stage II crack propagation normal to the tensile stress described by Brown and Miller (1979). Thus, assuming that the dominant stage will coincide with the mode producing the maximum plastic strain in the crack tip region, the transition from one stage to the next would occur at values of σ_P/τ in the region of 1.85. This prediction is not far out from the results of experiments on square section bars of mild steel by Cox and Field (1952) who found that the transition from Stage I to Stage II occurred at values of σ_P/τ varying from 1.5 to 1.7 or more but that the transition from Stage II to Stage I occurred at the lower value of 1.5. Note also the experiments by Gough et al. (1951) on the fatigue limits of three steels under combined bending and torsional alternating stresses. The results are reproduced by Frost et al. (1974) in their Fig. 3.8. They show that the ratios of the fatigue limit under pure bending stress to the fatigue limit under pure shear due to torsion are approximately equal to 1.82, 1.76 and 1.53 for 0.1% C steel, 3% Ni steel and 3.5% Ni-Cr steel, respectively. Frost *et al.* comment further that for those metals in which failure is initiated at a free surface, the ratio of the fatigue limit (or strength at very long endurance) in torsion to that in uniaxial loading has an average value not far removed from that predicted by the Von Mises criterion, the value for a particular material depending on the endurance at which the fatigue strength is estimated and on metallurgical structure.

It is probable that a large negative value of σ_T combined with mixed or pure Mode II loading would result in increased crack tip plasticity and it may be worthwhile repeating the analyses with compressive values of σ_Q equal to $-0.5\sigma_P$ and $-\sigma_P$, respectively.

9. Conclusions

(1) Crack separation energy rates G_I^Δ and G_{II}^Δ for coplanar extensions were calculated. Using quadrature coefficients $K_{ij}(\alpha)$, estimates of $G_I^\Delta(\alpha)$ and $G_{II}^\Delta(\alpha)$ for nonplanar crack extensions in the direction of α were made for cases involving only moderate plastic flow.

(2) An elastic-plastic fracture criterion for the onset of crack propagation was proposed. The criterion is based on the maximum crack separation energy rate $G^\Delta(\alpha_M) = G_c$ when the material is elastic-plastic, where α_M is the optimal fracture angle. When the material is linear-elastic the $G^\Delta(\alpha_M)$ criterion is equivalent to the maximum Griffith energy rate criterion, $G(\alpha_M) = G_c$.

(3) Predictions of the fracture angle α_M for different inclinations were made using the $G^\Delta(\alpha_M)$ criterion, for cases involving moderate plastic flow. The values of α_M obtained did not differ much from the values of the fracture angles predicted by some of the criteria based on linear elastic analyses.

(4) For a given uniaxial applied stress σ_P, the most dangerous inclination was found to be in the region of $\theta \simeq 22.5°$ with $\alpha_M \simeq -35°$ in the elastic-plastic material, using the $G^\Delta(\alpha_M)$ criterion. Approximately similar results are predicted for linear-elastic materials by the maximum energy criterion or the vanishing k_2 criterion.

(5) For equal intensities of plastic flow in the crack tip region under pure Mode I (σ_P) or pure shear (τ) loading, the ratio of σ_P/τ was found to be approximately 1.85. This ratio may have some bearing (a) on the transition from Stage I fatigue-crack propagation (on a plane of maximum shear) to Stage II crack propagation (normal to the tensile stress) and (b), on the ratios of the fatigue limits of some metals under pure Mode I and pure Mode II alternating stresses.

Acknowledgment

This work is part of a continuing program funded by the Nuclear Regulatory Commission (NRC) concerned with crack growth studies in complex stress-strain fields.

References

Bilby, B. A. and Cardew, G. E. (1975), "The Crack With a Kinked Tip," *Int. J. Fract.*, **2**, 708.

Bilby, B. A., Cardew, G. E. and Howard, I. C. (1978), "Stress Intensity Factors at the Tips of Kinked and Forked Cracks," *Fracture 1977, Advances in Research on the Strength and Fracture of Materials*, vol. 3, D. M. R. Taplin (ed.), Pergamon Press, New York, 197.

Brown, M. W. and Miller, K. J. (1979), "Initiation and Growth of Cracks in Biaxial Fatigue," *Fatigue Eng. Mats. Structs.*, **1**, 231.

Chiang, W. T. (1978), "Fracture Criteria for Combined Mode Cracks," *Fracture 1977, Advances in Research on the Strength and Fracture of Materials*, vol. 4, D. M. R. Taplin, (ed). Pergamon Press, New York, 135.

Corlett, P. E. G. (1977), "Crack Paths and Crack Stability," *M.Sc. dissertation, Department of Theory of Materials, University of Sheffield*.

Cox, H. L. and Field, J. E. (1952), "The Initiation and Propagation of Fatigue Cracks in Mild Steel Pieces of Square Section," *The Aeronautical Quarterly*, **4**, 1.

Eftis, J., Subramonian, N. and Liebowitz, H. (1977), "Crack Border Stress and Displacement Equations Revisited," *Eng. Fract. Mech.* **9**, 189.

Erdogan, F. and Sih, G. C. (1963), "On the Crack Extension in Plates Under Plane Loading and Transverse Shear," *J. Basic Eng.*, **85D**, 519.

Eshelby, J. D. (1951), "The Force on an Elastic Singularity," *Phil. Trans.* **A244**, 87.

Ewing, P. D. and Williams, J. G. (1974a), "Further Observations of the Angled Crack Problem," *Int. J. Fract.*, **10**, 135.

Ewing, P. D. and Williams, J. G. (1974b), "The Fracture of Spherical Shells under Pressure and Circular Tubes with Angled Cracks in Torsion," *Int. J. Fract.*, **10**, 537.

Finnie, I. and Saith, A. (1973), "A Note on the Angled Crack Problem and the Directional Stability of Cracks," *Int. J. Fract.*, **9**, 484.

Frost, N. E., Marsh, K. J. and Pook, L. P. (1974), *Metal Fatigue*, Clarendon Press, Oxford.

Gough, H. J., Pollard, H. V. and Clenshaw, W. (1951), "Some Experiments on the Resistance of Metals to Fatigue Under Combined Stresses," *Aero Research Council, R and M, 2522*, H.M.S.O.

Howard, I. C. (1978), "Simple Approximate Results for the Stress Intensity Factors at the Tip of a Kinked Crack," *Int. J. Fract.*, **14**, R307.

Iron and Steel Institute (1968), *Fracture* I.S.I. Publication 121, Iron and Steel Inst., London.

Kfouri, A. P. (1979a), "Crack Separation Energy-Rates for the DBCS Model Under Biaxial Modes of Loading," *J. Mech. Phys. Solids*, **27**, 135.

Kfouri, A. P. (1979b), "Plane Strain Elastic-Plastic Finite Element Analyses of a Plate Containing a Central Inclined Crack," Report, Department of Mechanical Engineering, University of Sheffield.

Kfouri, A. P. and Miller, K. J. (1974), "Stress, Displacement, Line Integral and Closure Energy Determinations of Crack Tip Stress Intensity Factors," *Int. J. Pressure Vessels Piping*, **2**, 179.

Kfouri, A. P. and Miller, K. J. (1976), "Crack Separation Energy Rates in Elastic-Plastic Fracture Mechanics," *Proc. Inst. Mech. Engrs. 190*, (Paper 48/76), 571.

Kfouri, A. P. and Miller, K. J. (1978), "The Effect of Load Biaxiality on the Fracture Toughness Parameters J and G^Δ," *Fracture 1977, Advances in Research on the Strength and Fracture of Materials*, D. M. R. Taplin (ed.), 3A Pergamon Press, New York 241.

Kfouri, A. P. and Rice, J. R. (1978), "Elastic-Plastic Separation Energy Rate for Crack Advance in Finite Growth Steps," *Fracture 1977, Advances in Research on the Strength and Fracture of Materials*, D. M. R. Taplin (ed.), 1, Pergamon Press, New York 43.

Khrapkov, A. A. (1971), "The First Basic Problem for a Notch at the Apex of an Infinite Wedge," *P.M.M.* 35, 677; *Int. J. Fract. Mech.*, 7, 373.

Lo, K. K. (1978), "Analysis of Branched Cracks," *J. Appl. Mech.*, 45, 797.

Miller, K. J. and Kfouri, A. P. (1974), "An Elastic-Plastic Finite Element Analysis of Crack Tip Fields Under Biaxial Loading Conditions," *Int. J. Fract.*, 10, 393.

Miller, K. J. and Kfouri, A. P. (1977), "A Comparison of Elastic-Plastic Fracture Parameters in Biaxial Stress States," *ASTM STP 688*, 214.

Palaniswamy, K. and Knauss, W. G. (1974), "On the Problem of Crack Extension in Brittle Solids under General Loading," Report SM 74-8, Grad. Aeronautical Labs., Calif. Inst. of Technology.

Rice, J. R. (1968a), "Mathematical Analysis in the Mechanics of Fracture," *Fracture: An Advanced Treatise* H. Liebowitz (ed.), Vol. II Mathematical Fundamentals, Academic Press, New York, Ch. 3, 191.

Rice, J. R. (1968b), "A Path Independent Integral and the Approximate Analysis of Strain Concentration by Notches and Cracks," *J. Appl. Mech.*, 35, 379.

Rice, J. R. (1975), "Elastic-Plastic Models for Stable Crack Growth," *Mechanics and Mechanisms of Crack Growth* M. J. May (ed.), British Steel Corp., Physical Metallurgy Center, Sheffield, 14.

Rice, J. R. and Johnson, M. A. (1970), "The Role of Large Crack Tip Geometry Changes in Plane Strain Fracture," *Inelastic Behavior of Solids* M. F. Kanninen, W. F. Adler, A. R. Rosenfield and R. I. Jaffee (eds.), McGraw-Hill, New York, 641.

Sih, G. C. (1974), "Strain-Energy-Density Factor Applied to Mixed Mode Crack Problems," *Int. J. Fract.*, 10, 305.

Sih, G. C. and Kipp, M. E. (1974), discussion on "Fracture Under Complex Stress" by J. G. Williams and P. D. Ewing, *Int. J. Fract.*, 10, 261.

Swedlow, J. L. (1976), "Criteria for Growth of the Angled Crack," *Cracks and Fracture*, ASTM STP 601, 506.

Tanaka, S. and Ichikawa, M. (1979), ICF4 discussion "Fracture Criteria for Combined Mode Cracks" by W. T. Chiang, *Int. J. Fract.*, 15, R49.

Williams, J. G. and Ewing, P. D. (1972), "Fracture Under Complex Stress—The Angled Crack Problem," *Int. J. Fract. Mech.*, 8, 441.

Williams, M. L. (1957), "On the Stress Distribution at the Base of a Stationary Crack," *J. Appl. Mech.*, 24, 109.

S. Nemat-Nasser, Editor
THREE-DIMENSIONAL CONSTITUTIVE RELATIONS AND DUCTILE FRACTURE
North-Holland Publishing Company (1981) 111–122

FATIGUE CRACK GROWTH THEORY FOR DUCTILE MATERIAL

Johannes WEERTMAN

Northwestern University, Evanston, IL, U.S.A.

In this paper it is shown that the recent fatigue crack growth theory of the author developed for crack advance in a cleavage mode also can be applied directly to the case in which crack advance occurs in a crack tip shear sliding mode. For the situation where a fatigue crack begins its advance when the cyclic stress has almost reached its maximum value the fatigue crack growth equation is $da/dN \approx (2/\pi\tau_0\mu)(1 - \eta^p)(\Delta K/2)^{p+2}/p(gK_{cb})^p$, where da/dN is the crack advance per cycle, τ_0 is the yield stress, μ is the shear modulus, η is the ratio of the stress at which crack advance starts to the maximum stress, ΔK is the cyclic stress intensity factor, g is a term that depends upon the ratio of the theoretical tensile strength σ_T to theoretical shear strength τ_T ($g=1$ for $\sigma_T/\tau_T \approx 7$ and $g=0.6$ for $\sigma_T/\tau_T \approx 20$), K_{cb} is the critical stress intensity factor for brittle fracture ($K_{cb} = \{4\mu\gamma/(1-\nu)\}^{1/2}$ where γ is the surface energy and ν is Poisson's ratio), and the exponent p can be expected to have values of the order of $p \approx 1$ to 2. When η is small compared to 1 this last equation is replaced with $da/dN \approx (\Delta K)^2/2\pi\tau_0\mu$.

The theory developed for crack advance through a cleavage mode cannot be used for crack advance in a crack tip shear sliding mode when geometric blunting of the crack tip is important. The fatigue crack growth rate when blunting determines the crack advance distance is given by the equation, $da/dN \approx b(\Delta K)^2/4g^2K_{cb}^2$, where b is the atomic distance.

For ductile materials the crack propagation rate should be determined by whichever of the equations above give the smallest crack propagation rate.

1. Introduction

We recently presented a theory (Weertman, 1978a, 1979) of fatigue crack growth that is based on an elastic crack tip enclave model of fracture (Thomson, 1976, 1978; Weertman, 1978b, 1980). In our fatigue crack growth theory the incremental advance of a fatigue crack is considered to occur in a brittle cleavage mode. The purpose of this paper is to show that that fatigue crack growth theory applies also to the situation in which crack tip advance occurs through the shear sliding mode of Laird and Smith (1962; Laird, 1979) and of Neumann (1974a, 1974b; Vehoff and Neumann, 1979).

Figure 1 shows (one way out of many) how fatigue crack growth can occur through a shear sliding mode. As seen in Fig. 1 alternate slip on two

J. Weertman

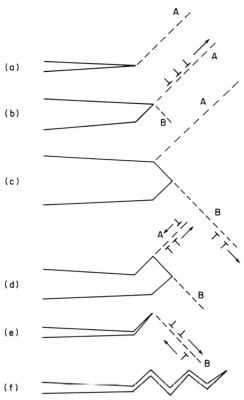

Fig. 1. Fatigue crack advance by crack tip shear sliding: (a) Initial crack condition; (b) Dislocation emission on slip plane A; (c) Dislocation emission on slip plane B; (d) Reversed slip on slip plane A by either emission of dislocations of opposite sign or return of the original dislocations; (e) Reversed slip on slip plane B leading to crack advance on inclined plane A. Note that if in (b) dislocation emission occurred first on plane B the crack advance would occur on inclined plane B; (f) Crack advance after many repetitions of the shear sliding process with alternative advances on planes A and B.

slip planes that have a common intersection with the crack front followed by alternate reverse slip produces an incremental crack advance. The crack advance distance per stress cycle is equal to the number of dislocations of like sign times a geometric factor.

  Although Fig. 1 shows how a fatigue crack can advance through a crack tip shear sliding process it provides no additional information than does an analogous drawing of a crack that propagates in a brittle mode in which the atoms are shown as being separated far enough apart to break the atomic bonds across the crack plane. In particular, Fig. 1 provides no answers to questions such as: At what value of the stress intensity factor does crack tip shear sliding take place? How many dislocations are emitted from the crack tip region? What effect does a plastic zone around the crack tip have on the shear sliding process? These are essentially the same questions that occur for the model of crack advance in a cleavage mode and, we believe, have essentially the same answers (not all of which are known at present).

It is shown in this paper that a theory developed for cleavage mode fatigue crack propagation applies equally well to a shear sliding mode fatigue crack propagation. However, our extended theory does not apply when geometric blunting of the crack tip is the most important mechanism that limits crack growth. For this latter situation, as will be seen, a fatigue crack propagation equation is easily obtained.

2. Theory

2.1. Review

Our fatigue crack propagation theory made use of the assumption that a small region around the crack tip always can be considered to be elastic. The tip of the crack propagates in a cleavage mode whenever the "true" stress intensity factor K_t within this elastic region, which is surrounded by a plastic zone, is equal to the critical stress intensity factor K_{cb} of a perfectly brittle solid. (The analysis is restricted to the case of small scale yield condition.) The value of K_{cb} is given by,

$$K_{cb} = \{4\mu\gamma/(1-\nu)\}^{1/2}, \tag{1}$$

where μ is the shear modulus, γ is the surface energy of the solid, and ν is Poisson's ratio.

For a virgin crack $K_t = K$ where K is the conventional stress intensity factor and is equal to $K = \sigma(\pi a)^{1/2}$ for a crack in an infinite medium. Here σ is the applied stress and a is the half length of the crack. By a virgin crack is meant one that exists in a material which had not previously been stressed and whose crack tips have not advanced under application of the stress σ. Thus, as shown in Fig. 2a, if the position of the crack tip remains at $x = a$ the value of $K_t = K$ as the conventional stress intensity factor is increased from its initial value of $K = 0$ to its final value of K. Were the position of the crack tip now to increase while K remains a constant the value of K_t would decrease as shown in Fig. 2a from the value $K_t = K$ towards a limiting value $K_t = K^*$ where K^* is given by the equation,

$$K^* = K\{1 - 2\beta/\pi(1-\nu)\}^{1/2}, \tag{2}$$

where β is a constant whose value depends upon the elastic-plastic stress strain curve of the solid (Weertman, 1978b, 1980).

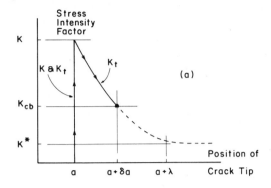

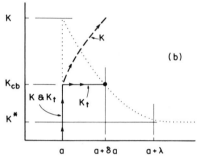

Fig. 2.   Plot for virgin crack of K_t and K versus crack tip position: (a) When crack tip moves only after K reaches its final value; (b) When crack tip moves when K_t reaches the value K_{cb}.

The crack tip will not advance whenever K_t is smaller than K_{cb}. Thus in Fig. 2a the crack advance must stop at $x = a + \delta a$ where $K_t = K_{cb}$. Moreover, because a crack will advance whenever $K_t = K_{cb}$ actually K_t cannot, as shown in Fig. 2a, exceed the value of K_{cb}. Hence as the conventional stress intensity factor is increased to a value greater than K_{cb} the true stress intensity factor K_t actually will change in the manner shown in Fig. 2b. Once $K \geqslant K_{cb}$ thereafter $K_t = K_{cb}$. The crack growth increment δa can be estimated by assuming that in Fig. 2a the value of K_t decreases linearly with crack advance out to a distance λ where λ is equal to (or to some fraction of) a characteristic length such as the crack opening displacement. Thus,

$$\delta a \approx \lambda (K - K_{cb})/(K - K^*). \tag{3}$$

Equation (3) requires that $K > K_{cb}$. Otherwise, if $K < K_{cb}$ then K_t is always smaller than K_{cb} and no crack advance can occur. Moreover Eq. (3) requires that $K < K_{Ic}$ where K_{Ic} is the critical value for catastrophic crack advance for a monotonically loaded specimen. K_{Ic} is found from Eq. (2) by setting $K^* = K_t = K_{cb}$ to give,

$$K_{Ic} = \{4\pi\mu\gamma / [\pi(1-\nu) - 2\beta]\}^{1/2}. \tag{4}$$

Now let a cyclic, rather than a monotonically increasing, stress be applied to a specimen containing a crack. The effect of stress reversal during the stress cycle is to make the advancing fatigue crack into a pseudo-virgin crack at the start of each forward portion of the stress cycle.

The true stress intensity factor K_t can no longer equal the conventional stress intensity factor K during the period that the stress is increasing but no crack advance occurs. However, K_t will reach the value K_{cb} before the stress has reached its maximum value, thus permitting crack advance during each stress cycle. The Fig. 2 analogue to the monotonically increasing stressed crack is shown in Fig. 3 for the cyclically stressed crack. When the crack is

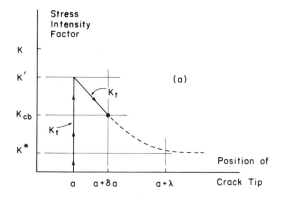

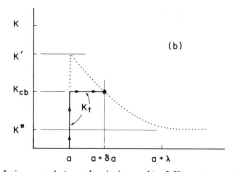

Fig. 3. Plot for fatigue crack (pseudo-virgin crack) of K_t versus crack tip position: (a) When crack tip moves only after K reaches its final value; (b) When crack tip moves whenever $K_t = K_{cb}$.

permitted to advance up to the moment the applied stress reaches its maximum value, K_t does not equal K, as it did in Fig. 2a, but rather is equal to $K_t = K'$ where $K' \neq K$. Thus the crack advance distance δa now is given by,

$$\delta a \approx \lambda (K' - K_{cb})/(K' - K^*), \tag{5}$$

and K^* again is given by Eq. (2).

The term K' can be expressed in terms of the ratio η defined by the relation,

$$\eta = K_{start}/K_{max}, \tag{6}$$

where K_{max} is the maximum value of the conventional stress intensity factor K during a stress cycle and K_{start} is the value of K when a fatigue crack begins to grow during a stress cycle. When $K = \eta K_{max}$ then $K_t = K_{cb}$. In the Weertman (1979) reference it is argued if a fatigue crack were prevented from advancing and K were increased to a value much greater than K_{max} then the fatigue crack becomes for all practical considerations a virgin crack. In this situation $K_t \approx K$. A rather general equation for K_t that satisfied the conditions that $K_t = K_{cb}$ when $K = \eta K_{max}$ and $K_t \approx K$ when $K \gg K_{max}$ is,

$$K_t^p \approx K^p + K_{cb}^p - (\eta K_{max})^p, \tag{7}$$

where p is any constant greater than 0. (Reasonable values of p are $p \approx 1$ to 2.) For K_t given by Eq. (7) the term K' must be equal to

$$K'^p = K_{cb}^p + (1 - \eta^p)K_{max}^p. \tag{8}$$

When $K'^p \approx K_{cb}$, which requires that $\eta \approx 1$ when $K_{max} \gg K_{cb}$, and $K_{cb} \gg K^*$ Eq. (5) reduces to,

$$\delta a \approx \lambda (1 - \eta^p)(K_{max}/K_{cb})^p/p. \tag{9}$$

Since λ can be expected to be proportional to K_{max}^2 this last equation is a Paris fatigue crack growth equation of power $2 + p$. When $p = 1$ it is a 3rd power equation and when $p = 2$ it is a 4th power equation.

When K_{max} is large compared with K_{cb} and the expression $(1 - \eta^p)$ is of the order of magnitude of 1 Eq. (5) reduces to,

$$\delta a \approx \lambda. \tag{10}$$

Equation (10) corresponds to a 2nd power Paris equation. Of course, Eq. (9) also is approximately a 2nd power Paris equation if $p \ll 1$. In this situation Eq. (5) becomes,

$$\delta a \approx \lambda (1 - \eta^p) \approx \lambda \log \eta^{-1}. \tag{11}$$

2.2. *Application to fatigue crack advance by shear sliding*

We have considered so far that a crack advances when $K_t = K_{cb}$ and the advance occurs in a cleavage type mode. But suppose the solid is one in which a crack tip shear sliding process of the kind shown in Fig. 1 occurs when $K_t < K_{cb}$. Crystalline material for which this can happen are those in which the ratio σ_T/τ_T has a large value. Here σ_T is the theoretical tensile strength of the solid and τ_T is its theoretical shear strength. The conditions under which dislocations are emitted from the crack tip region has been analyzed by Rice and Thomson (1974) and more recently by us (Weertman, 1981). We found that when $\sigma_T/\tau_T > 7$ a crack tip shear sliding process will occur when $K_t < K_{cb}$. In particular it will occur at $K_t = K_{cs}$ where K_{cs} is given by,

$$K_t = K_{cs} = gK_{cb},$$

where g is a constant whose value depends upon the ratio σ_T/τ_T. It has the values (Weertman, 1981) $g \approx 1$ for $\sigma_T/\tau_T = 7$, $g \approx 0.6$ for $\sigma_T/\tau_T - 20$, and $g \approx (\tau_T/\sigma_T)^{1/3}$ for values of the ratio σ_T/τ_T that are appreciably larger than 20.

The critical stress intensity constant K_{cs} is the critical value of K_t for the onset of shear sliding at the crack tip. It plays the analogous role as the critical stress intensity constant K_{cb} does for the onset of cleavage separation ahead of the crack tip.

When $K_{cs} < K_{cb}$ fatigue crack advance should occur by the shear sliding process of the type shown in Fig.1, at least for those cracks whose fronts lie on an intersection with a slip plane. For the virgin crack shear sliding as shown in Fig. 1b and 1c should start once K reaches the value $K = K_{cs} = K_t$. For the pseudo-virgin fatigue crack the crack advance will commence when $K_t = K_{cs}$ and $K = \eta K_{max}$. As before, the crack will continue to advance as long as $K_t = K_{cs}$. This condition requires that K increases in value as the crack advances. Thus in the equations of the previous section it is only necessary to replace K_{cb} with K_{cs} to obtain the fatigue crack growth equations. Equation (8) becomes,

$$K'^p = K_{cs}^p + (1 - \eta^p)K_{max}^p, \tag{12}$$

and Eq. (9) becomes,

$$\delta a \approx \lambda(1 - \eta^p)(K_{max}/K_{cs})^p/p. \tag{13}$$

Equations (10) and (11) remain unaltered.

There is, however, one complication in this rather direct analogy between the shear sliding mode and cleavage mode of fatigue crack advance. Consider Fig. 4 and the situation in which K_{max} is so large that a huge

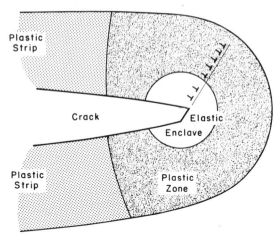

Fig. 4. Schematic figure of crack tip region when dislocations emitted from the crack tip region move out of crack tip elastic enclave into the plastic zone.

number of dislocations must be emitted from the crack tip region in order to account for a large value of the crack tip advance distance δa per cycle. Many dislocations must move from the elastic enclave into the plastic zone as shown in Fig. 4. (The plastic deformation in the plastic zone normally occurs through the motion of dislocations created within this zone and not from dislocations created at the crack tip. The effect of this plastic deformation on the growing crack is to reduce the value of K_t from the value $K_t - K$ for the virgin crack to a value $K_t < K$.) The dislocations within the work hardened plastic zone will resist the motion of dislocations emitted from the crack tip region that move into the plastic zone. Consequently, it will become more and more difficult for dislocations to be emitted from the crack tip region. The value of K_{cs} effectively is increased. If it is increased to the level of $K_{cs} = K_{cb}$ a cleavage mode of crack propagation will take over. Thus a mixed mode mechanism of fatigue crack propagation might be expected to occur as K_{max} is increased.

3. Geometric blunting

The equations of the last section were developed under the assumption that the plastic deformation around a growing crack is the physical process that ultimately stops the growth of a crack. But when shear sliding occurs at a crack tip it is possible to stop crack growth through geometric blunting. No plastic zone need form to accomplish this. If Figs. 1a through 1c are examined it is seen that slip on two intersecting slip planes has caused the atomically sharp crack of Fig. 1a to become the blunter appearing crack of Fig. 1c. And if slip also occurred on planes parallel to the two planes shown in Fig. 1 the crack would appear even blunter. The crack tip can be considered to be rounded out to a radius of curvature ρ, which is of the order of the crack advance distance δa. The true stress intensity factor K_t required to have further crack advance and crack tip shear sliding must be of the order of,

$$K_t \approx K_{cs}(\rho/b)^{1/2} = g K_{cb}(\rho/b)^{1/2}, \tag{14}$$

rather than $K_t = K_{cs}$. Here b is the interatomic distance.

If geometric blunting rather than plastic zone deformation causes a growing crack to stop the plastic zone deformation must be ineffectual in reducing appreciably the value of K_t from its virgin crack value of $K_t = K$. Hence in Eq. (14) $K_t \approx K$ and $\delta a \approx \rho$. Thus the crack advance distance is given by,

$$\delta a \approx b K_{max}^2 / g^2 K_{cb}^2. \tag{15}$$

This equation also applies to the cyclically stressed fatigue crack. The incremental crack advance per cycle should be given by Eq. (15) if it predicts the smallest δa value.

4. Discussion and summary

The starting point of our previous fatigue crack growth theory was the fact that if an elastic enclave is considered to exist at a crack tip then a virgin crack will begin to grow when the conventional stress intensity factor K is equal to the critical value K_{cb} for a crack in a perfectly brittle material. The true stress intensity factor K_t at the crack tip is also equal to K at this instant. The plastic zone that is created around the virgin crack as K is increased from the value $K=0$ to the value $K=K_{cb}$ does not change the relationship $K_t = K$ for the virgin crack. The stress-strain field of the plastic zone for a growing crack is different than that of the stationary crack and consequently when the crack grows $K_t < K$. The crack will grow as long as $K_t = K_{cb}$. This condition requires that K increase during crack growth. (Of course, catastrophic growth occurs once $K = K_{Ic}$.)

A fatigue crack growth theory can be built on these considerations through the assumption that reverse loading has the effect on a growing crack of at least partially restoring the crack to condition of a virgin crack. Thus each stress cycle K_t reaches the level K_{cb} before the cyclic stress has reached its maximum value. Consequently incremental crack growth can occur each stress cycle.

This picture is not changed in any essential way for more ductile material in which a crack tip shear sliding process occurs when $K_t = K_{cs} < K_{cb}$. Here K_{cs} represents the critical stress intensity factor for the beginning of emission of dislocations from the crack tip region. A crack will advance through the emission of dislocations from the crack tip region. The crack will stop advancing when, as a consequence of crack growth, K_t falls below the level K_{cs}. Thus the same equations that were derived previously for the fatigue crack growth also apply after the substitution is made in them of $K_{cs} = gK_{cb}$ for K_{cb}.

When a fatigue crack begins to grow in each stress cycle only when the applied stress almost reaches its maximum value the fatigue crack growth equation is,

$$da/dN \approx (2/\pi\tau_0\mu)(1-\eta^p)(\Delta K/2)^{p+2}/p(gK_{cb})^p, \tag{16}$$

where N is the number of stress cycles and the characteristic distance λ has been set equal to the crack opening displacement ($\lambda \approx 2K_{max}^2/\pi\tau_0\mu$ where τ_0 is the yield stress and μ is the shear modulus). In Eq. (16) the cyclic stress intensity factor ΔK is equal to $\Delta K = 2K_{max}$. The term η in this equation represents the ratio of stress at which a fatigue crack begins to grow to the maximum stress; g is a constant whose value is smaller the larger is the ratio of the theoretical tensile strength to the theoretical shear strength. Reasonable values of the exponent p are $p \approx 1$ to 2. However, if the exponent p is almost equal to zero Eq. (16) reduces to,

$$da/dN \approx (\log \eta^{-1})(\Delta K)^2/2\pi\tau_0\mu. \tag{17}$$

When the crack begins its advance each cycle well before the stress approaches its maximum value the fatigue crack growth equation is,

$$da/dN \approx (\Delta K)^2/2\pi\tau_0\mu. \tag{18}$$

There are two complicating factors that should be noted. When large numbers of dislocations are emitted from the crack tip region most of these will leave the elastic enclave and enter the plastic zone. The effect of the resistance to dislocation motion within the plastic zone is to make dislocation emission from the crack tip region more difficult. Consequently the value of the critical stress intensity factor K_{cs} is increased. Increasing the value of K_{cs} will decrease the crack growth rate if other factors remain unchanged.

The second complication to the theory arises from the change in the crack tip geometry produced by the shear sliding process. The crack tip is more blunted (compare Figs. 1a and 1c). When crack tip blunting determines the propagation rate the fatigue crack growth equation is,

$$da/dN \approx b(\Delta K)^2/4g^2K_{cb}^2, \tag{19}$$

where b is the interatomic distance. Since the surface energy of a solid is of the order of $\mu b/h$ where $h \approx 3$ this last equation can be written as,

$$da/dN \approx h(\Delta K)^2/16g^2\mu^2(1-\nu).$$ (20)

Equation (20) predicts a smaller crack growth rate than Eq. (18) and presumably Eq. (20) would apply whenever the term η has a value that is not approximately equal to 1. Equation (20) gives a reasonable account of fatigue crack growth data considered by Donahue *et al.* (1972). Donahue *et al.* showed that data reviewed by them that was obtained in a nonaggressive environment obeyed the second power relationship $da/dN \approx (8/\pi)K_{max}^2/E^2$ where E is Young's modulus. Since $E = 2\mu(1+\nu)$ and $\Delta K = 2K_{max}$ this relationship is $da/dN \approx (\Delta K)^2/2\pi(1+\nu)^2\mu^2$ and is approximately the same as that given in Eq. (20). (The fatigue crack growth rate found in aggressive environments was larger by up to a factor of 10. This increase might be explained by a reduction of the surface energy γ that appears in Eq. (19).) Equation (20) predicts that when the yield stress $\tau_0 \ll \mu$ the crack advance distance per cycle is only a very small fraction of the crack opening displacement calculated from Eq. (18). Thus the geometric blunting model offers a successful explanation of the puzzling experimental result why, when a second power fatigue crack growth equation is obeyed, the crack advance distance is so much smaller than the crack opening displacement that is calculated using the small scale plastic yield condition.

It should be noted that because the value of K_{cb} is altered when an active environment is present environmental effects are predicted for the equations that contain K_{cb}. It should also be noted that these theories and models break down if the crack advance distance becomes smaller than a few atomic distances. Thus the theory also predicts a threshold effect.

Acknowledgment

This work was supported under the NSF-MRL program through the Materials Research Center of Northwestern University (Grant DMR76-8087).

References

Donahue, R. J., Clark, H. M., Atanmo, P., Kumble, R. and McEvily, A. J. (1972), "Crack Opening Displacement and the Rate of Fatigue Crack Growth," *Int. J. Fract. Mech.*, **8**, 209.
Laird, C. and Smith, G. C. (1962), "Crack Propagation in High Stress Fatigue," *Phil. Mag.*, 7, 847.
Laird, C. (1979), "Mechanisms and Theories of Fatigue," *Fatigue and Microstructure*, American Society for Metals, Metals Park, Ohio, 149.

Neumann, P. (1974a), "New Experiments Concerning the Slip Processes at Propagating Fatigue Cracks—I," *Acta Met.*, **22**, 1155.

Neumann, P. (1974b), "The Geometry of Slip Processes at a Propagating Fatigue Crack—II," *Acta Met.*, **22**, 1167.

Rice, J. R. and Thomson, R. (1974), "Ductile Versus Brittle Behavior of Crystals," *Phil. Mag.*, **29**, 73.

Thomson, R. (1976), "Some Microscopic and Atomic Aspects of Fracture," *The Mechanics of Fracture*, AMD-Vol. 19, F. Erdogan (ed.), American Society of Mechanical Engineers, New York, 1.

Thomson, R. (1978), "Brittle Fracture in a Ductile Material with Application to Hydrogen Embrittlement," *J. Mater. Sci.*, **13**, 128.

Vehoff, H. and Neumann, P. (1979), "*In situ* SEM Experiments Concerning the Mechanism of Ductile Crack Growth," *Acta Met.*, **27**, 915.

Weertman, J. (1978a), "True Stress Intensity Factor Rationalization of a 2nd and 4th Power Paris Fatigue Crack Growth Equation," *Fracture Mechanics*, N. Perrone, H. Liebowitz, D. Mulville and W. Pilkey (eds.), University of Virginia Press, Charlottesville, 193.

Weertman, J. (1978b), "Fracture mechanics: A Unified View for Griffith-Irwin-Orowan Cracks," *Acta Met.*, **26**, 1731.

Weertman, J. (1979), "Fatigue Crack Propagation Theories," *Fatigue and Microstructure*, American Society for Metals, Metals Park, Ohio, 279.

Weertman, J. (1980), "Fracture Stress Obtained from the Elastic Crack Tip Enclave Model," *J. Mater. Sci.*, **15**, 1306.

Weertman, J. (1981), "Crack Tip Blunting by Dislocation Pair Creation and Separation," *Phil. Mag.*, (in press).

S. Nemat-Nasser, Editor
THREE-DIMENSIONAL CONSTITUTIVE RELATIONS AND DUCTILE FRACTURE
North-Holland Publishing Company (1981) 123–145

PROBLEMS IN ENVIRONMENTALLY-AFFECTED CREEP CRACK GROWTH

Frank A. McCLINTOCK and John L. BASSANI*

Massachusetts Institute of Technology, Cambridge, MA, U.S.A.

Two idealized models are proposed for creep cracking due to environmental effects: a steady, atom-by-atom cracking due to the saturation of the grain boundary by a weakening species, and an intermittent crack advance cleaving between grains of size d_g when the strength at a distance d_g from the crack tip falls to the level of the local stress. With either criterion crack meandering does not appear critical, but multiple branching with statistical effects remains unsolved.

The fracture criteria are introduced into the asymptotic flow fields for elastic, plastic, and primary, secondary, and tertiary creep fields around stationary cracks and secondary creep fields around growing cracks. The results are illustrated and compared with data and service conditions, showing the difficulties of predicting long-time service from short-time tests. Micrographs indicate the limiting scales to which continuum mechanics can be applied.

Needed improvements in the flow relations are discussed, especially when nonradial or reversed loading gives important primary creep effects, either due to hardening or to polarity of the dislocation structure.

For flexibility in calculating strain fields for such flow relations, we describe a cost effective, explicit time integration scheme, with initial strain finite element solutions. Short time steps for stability are alternated with far larger steps to reduce costs.

1. Introduction

Floreen (1980) has shown that in superalloys, an oxidizing atmosphere can accelerate the growth of creep cracks by an order of magnitude or more. Oxidation in the grain boundary has been shown metallurgically by McMahon (1974), Chaku and McMahon (1974), and McMahon and Coffin (1970). Continuous cracking is shown in Fig. 1, and cracking due to deeper oxidation in Fig. 2. On the other hand, oxidation can inhibit creep near a surface or a crack (Sessions, McMahon, and Walker (1977), Kennedy (1953), and Widmer and Grant (1960)). Since most of the data have been obtained in air and since superalloys are used in an oxidizing environment, it is appropriate to consider the corresponding crack growth criteria. Although environmental effects are much less in austenitic stainless steels, extrapolation of the relatively short time data available indicates that for the

*Now at University of Pennsylvania, Philadelphia, PA, U.S.A.

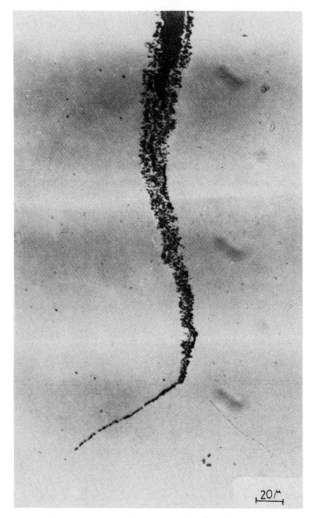

Fig. 1. Crack in Udimet 700. Note internal grain boundary oxidation. Unetched. From McMahon (1974).

30 year (10^9 s) times expected in service, crack growth rates may be seriously affected by the atmosphere in these materials as well (Sadananda and Shahinian, 1980).

A crack growth criterion must be embedded in an appropriate stress analysis. Previous work has been based rather empirically on overall static fracture analyses or modifications of them (Floreen (1975), Haigh (1975), Gooch et al. (1977), Sadananda and Shahinian (1978)). Finite element

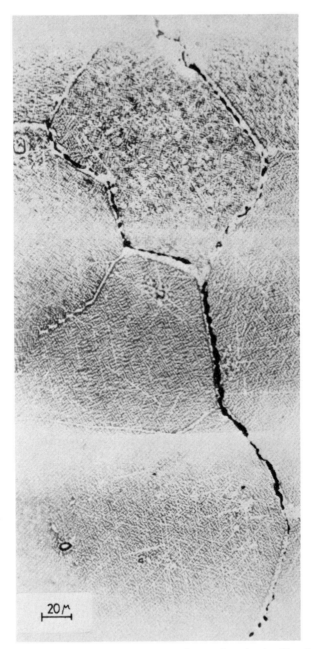

Fig. 2. Etched microstructure from a grain boundary region ahead of another crack in Udimet 700 showing internal oxidation. From McMahon (1974).

analysis has been carried out, for example by Bassani and McClintock (1980), but the studies have not shown details at the crack tip, nor the regimes of dominance of the various modes of creep.

2. Fracture criteria

Environmentally affected cracking usually progresses along grain boundaries. Consider two limiting cases, the first steady-state cracking at the atom-by-atom scale, and the second a repeated grain boundary cracking controlled by diffusion to a point one grain diameter d_g ahead of the main crack, starting from the clean surface left by the previous crack jump (e.g. Figs. 1 and 2).

2.1. Steady-state cracking

Assume the diffusion has a coefficient D_g along a grain boundary of constant width, with the crack front progressing at a constant velocity $\dot{a}$. The rate of change of concentration per unit volume c is proportional to the rate of change of concentration gradient along the crack:

$$-\dot{a}\frac{dc}{dx} = \frac{dc}{dt} = D_g\frac{d}{dx}\frac{dc}{dx}. \tag{1}$$

For a concentration c_1 at the crack tip ($x=0$), the solution to Eq. (1) is

$$c = c_1 e^{-\dot{a}x/D_g}. \tag{2}$$

The mass flow rate dm/dt into the grain boundary, through the cross-sectional area A_g, is proportional to the concentration gradient at the tip:

$$\frac{1}{A_g}\frac{dm}{dt} = -D_g\left(\frac{dc}{dx}\right)_1 = c_1\dot{a}. \tag{3}$$

The difference between the far field and crack tip concentrations, $c_0 - c_1$, drives the diffusion through a resistance Ω_s:

$$\frac{1}{A_g}\frac{dm}{dt} = \frac{c_0 - c_1}{\Omega_s} = c_1\dot{a}. \tag{4}$$

Eliminating c_1 with Eq. (4) gives the concentration along the crack in terms of the concentration c_0 outside the surface film:

$$c = \frac{c_0 e^{-\dot{a}x/D_g}}{1 + \dot{a}\Omega_s}. \tag{5}$$

Assume that fracture occurs at a stress reduced from the pure grain boundary fracture strength σ_f by the local concentration in the grain boundary:

$$\frac{\sigma_{fs}}{\sigma_f} = 1 - \frac{c}{c_0} = 1 - \frac{e^{-\dot{a}x/D_g}}{1 + \dot{a}\Omega_s} \xrightarrow[\dot{a}\to 0]{} \dot{a}\left(\frac{x}{D_g} + \Omega_s\right). \tag{6}$$

For continuous cracking, take x to be the Burgers vector b.

The local stress to be compared to the fracture strength of Eq. (6) is limited by creep in the surrounding zone. Furthermore, the stress does not increase without limit as the crack tip is approached, but rather becomes relatively constant, on the average, within regions of the order of the dislocation spacing. This spacing can be approximated by the commonly appearing equation of dislocation mechanics relating the flow strength σ, the modulus of elasticity E, the Burgers vector b, and the spacing ℓ_d between dislocations or pinning points:

$$\ell_d = bE/\sigma. \tag{7}$$

Within this region, the stress could conceivably rise by the ordinary inverse square root singularity in an elastic region. The crack would advance very quickly in such cases, until it reached a dislocation or slip band whose motion relieved the stress. Then the crack would have to be reactivated at the flow strength σ. Therefore it seems reasonable to truncate the stress–radius relation around the crack tip (set by continuum mechanics as discussed in Section 4 below) at the stress corresponding to Eq. (7) with ℓ_d roughly the radius to the crack tip. The resulting creep stress is to be equated to σ_{fs} of Eq. (6) to determine the crack velocity.

2.2. Intermittent crack growth

The transient diffusion along an initially uncontaminated grain boundary is a nonanalytic function of time and distance, typically presented in graphical form as the Gurney–Lurie charts (e.g. McAdams, 1933). A close approximation in analytical form can be obtained by considering a lumped parameter system in which the contamination in the grain boundary is a constant value c over an instantaneous length x at time t. The concentration gradient controlling diffusion is approximated by the difference between that at the crack tip with no film, c_0, and c. Equating the mass flux rate into the element to its rate of increase within the element gives

$$\frac{D_g(c_0 - c)}{x} = \frac{cx}{t}; \qquad 1 - \frac{c}{c_0} = 1 - \frac{1}{1 + x^2/D_g t} = \frac{1}{1 + D_g t/x^2}. \tag{8}$$

Equation (8) represents the Y of the Gurney–Lurie charts. For $Dt/x^2 = 1$, Eq. (8) gives 0.5 compared to the 0.53 of the charts; for $Dt/x^2 = 4$, 0.2 compared to 0.28. Thus the single lumped parameter approximation may be quite adequate for describing an initial film. Again assume the strength drops linearly as the concentration rises to its saturation value c_0. An average crack velocity $\dot{a}$ can be obtained by taking it to be the ratio of the grain boundary size d_g to the cracking time t. Equation (8) then gives, for the local contaminated fracture strength for intermittent cracking,

$$\frac{\sigma_{fi}}{\sigma_f} = 1 - \frac{c}{c_0} = \frac{1}{1 + D_g/\dot{a}\,d_g} \xrightarrow[\dot{a} \to 0]{} \frac{\dot{a}\,d_g}{D_g}. \tag{9}$$

The fracture criterion given by Eq. (9) applies to strengths determined over a region of the order of the grain size d_g, so the stress-radius relation of continuum mechanics should be truncated at that value, with the creep strength equated to σ_{fi}.

2.3. Crack meandering and branching

For the mechanisms considered here, fracture is assumed to depend on the normal stress across a grain boundary. The deviation of cracks from one grain boundary to the next can be modeled roughly by a dog-leg crack in which the tip of the crack is deflected by an angle θ_d.

Elastic fields. For dog-legs of both finite and infinitesimal lengths under Mode I loading, as well as for cracks with branches on one or both ends, Lo (1978) has given the local stress intensity factors $K_{I\ell}$, $K_{II\ell}$ for several dog-leg angles, θ_d, by formulating a singular integral equation that could be readily evaluated using the quadrature formula developed by Erdogan et al. (1973). Using a less direct method, Chiang (1978) has given the local stress intensity factors for infinitesimal dog-leg for $\theta_d = 0°$ (5°) 90° under global loading in both Mode I and Mode II. Because the results are inconsistent for large θ_d, we have presented both sets in Table 1, along with the simple approximation that the local stress intensities $K_{I\ell}$ are the global ones K_{Ig} multiplied by the same trigonometric functions that apply to the circumferential normal and shear stress components around the tip of a straight crack:

$$K_{I\ell} = K_{Ig}(0.5)(1 + \cos\theta)\cos\frac{\theta}{2} + K_{IIg}(1.5)(\sin\theta)\cos\frac{\theta}{2},$$

$$K_{II\ell} = K_{Ig}(0.5)(\sin\theta)\cos\frac{\theta}{2} + K_{IIg}(1.5\cos\theta - 0.5)\cos\frac{\theta}{2}. \tag{10}$$

Table 1 also presents the resulting orientation and magnitude of the maximum circumferential normal stress intensity[1]. In spite of the disagreement between the two solutions, the general conclusion that can be drawn is that for dog-legs up to 60°, there is only a 10% reduction in stress intensity

---

[1] Note added in proof: Recent results by Nemat-Nasser should have been included. They are in close agreement with Lo's work, but controversy still exists.

Table 1
Local stress in intensity factors $K_{I\ell}$ and $K_{II\ell}$ for global Mode I loading of a crack with an infinitesimal dog-leg at θ_d, and corresponding local normal stress singularity, $\sigma_{\theta\theta}\sqrt{2\pi r}/K_{Ig}$

Dog-leg θ_d	0	-15	-30	-45	-60	-75	-90
				Local stress intensities			
Lo (1978)							
$K_{I\ell}$	1.0	0.974		0.789		0.505	
$K_{II\ell}$	0	-0.129		-0.324		-0.376	
Approx. (Eq. (10))							
$K_{I\ell}$	1.0	0.975	0.901	0.789	0.650	0.499	0.354
$K_{II\ell}$	0	-0.128	-0.241	-0.327	-0.375	-0.383	-0.354
Chiang (1978)							
$K_{I\ell}$	1.0	0.974	0.893	0.748	0.532	0.251	-0.058
$K_{II\ell}$	0	-0.129	-0.257	-0.373	-0.465	-0.510	-0.490
				Local circumferential normal stress singularities			
Approx.							
$\sigma_{\theta\theta\,mx}\sqrt{2\pi r}/K_{Ig}$	1.0	0.9991	0.988	0.952	0.880	0.771	0.632
$\theta_{\sigma mx}$	0	14.5	27	36	43	43	53
Chiang							
$\sigma_{\theta\theta\,mx}\sqrt{2\pi r}/K_{Ig}$	1.0	0.9994	0.991	0.959	0.880	0.738	0.535
$\theta_{\sigma mx}$	0	15	28	41	51	62	73

factor, and the subsequent direction of the crack tip should be within 15° of the direction of the main crack. The local stress intensity factors themselves show less agreement.

Plastic fields. In the plane strain plastic regime with full constraint, the slip line field of Fig. 3, adapted from Carson (1970), indicates that the maximum normal stress at the tip should be $\sigma_{\theta\theta} = 2k(1 + \pi/2 - \theta_d/2)$, at an angle of $\theta_d/2$ to the direction of the main crack. Thus for a 60° dog-leg the maximum circumferential stress is 30° outward from the main crack, which is somewhat more deviation than in the elastic case.

When grain boundary sliding (strain) is important, one would expect the crack to return to the general straight ahead direction, reducing meandering, because of the high shear strain in the fan along slip lines that lead across the ligament. The tendency for meandering is then reduced, compared to a pure stress criterion for fracture.

Statistical prenucleation and branching. McMahon (1974) has observed precracking ahead of the main crack even when oxidation is important. This introduces essential questions of statistics and three-dimensionality. Approximate analyses have been made by Dvorak (1976), Batdorf and Crose (1973), McClintock and Zaverl (1974), and McClintock and Mayson (1976).

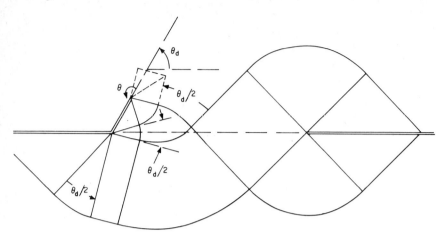

Fig. 3.   Slipline field for interaction of a single dog-leg with a straight crack.

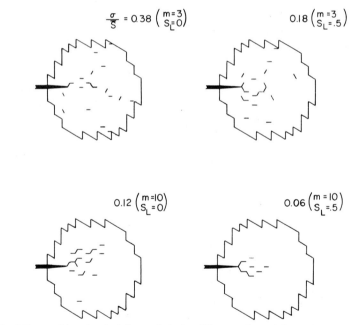

Fig. 4.   Microcracking ahead of the crack tip for different values of the coefficient of variation $\sigma/\bar{S}$.

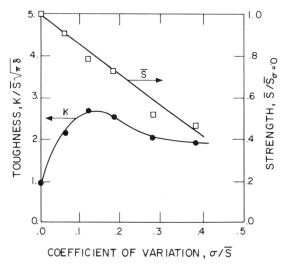

Fig. 5. The effect of variability on strength and crack toughness. (Strength data for a part size of 100 segments.)

Some numerical studies of crack growth have been reported by McClintock and Zaverl (1974) with the results shown in Fig. 4 for a variety of coefficients of variation. As shown in Fig. 5, a maximum toughness appeared possible, at a coefficient of variation of grain boundary strength of approximately 0.12. No analytical model has been obtained.

3. Asymptotic stress and flow fields for creep cracks

3.1. Stationary cracks

Riedel and Rice (1979) and Riedel (1979) have derived various singular stress fields that exist around a stationary crack tip under creep conditions when one deformation mode dominates in the far field and perhaps another at the crack tip. In the spirit of small-scale-yielding approximations, Riedel's (1979) results can be pieced together to demonstrate the evolution and disappearance of regions surrounding the crack tip that are dominated each by either elastic, time-independent plastic, primary creep, or secondary creep deformations. The uniaxial tension, stress–strain-rate (σ–$\dot{\varepsilon}$) relation is modeled in terms of Young's modulus E, plasticity exponent N, primary creep exponent n_p and hardening exponent p, secondary creep exponent n_s, and several normalizing quantities $(\ )_0$ as

$$\dot{\varepsilon} = \dot{\sigma}/E + (1/N)\varepsilon_0(\sigma/\sigma_0)^{(1/N)-1}(\dot{\sigma}/\sigma_0)$$
$$+ \dot{\varepsilon}_{p0}(\sigma/\sigma_0)^{n_p}(\varepsilon/\varepsilon_{p0})^{-p} + \dot{\varepsilon}_{s0}(\sigma/\sigma_0)^{n_s}. \qquad (11)$$

The normalizing quantities in Eq. (11) are not lumped together, to emphasize dimensional consistency throughout. For primary creep, $p > 0$. An isotropic, multiaxial generalization of Eq. (11) based on the Mises effective stress is obtained in the usual manner.

Initial loading of a crack in a material that obeys Eq. (11) may give rise to a nominal stress region surrounding an elastic singularity, a plastic singularity, and a fracture process zone. The two-dimensional elastic crack tip stresses are given in terms of the crack tip polar coordinates r and θ, the stress intensity factor K, and known nondimensional functions $f_{ij}(\theta)$ as

$$\sigma_{ij} = \frac{K}{\sqrt{2\pi r}} f_{ij}(\theta). \tag{12}$$

The singular fields associated with time-independent "plastic" and creep deformations are each of the Hutchinson (1968) or Rice and Rosengren (1968) (HRR) type. For example, a singular plastic field, that may be embedded within or engulf the singular elastic field, is given in terms of a J integral value that sets its amplitude, the crack length a, and the nondimensional functions $I_{1/N}$ and $\tilde{\sigma}_{ij}(\theta; 1/N)$ as

$$\frac{\sigma_{ij}}{\sigma_0} = \left(\frac{J}{\sigma_0 \varepsilon_0 a I_{1/N}} \frac{a}{r} \right)^{N/(N+1)} \tilde{\sigma}_{ij}(\theta; 1/N). \tag{13}$$

To within 2% accuracy, for plane strain, McClintock (1977) gives the approximation

$$I_n \approx 10.3\sqrt{0.13 + 1/n} - 4.8/n. \tag{14}$$

As time progresses, creep deformation can lead to crack tip regions governed by other singular stress fields. At a crack tip where the primary creep rates dominate the total strain rates, the stresses are given in terms of a time-dependent amplitude $C_p(t)$ as

$$\frac{\sigma_{ij}}{\sigma_0} = \left(\frac{C_p(t)}{\left[(p+1)\dot{\varepsilon}_{p0}/\varepsilon_{p0} \right]^{1/(p+1)} \sigma_0 \varepsilon_{p0} a I_m} \frac{a}{r} \right)^{1/(m+1)} \tilde{\sigma}_{ij}(\theta; m), \tag{15}$$

where

$$m = n_p/(p+1). \tag{16}$$

If secondary creep rates dominate at the crack tip, then the stresses are given in terms of a time-dependent amplitude $C_s(t)$ as

$$\frac{\sigma_{ij}}{\sigma_0} = \left(\frac{C_s(t)}{\sigma_0 \dot{\varepsilon}_{s0} a I_{n_s}} \frac{a}{r} \right)^{1/(n_s+1)} \tilde{\sigma}_{ij}(\theta; n_s). \tag{17}$$

The approximate field matching of Riedel and Rice (1979) and Riedel (1979) can lead to simultaneous existence, albeit in separate regions, of the stress fields in Eqs. (12), (13), (15), and (17). This matching assumes approximate path independence of the J integral. Proceeding from this innermost crack tip field out to the nominal field we will calculate the amplitude of each field in terms of the amplitude of its surrounding field. Certain restrictions will arise, and the reader is referred to Riedel (1979) for a more complete discussion. The plausibility of each restriction will be demonstrated in terms of a case study on 304 stainless steel.

(a) *Secondary creep region embedded in a primary creep region.* For $n_s > n_p/(p+1)$, the amplitude $C_s(t)$ of the secondary creep field in Eq. (17) is given in terms of the amplitude $C_p(t)$ of the surrounding primary creep field in Eq. (15) as

$$C_s(t) = \frac{n_s + p + 1}{(p+1)(n_s + 1)} \frac{C_p(t)}{t^{p/(p+1)}}. \tag{18}$$

(b) *Primary creep region embedded in a plastic region.* For $n_p/(p+1) > 1/N$,

$$C_p(t) = J \Big/ \left(\frac{n_p + p + 1}{p+1} t \right)^{1/(p+1)}, \tag{19}$$

where J is the amplitude of the plastic field given in Eq. (13).

(c) *Plastic region embedded in an elastic region.* For $1/N > 1$,

$$J = (1 - \nu^2) K^2 / E, \tag{20}$$

where ν is Poisson's ratio and K is the elastic stress intensity factor appearing in Eq. (12).

(d) *Elastic region embedded in a nominal stress field.* With σ_∞ denoting the nominal far field stress and F_K a number depending on geometry and loading that is catalogued for several cases in the handbook by Tada *et al.* (1973),

$$K = F_K \sigma_\infty \sqrt{\pi a}. \tag{21}$$

(e) *Elimination of fields.* Any one of the fields described above may, at some time, disappear. Furthermore, for high initial loading, the plastic field will dominate over the elastic one. If the secondary creep region wipes out the primary creep region but is within a plastic region, then

$$C_s(t) = \frac{J}{(n_s + 1)t}. \tag{22}$$

If this secondary creep region also wipes out the plastic region but is within the elastic region, then Eq. (22) still holds with Eq. (20) substituted for J.

This also applies if the primary region of Eqs. (15) and (19) engulfs the plastic but is within the elastic.

For high initial loading or long times, in which case the elastic region given by Eq. (12) does not exist, the amplitude of the outer most HRR field can be given in terms of the fully plastic solutions of Goldman and Hutchinson (1975), Parks (1978), Ranaweera and Leckie (1978), and Hutchinson, Needleman, and Shih (1978), for various geometries and loadings.

When the plastic field is embedded in the nominal field,

$$J = F_J \sigma_0 \varepsilon_0 a (\sigma_\infty / \sigma_0)^{(N+1)/N}. \tag{23}$$

For the plane strain, center-cracked panel in tension, for instance, the nondimensional number F_J is of order unity for moderate ratios of crack length to panel width. When the primary creep field is embedded in the nominal field, $C_p(t)$ approaches its steady-state value $C_p{}^*$ given by

$$C_p{}^* = F_p \sigma_0 \left[(p+1) \dot{\varepsilon}_{p0} / \varepsilon_{p0} \right]^{1/(p+1)} \varepsilon_{p0} a (\sigma_\infty / \sigma_0)^{(n_p + p + 1)/(p+1)}, \tag{24}$$

where, once again, F_p is a nondimensional number that depends only on geometry and the far field loading configuration. Finally, when the secondary creep field is embedded in the nominal field, $C_s(t)$ approaches its steady-state value $C_s{}^*$ given by

$$C_s{}^* = F_s \sigma_0 \dot{\varepsilon}_{s0} a (\sigma_\infty / \sigma_0)^{n_s + 1}. \tag{25}$$

The order suggested above for embedding the various crack tip solutions depends on the restrictions stated for the exponents N, n_p, p, and n_s. If, for instance, $1/N > n_p/(p+1)$ or n_s, then time-independent plasticity may persist at the crack tip within the creep zones. The limiting case of nonhardening plasticity $N = 0$ has been considered by Leckie and Mc-Meeking (1980). In that case, the nonsingular stresses at the crack tip are limited by the Prandtl solution.

3.2. Growing cracks

For the steady growth of cracks in a material that deforms both elastically and in secondary creep, Hui and Riedel (1980) have found an asymptotic solution that surprisingly depends only on crack velocity and not on any far-field boundary condition. With $\dot{a}$ denoting the crack velocity, in terms of the nondimensional function $\beta(n_s)$ (which is of order unity for plane stress or strain) and $\hat{\sigma}_{ij}(\theta; n_s)$,

$$\frac{\sigma_{ij}}{\sigma_0} = \beta(n_s) \left(\frac{\sigma_0 \dot{a}}{E \dot{\varepsilon}_{s0} r} \right)^{1/(n_s - 1)} \hat{\sigma}_{ij}(\theta; n_s). \tag{26}$$

As Hui and Riedel have discussed, $\dot{a}$ can be related to the far field geometry and loading through a nonlocal or history dependent fracture criterion. If the environmental fracture criterion considered here is a local one with no crack tip impedance to foreign species, then no coupling between $\dot{a}$ and σ_∞ is immediately obvious. If, on the other hand, the diffusion of a foreign species is limited by overall deformation such as the crack opening, then a coupling is possible.

4. Comparison with data and service conditions

Complete material data under creep conditions are readily available for only a few structural alloys. In particular the short time, primary creep response, that can lead to over half of the allowable total strain in some cases, is not well documented. As an exception, the data of Blackburn (1972) (also in the Nuclear Systems Materials Handbook) and Bynum and Roberts (1974) are complete enough to choose the constants that appear in Eq. (11) for annealed 304 stainless steel. Curve fitting to their data at 600° C has revealed limitations in the various power-law terms. The exponents and normalizing quantities are found to be dependent on stress level, as summarized in Table 2. The primary creep data are fitted at roughly one tenth the time it takes for primary creep to be exhausted. Our survey at other times indicates that n_p, p, and $\dot{\varepsilon}_{p0}$ are also functions of time. Note that the fit for plastic flow and secondary creep are relatively good over a factor of three in stress, but the coefficient $\dot{\varepsilon}_{p0}$ rises by a factor of 50. This emphasizes a need for other stress-strain relations, especially when analyzing the behavior at a crack tip. The stress levels range from the 50 MN/m² required for 1% creep in 30 years (10^{-11}/s) to the high stress levels which cause fracture in a few hours (10^4 s), which is often the case for tests in a

Table 2
Fit of Eq. (11) to data of Blackburn (1972)
to annealed 304 stainless steel at 600° C

Fix $\varepsilon_{p0}=0.01$, $\dot{\varepsilon}_{s0}=0.01/10^9\,\text{s}=10^{-11}\,\text{s}^{-1}$, $E=145\,500$ MN/m²			
σ_∞ (MN/m²)	50	100	150	Sample
n_s	6.32	7.22	8.63	7
σ_0 (MN/m²)	43.5	46.7	54.9	50
n_p	17.3	10.6	10.85	13
p	1.84	2.14	1.77	2
$\dot{\varepsilon}_{p0}(10^{-12}\,\text{s}^{-1})$	0.089	0.82	4.8	0.1
N	0.3	0.3	0.3	0.3
$\varepsilon_0(10^{-5})$	3.0	4.3	9.6	5

research laboratory. Round numbers were chosen for the material parameters and are given in the last column of Table 2. As an example we will consider the center-cracked panel, with crack length-to-width ratio equal to 1/2, under plane-strain, Mode I tension.

4.1. Laboratory stress levels

For a nominal far-field stress of 150 MN/m² ($\sigma_y/E = 0.001$), the resulting nested dominant singularities of stress are presented in Fig. 6. This stress level is perhaps typical of accelerated tests in a laboratory. For illustrative purposes we consider a short time of 10^4 s with a typical crack growth of 10 mm, giving a characteristic crack velocity of 10^{-6} m/s.

On first loading, the nominal stress reaches the yield regime, and the strain hardening singularity governed by the J factor of Eq. (23) dominates. After only 10 s the primary creep zone, governed by the coefficient C_p of Eq. (19) has cut into the strain hardening zone significantly. At perhaps 100 s the primary creep zone reaches the nominal regime. (Even less time would have been required had we fitted $\dot{\varepsilon}_{p0}$ at this stress.) Soon after 1000 s the secondary creep zone, governed by C_s of Eq. (18), begins to appear. At

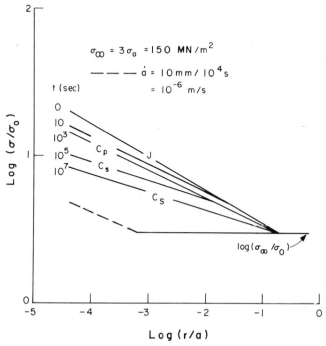

Fig. 6. Crack tip stresses in typical laboratory conditions.

about 10^7 s (3 months) the secondary zone reaches the nominal far field, governed by C_s of Eq. (25). From then on, steady-state creep persists, except for possible crack growth.

The stress singularity due to steady-state crack growth, given by Eq. (26), is shown by the dashed line of Fig. 6. Even for cracks growing at the relatively fast rate of 10 mm in 10^4 s, the singularity for growth is swamped by the secondary-creep singularity of a stationary crack. Only for rates of the order of 10^{-4} m/s, which would be attained toward the end of a constant load test, does the steady-growth singularity dominate over any substantial region. (Note that the differences in *strain* dominance between the different regimes are much greater than the differences in *stress* dominance, shown here.)

4.2.   Service stress levels

The nominal far-field stress of 50 MN/mm^2 shown in Fig. 7 is more typical of the stress levels for long time service. On first loading there is now a K-field surrounding the non-linear J-field. In perhaps 100 s the J-field has been squeezed out between the primary (C_p) and the elastic (K) fields. The

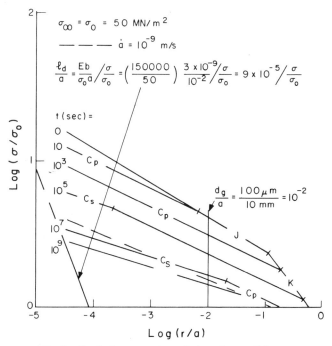

Fig. 7.   Crack tip stresses in typical service conditions.

secondary creep field C_s does not begin to play an important role until after 10^7 s (4 months). Even after 10^9 s (30 years) the secondary regime is still dominated by the primary zone, before the nominal applied stress is reached. The effect of a growing crack is more important at these service stress levels; its stress field dominates the static fields if the crack is growing at the rate of 10 mm in 4 months (10^{-9} m/s). This would be marginally acceptable for monthly inspection periods. With annual inspections, the crack growth would again have to be so slow that the stress fields could be estimated as being due to a succession of static states.

4.3. Microstructural limits to the singular fields

The first possible limit to the validity of the homogeneous continuum singularities is the grain size, typically 100 μm in austenitic stainless steels. With a crack length of 10 mm, this gives the vertical line of Fig. 7.

With 12 slip systems, individual crystals are at most $\pm 25\%$ from being plastically isotropic, however, so one may regard the singular solution as a reasonable approximation to a finer scale. The ultimate limit is set by the spacing of dislocation sources or dislocations themselves. An estimate of this limit can be found by adapting the usual equation for the shear strength required to replicate pinned dislocations by bowing out, or to force parallel dislocations past each other:

$$\tau = Gb/\ell_d \rightarrow \ell_d/a = (E/\sigma_0)(b/a)/(\sigma/\sigma_0). \tag{27}$$

The line corresponding to Eq. (27) for $E = 150\,000$ MN/mm^2, $b = 0.3 \times 10^{-9}$ m is shown in Fig. 7. Dislocation cell structures would give sizes intermediate between the dislocation spacing and the grain size.

Note that the above rules limit the stress concentration under long time creep, even under laboratory conditions, to about three times the nominal stress, in contrast to the stress concentration of a factor of 10 that occurs on rapid loading. This high stress itself indicates severe effects due to periodic unloading and reloading, and also the importance of changing the load slowly.

4.4. Possible corrosion crack growth rates

A very rough estimate of the grain boundary diffusion rate of oxygen in stainless steel can be obtained from the dimensional equation for the diffusion coefficient:

$$D_g = \nu b^2 e^{-U/kT}. \tag{28}$$

Take the Debye frequency to be $\nu = 10^{13}$/s, the atomic dimension to be the same as the Burgers vector at $b = 0.3 \times 10^{-9}$ m, the activation energy at 600–700° C to be 1 eV, and kT at 600–700° C to be $3/40$ eV. Then,

$$D_g \approx 10^{13}/s \, (0.3 \times 10^{-9} \, m)^2 \, e^{-1(3/40)} = 1.5 \times 10^{-12} \, m^2/s \qquad (29)$$

(perhaps a factor of 10 less than Gjostein's (1974) data for metallic grain boundaries).

Steady state crack growth. In Eq. (6), take the local stress at fracture to be $\sigma_{fs} = 3\sigma_0 = 150$ MN/m², σ_f to be $E/10$; with other numbers as taken above. Neglecting surface resistance, the crack growth rate is then

$$\dot{a} \approx \frac{\sigma_{fs}}{\sigma_f} \frac{D_g}{b} = \left(\frac{150 \text{ MN/m}^2}{150\,000/10 \text{ MN/m}^2} \right) \left(\frac{1.5 \times 10^{-12} \text{ m}^2/s}{0.3 \times 10^{-9} \text{ m}} \right)$$

$$= 5 \times 10^{-5} \text{ m/s}. \qquad (30)$$

This growth rate would certainly be great enough to be of practical importance.

Direct experimental evidence for the contribution of oxidation to grain boundary cracking is hard to find for stainless steel. In super-alloys, however, McMahon (1974) has found abundant evidence as shown, for example, in Fig. 1. Evidently in such cases, oxide plugging of the crack markedly reduces the rate of crack advance. The interaction of plugging and crack opening would also allow the far-field loading to play a role, even though it does not dominate the field of the growing crack itself. It is perhaps partly for this reason, as well as because of the high applied stresses, that a plateau region of constant crack growth rate at varying loading is not observed in growing creep cracks, although it does occur in stress corrosion cracking at room temperature.

Grain boundary cracking. As shown in Fig. 2, McMahon has also observed, in nickel-based superalloys, a penetration of the oxide film over several grains, with some grain boundaries cracking ahead of the main crack. For this mode, the intermittent cracking model of Eq. (9) is appropriate. At slow growth rates, the intermittent crack growth rate is slower than the continuous value simply by the ratio of interatomic spacing b to grain diameter d_g:

$$\dot{a} \approx 5 \times 10^{-5} \, m/s \left(\frac{0.3 \times 10^{-9} \text{ m}}{100 \times 10^{-6} \text{ m}} \right) \approx 1.5 \times 10^{-10} \text{ m/s}. \qquad (31)$$

This is at the upper limit of the crack growth rates of practical interest, which might be growth by 100 mm in 30 years (10^9 s), or 10^{-10} m/s.

5. Needed developments in plastic flow relations

As pointed out by Parks (1980), the stress history near a crack is very nonradial as a load sweeps by, so that the HRR strain singularity may be significantly in error. Furthermore, the primary creep may be due not only to hardening as suggested by Eq. (11), but also to a polarity in dislocation structure. As shown by an example of a hexagonal grid of screw dislocations in Fig. 8, the polarity may be apparent only from the Burgers vectors of the dislocations, and not from the configuration of the net itself. Transmission micrographs of the structure of René 95 illustrate the difficulty of representing the microscopic features of flow and fracture with a small number of variables. Fracture in the forged René 95 of Fig. 9 is associated with occasional large carbide particles. The $Ni_3(Al, Ti)$ precipitates may be tri-modal: very large precipitates in the grain boundaries, micron-sized cubic precipitates within the grain, and fractional micron-sized precipitates in the matrix between the micron-sized precipitates. The dislocation structures seem to sweep through the grains, around the micron-sized particles. Cell formation seems to be inhibited by the micron-sized or smaller precipitates. In some cases, these particles are dense enough so that their alignment with the cubic structure leads to very different flow strengths against shear and normal distortion, as referred to the cube axes (e.g. McClintock and Argon, 1966). The hot isostatically pressed (HIP) René 95 of Fig. 10 has a much more irregular structure, with multiple sizes of precipitates.

Steady state secondary and tertiary creep may be associated with (i) softening due to climb of dislocations past obstacles (with stress activation

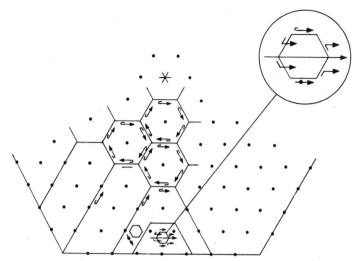

Fig. 8.   Hexagonal network of screw dislocations.

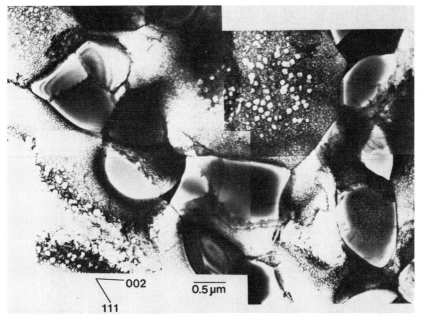

Fig. 9. Transmission micrograph of forged René 95.

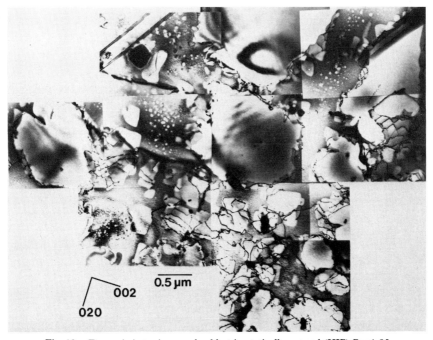

Fig. 10. Transmission micrograph of hot isostatically pressed (HIP) René 95.

still playing a role), (ii) overaging, (iii) dislocation cell formation, and (iv) the development of fracture. Each of these mechanisms would require a different flow formulation, all different from the mechanically convenient terms of Eq. (11).

6. Integration techniques

Improved plastic flow relations will probably require numerical solutions. As a step in this direction, we have done numerical calculations for the transient singularity for secondary creep in a far-field elastic regime, to verify the numerical methods against the known numerical results. For flexibility in calculating strain fields for such flow relations, we have used a cost effective, explicit time integration scheme, with initial strain finite element solutions. Short time steps for stability were taken from the Irons–Treharne (1971) and Cormeau (1976) estimates, but occasionally far larger steps are taken with the short steps used for re-stabilization.

7. Conclusions

(1) Equation (6) is given for cracking due to continuous grain boundary oxidation at the atomic scale. It includes a factor for oxide plugging at the crack tip, which is required to reduce the crack growth rates to reasonable values. Even cracking at oxides that have penetrated to a grain diameter beyond the crack tip can occur fast enough to be of practical concern (Eq. (31)).

(2) Riedel's regimes of dominance of singularities of elastic, strain hardening, and primary and secondary creep deformation around stationary cracks have been illustrated for stainless steel, in the regimes of interest in laboratory tests and in service. Important differences are noted in the early stages, but for the longest typical times in each case the steady-state creep singularity is embedded in a primary creep zone which is in turn imbedded in the nominal field. In the early stages, an elastic K-field is present for the low service stresses, but not for the high nominal stresses often encountered in laboratory tests.

(3) Microstructural considerations limit the effective stress concentration factor under long time conditions to a value of about 3, in contrast to the value of 10 for rapid initial loading.

(4) Needed improvements in stress–strain relations are discussed. These are particularly important in cases of nonradial loading due to out-of-phase loading or to the growth of a crack, and due to the possibility of repeated primary creep on unloading and reloading.

(5) The use of variable step sizes to accelerate numerical finite element calculations, with repeated Irons–Treharne–Cormeau small steps to restabilize the calculations, appears practical for strain rate exponents less than 10, even for the inhomogeneous fields associated with sharp cracks.

Acknowledgments

It is a pleasure to acknowledge the support of the National Science Foundation under Grant DMR-78-24185 to the M.I.T. Center for Materials Science and Engineering, which led to the ideas on fracture criteria and flow relations discussed here and also to the transmission micrographs of René 95 taken by Dr. Wm. M. Sherry. Likewise thanks are due to the Department of Energy under Contract EG-77-S-02-4461 for the application of the fracture criteria to the asymptotic stress and strain distributions in the different regimes, and for the discussion of numerical methods needed to treat intermediate regimes and more realistic flow relations.

References

Bassani, J. L. and McClintock, F. A. (1980), "Creep Relaxation of Stress Around a Crack Tip," accepted by *Int. J. Solids Struct.*

Batdorf, S. B. and Crose, J. G. (1973), "A Statistical Theory for the Fracture of Brittle Structures Subjected to Non-uniform Polyaxial Stresses," Aerospace Corp. Rep. TR-0073(3450-76)-2.

Blackburn, L. D. (1972), "Isochronous Stress–Strain Curves for Austenitic Stainless Steels," *The Generation of Isochronous Stress–Strain Curves*, A. O. Schaefer (ed.), Am. Soc. Mech. Eng., New York, 15.

Bynum, J. E. and Roberts, B. W. (1974), "Primary Creep Behavior of Type 304 Austenitic Steel," *Elevated Temperature Properties of Austenitic Stainless Steels*, A. O. Schaefer (ed.), Am. Soc. Mech. Eng., New York, 49.

Carson, J. W. (1970), "A Study of Plane Strain Ductile Fracture," Ph.D. Thesis, Dept. of Mech. Eng., Mass. Inst. of Tech., Cambridge, MA.

Chaku, P. N. and McMahon, C. J., Jr. (1974), "The Effect of an Air Environment on the Creep and Rupture Behavior of a Nickel-Base High Temperature Alloy," *Met. Trans.*, **5**, 441.

Chiang, W. T. (1978), "Fracture Criteria for Combined Mode Cracks," *Fracture 1977*, **4**, D. M. R. Taplin (ed.), Pergamon, New York, 135.

Cormeau, I. (1975), "Numerical Stability in Quasi-Static Elasto/Visco-Plasticity," *Int. J. Num. Meth. Eng.*, **9**, 109.

Dvorak, G. J. (1976), "The Influence of Stress States on the Fracture Toughness of bcc Metals," *Recent Advances in Engineering Science*, **7**, No. 2, T. S. Chang (ed.), 75.

Erdogan, F., Gupta, G. D. and Cook, T. S. (1973), "Numerical Solution of Singular Integral Equations," *Mechanics of Fracture 1, Methods of Analysis and Solutions of Crack Problems*, G. C. Sih (ed.), Noordhoff, Leyden, 368.

Floreen, S. (1975), "The Creep Fracture of Wrought Nickel-Base Alloys by a Fracture Mechanics Approach," *Met. Trans.*, **6A**, 1741.

Floreen, S. (1980), "High Temperature Crack Growth Structure–Property Relationships in Nickel-Base Superalloys," AIME Seminar, Sept. 1979, to be published in *Creep-Fatigue–Environment Interactions*, AIME.

Gjostein, N. A. (1974), "Short Circuit Diffusion," *Diffusion Seminar*, Am. Soc. Met., 241.

Goldman, N. L. and Hutchinson, J. W. (1975), "Fully Plastic Crack Problems: the Center-Cracked Strip Under Plane Strain," *Int. J. Solids Struct.*, **11**, 575.

Gooch, D. J., Haigh, J. R. and King, D. L. (1977), "Relationship Between Engineering and Metallurgical Factors in Creep Crack Growth," *Met. Sci.*, **11**, 545.

Haigh, J. R. (1975), "The Mechanisms of Macroscopic High Temperature Crack Growth, Part I: Experiments on Tempered Cr–Mo–V Steels," *Mater. Sci. Eng.*, **20**, 213.

Haigh, J. R. (1975), "The Mechanisms of Macroscopic High Temperature Crack Growth, Part II: Review and Re-analysis of Previous Work," *Mater. Sci. Eng.*, **20**, 225.

Hui, H. and Riedel, H. (1980), "The Asymptotic Stress and Strain Field Near the Tip of a Growing Crack under Creep Conditions," submitted to *Int. J. Fract.*

Hutchinson, J. W. (1968), "Singular Behaviour at the End of a Tensile Crack in a Hardening Material," *J. Mech. Phys. Solids*, **16**, 13.

Hutchinson, J. W., Needleman, A., and Shih, C. F. (1978), "Fully Plastic Crack Problems in Bending and Tension," *Fracture Mechanics*, N. Perrone et al. (eds.), Univ. Press of Virginia, Charlottesville, 515.

Irons, B. M. and Treharne, G. (1971), "A Bound Theorem in Eigenvalues and its Practical Applications," *Proc. 3rd. Conf. on Matrix Methods in Structural Mechanics*, R. M. Bader, et al. (eds.), AFFDL-DR-71-160.

Kennedy, A. J. (1953), "Creep and Recovery in Metals," *Brit. J. Appl. Phys.*, **4**, 225.

Leckie, F. A. and McMeeking, R. M. (1980), "Stress and Strain Fields at the Tip of a Stationary Tensile Crack in a Creeping Material," to appear in *Int. J. Fract.*

Lo, K. K. (1978), "Analysis of Branched Cracks," *J. Appl. Mech.*, **45**, 797.

McAdams, W. H. (1933), *Heat Transmission*, McGraw-Hill, New York, 35.

McClintock, F. A. (1977), "Mechanics in Alloy Design," *Fundamental Aspects of Structural Alloy Design*, R. I. Jaffee and B. A. Wilcox (eds.), Plenum, New York, 147.

McClintock, F. A. and Argon, A. S. (1966), *Mechanical Behavior of Materials*, Addison-Wesley, Reading, MA, 635.

McClintock, F. A. and Mayson, H. J. (1976), "Principal Stress Effects on Brittle Crack Statistics," *Effects of Voids on Material Deformation*, ASME Appl. Mech. Div., **16**, 31.

McClintock, F. A. and Zaverl, F., Jr. (1974), *Proc. 11th Ann. Meeting Soc. Eng. Sci.*, Duke Univ., 6.

McClintock, F. A. and Zaverl, F., Jr. (1979), "An Analysis of the Mechanics and Statistics of Brittle Crack Initiation," *Int. J. Fract.*, **15**, 107.

McMahon, C. J. (1974), "On the Mechanism of Premature In-Service Failure of Nickel-Base Superalloy Gas Turbine Blades," *Mater. Sci. Eng.*, **13**, 295.

McMahon, C. J., Jr. and Coffin, L. F., Jr. (1970), "Mechanisms of Damage and Fracture in High-Temperature, Low-Cycle Fatigue of a Cast Nickel-Based Superalloy," *Met. Trans.*, **1**, 3443.

Parks, D. M., (1978) "Virtual Crack Extension: a General Finite Element Technique for *J*-Integral Evaluation," *Numerical Methods in Fracture Mechanics*, A. R. Luxmoore and D. R. J. Owen (eds.), University College, Swansea, 464.

Parks, D. M. (1980), "The Dominance of the Crack Tip Fields of Inelastic Continuum Mechanics," *2nd Int. Conf. on Numerical Methods in Fracture Mechanics*, Pineridge Press, Swansea, 239.

Ranaweera, M. P. and Leckie, F. A. (1978), "Solution of Nonlinear Elastic Fracture Problems by Direct Optimization," *Numerical Methods in Fracture Mechanics*, A. R. Luxmoore and D. R. J. Owen (eds.), Swansea.

Rice, J. R. and Rosengren, G. F. (1968), "Plane Strain Deformation near a Crack Tip in a Power-Law Hardening Material," *J. Mech. Phys. Solids*, **16**, 1.

Riedel, H. (1979), "Creep Deformation at Crack Tips in Elastic-Visco-plastic Solids," Brown University Report MRL E-114.

Riedel, H. and Rice, J. R. (1979), "Tensile Cracks in Creeping Solids," Brown University Report E(11-1), 3084-64, submitted for publication to ASTM.

Sadananda, K. and Shahinian, P. (1978), "Application of Fracture Mechanics Techniques to High Temperature Crack Growth," *Fracture Mechanics*, N. Perrone, H. Liebowitz, D. Mulville, and W. Pilkey (eds.), University Press of Virginia, Charlottesville, 685.

Sadananda, K. and Shahinian, P. (1980), "Effect of Environment on Crack Growth Behavior in Austenitic Stainless Steels Under Creep and Fatigue Conditions," *Met. Trans.* **11A**, 267.

Sessions, M. L., McMahon, C. J., Jr. and Walker, J. L. (1977), "Further Observations on the Effect of Environment on the Creep Rupture Behavior of a Nickel-Base High Temperature Alloy: Grain Size Effects," *Mater. Sci. Eng.*, **27**, 17.

Tada, H., Paris, P. C. and Irwin, G. R. (1973), *The Stress Analysis of Cracks*, Del Research Corp., Hellertown, PA.

Widmer, R. and Grant, N. J. (1960), "The Creep Rupture Properties of 80Ni–20Cr Alloys," Trans. ASME, *J. Basic Eng.*, Ser D **82**, pp. 829 and 882.

S. Nemat-Nasser, Editor
THREE-DIMENSIONAL CONSTITUTIVE RELATIONS AND DUCTILE FRACTURE
North-Holland Publishing Company (1981) 147–153

DISCUSSION ON SESSION 2

T. MURA

Northwestern University, Evanston, IL, U.S.A.

Energy release rate and the *J*-integral

A. F. Kfouri and K. J. Miller have investigated the crack branching angle and the associated crack separation energy release rate G^Δ under a biaxial stress state by the use of plane strain elastic-plastic finite element method. They have assumed a simple relation (10), among stress intensity factors of the main crack and the branch crack, and found that the branching behavior of the elastic-plastic material is almost similar to that of the linear elastic material. This conclusion seems to be extremely useful, if it is true. Since the relation (10) is proved only for the linear elastic material, a further theoretical investigation is necessary to study its validity for elastic-plastic materials.

J. Weertman has theoretically derived the fatigue crack growth rate in terms of the stress amplitude intensity factor, the yield stress, the shear modulus, the ratio of the stress at which crack advance starts to the maximum stress, the surface energy, and a term that depends upon the ratio of the theoretical tensile strength to the theoretical shear strength. He has also derived an equation of crack growth when blunting determines the crack advance. Although some parts of his theory are not clear to me, his results are interesting because this is the first paper to deal with ductile versus brittle and blunting in connection with the fatigue crack growth. The basic assumption made by Weertman is that a small region around the crack tip always can be considered to be elastic and this region is surrounded by a plastic zone. The stress intensity factor at the crack tip is called the "true" stress intensity factor, K_t, which is smaller than that of the Griffith crack. In order to have such a small elastic enclave at the crack tip, the stress-strain curve of the material must have the elastic tangent modulus at large strains. I suspect that Weertman's theory may not be valid when a stress strain curve of material becomes flat at large strains.

*This work was supported by the National Science Foundation through the Materials Research Center of Northwestern University under Grant No. 7601057, and by the U.S. Army Research Office under Grant No. DAAG 29-77-G-0042.

Although these two papers, one by Kfouri and Miller and the other by Weertman, have different objectives, coincidentally they have a common background of motivation using the concept of the crack separation energy release rate G^Δ and the concept of the elastic enclave at the crack tip.

I would like to discuss this background in terms of the energy release rate of the crack extension and the J-integral.

Let us consider a quasi-static growth of a crack A in a body D subjected to an applied force F_i on the surface S (Fig. 1). The displacement and stress field are denoted, respectively, by u_i and σ_{ij}, when the crack surface is A. We are interested in evaluating the change of energy associated with a crack growth δA (Fig. 1) in order to define the generalized force at the crack tip necessary for the crack extension. Changes of the displacement and stress fields due to δA are denoted by δu_i and $\delta \sigma_{ij}$, we have

$$\int_S F_i \, \delta u_i \, \mathrm{d}S = \int_D \sigma_{ij} \, \delta u_{i,j} \, \mathrm{d}D - \int_B \sigma_{ij} n_j [\delta u_i] \, \mathrm{d}S, \qquad (1)$$

where B is a surface of discontinuity of δu_i. For most cases, B is δA. For the BCS model, B includes the plastic zone $[\delta u_i] = \delta u_i$ (on B^+) $- \delta u_i$ (on B^-).

Equation (1) has been derived from $\int_S F_i \, \delta u_i \, \mathrm{d}S = \int_S \sigma_{ij} n_j \, \delta u_i \, \mathrm{d}S$ by applying Gauss' theorem of integration, where the boundaries of D are S, A^+, A^-, B^+, and B^- (see Fig. 1) and $\sigma_{ij,j} = 0$ in D, and $\sigma_{ij} n_j = 0$ on $A^+ \equiv A$. The strain, $(1/2)(u_{i,j} + u_{j,i})$, is the sum of elastic strain, e_{ij}, and

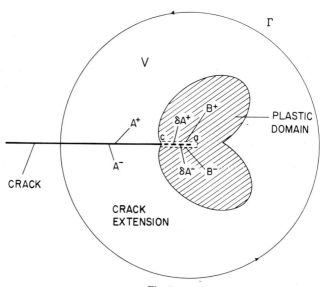

Fig. 1.

plastic strain, ε_{ij}^p, i.e.

$$(1/2)(u_{i,j}+u_{j,i})=e_{ij}+\varepsilon_{ij}^p. \tag{2}$$

Therefore, (1) can be written,

$$\int_S F_i\,\delta u_i\,\mathrm{d}S-\int_D\sigma_{ij}\,\delta e_{ij}\,\mathrm{d}D=\int_D\sigma_{ij}\,\delta\varepsilon_{ij}^p\,\mathrm{d}D-\int_B\sigma_{ij}n_j[\delta u_i]\,\mathrm{d}S. \tag{3}$$

Here, $\int_S F_i\,\delta u_i\,\mathrm{d}S$ is the work done by applied force F_i during the change of crack surface δA and can be written as $\delta\int_S F_i u_i\,\mathrm{d}S$ under the assumption that F does not change during the infinitesimal crack growth. $\int_D\sigma_{ij}\,\delta e_{ij}\,\mathrm{d}D=\delta[(1/2)\int_D\sigma_{ij}e_{ij}\,\mathrm{d}D]$ is the change of elastic strain energy. $\int_D\sigma_{ij}\,\delta\varepsilon_{ij}^p\,\mathrm{d}D$ is the plastic work done during the crack growth. Finally, $-\int_B\sigma_{ij}n_j[\delta u_i]\,\mathrm{d}S$ is the positive work done at the crack tip associated with the crack opening, $[\delta u_i]$.

The mechanical energy,

$$\Pi=(1/2)\int_D\sigma_{ij}e_{ij}\,\mathrm{d}D-\int_S F_i u_i\,\mathrm{d}S, \tag{4}$$

is a part of the Gibbs free energy. The extended Griffith fracture criterion can be written as

$$\delta G=\delta(\Pi+2A\gamma)=0, \tag{5}$$

where G is the Gibbs free energy and γ is the surface energy per unit area of the crack surface. Since the right-hand side in (3) expresses irreversible energies, $-\delta\Pi$ can be interpreted as the energy release during the crack growth δA. When the energy release rate is denoted by $\mathfrak{G}_l$, (3) can be written as

$$\mathfrak{G}_l\,\delta\xi_l=-\delta\Pi=\int_D\sigma_{ij}\,\delta\varepsilon_{ij}^p\,\mathrm{d}D-\int_B\sigma_{ij}n_j[\delta u_i]\,\mathrm{d}S, \tag{6}$$

where $\delta\xi_i$ is the displacement (change of position) of the crack tip due to δA. $\mathfrak{G}_l$ can also be called the crack extension force.

Irwin (1958) first evaluated,

$$-\int_{\delta A}\sigma_{ij}n_j[\delta u_i]\,\mathrm{d}S=\frac{1-\nu}{\mu}k_I^2\,\delta\xi_1, \tag{7}$$

for the Griffith elastic crack with length $2c$ under tension σ_{22}^o, where $k_I=\sigma_{22}^o(\pi c)^{1/2}$ is the stress intensity factor. As demonstrated by Yokobori and his coworker (1966, 1968), Rice (1966), Kfouri and Rice (1977), the BCS model yields

$$-\int_{\delta A}\sigma_{ij}n_j[\delta u_i]\,\mathrm{d}S=\frac{1-\nu}{2\mu}k_I^2(\delta\xi_1)^2/c, \tag{8}$$

which is one order smaller than (7).

Unless a stress singularity is considered as done for the elastic crack, the work of crack opening for a unit crack advance, $(-\int_{8A}\sigma_{ij}n_j[\delta u_i]\,dS)/\delta\xi_1$, becomes zero. This can be one of the reasons why Weertman considers the small elastic enclave. An alternative way to avoid this difficulty of zero work of crack opening may be to introduce the concept of the crack separation energy rate G^Δ as adopted by Kfouri and Miller.

First, the following assumption by Eshelby (1951, 1970) will be explained. After the crack tip has moved by $\delta\xi$, the change of displacement can be written as

$$\delta u_i = u_i(x-\delta\xi)-u_i(x)+\Delta u_i, \tag{9}$$

where Δu_i is a single-valued function. This equation implies that the displacement at point x after the crack advance $\delta\xi$, $u_i(x)+\delta u_i$, is almost equal to the displacement at point $x-\delta\xi$ before the crack advance, $u_i(x-\delta\xi)$. The difference is denoted by Δu_i, which is assumed to be small and single-valued. Then, $[\Delta u_i]=0$ on B, where B is defined in (1). Taylor's expansion of $u_i(x-\delta\xi)$ about point x is

$$u_i(x-\delta\xi)=u_i(x)-u_{i,l}\delta\xi_l. \tag{10}$$

Then, (9) becomes

$$\delta u_i = -u_{i,l}\delta\xi_l+\Delta u_i \tag{11}$$

or

$$[\delta u_i]=-[u_{i,l}]\,\delta\xi_l \tag{12}$$

on B, since Δu_i is a single-valued function.

Now we discuss the second term in the right-hand side in (6). Wnuk (1972, 1974) postulated that a part of the integral domain in the second term is a material constant (process zone) and derived a quasi-static extension of a tensile crack. He defined the crack separation energy by

$$G_l^\Delta \delta\xi_l = -\int_\Delta \sigma_{ij}n_j[\delta u_i]\,dS, \tag{13}$$

where G^Δ is the crack separation energy rate and Δ is the process zone size. For the BCS model, (13) becomes

$$G^\Delta\delta\xi_1=\int_c^{c+\Delta}k_0[\delta u_2]\,dx_1, \tag{14}$$

where $G^\Delta=G_1^\Delta$ and k_0 is the yield stress. From (12) $[\delta u_2]=-[u_{2,1}]\delta\xi_1$ and therefore

$$G^\Delta=-k_0\int_c^{c+\Delta}[u_{2,1}]\,dx_1=-2k_0\{u_2(c+\Delta,c)-u_2(c,c)\}$$
$$\simeq-2k_0\{u_2(c+\Delta,c)-u_2(c+\Delta,c+\Delta)\}. \tag{15}$$

For a small scale yielding,

$$u_2(x_1, c) = \frac{4(1-v^2)k_0}{\pi E} \{R - \xi/2 - (\xi/2) \log(4R/\xi)\}, \qquad (16)$$

where $a - c = R$, $x_1 - c = \xi$, and a is the x_1-coordinate of the plastic zone tip. Then, (15) yields

$$G^\Delta = \frac{8(1-v^2)k_0^2\Delta}{\pi E} \left(\frac{dR}{dc} + \frac{1}{2} + \frac{1}{2} \log \frac{4R}{\Delta} \right). \qquad (17)$$

The above equation was obtained by Wnuk (1972) and a similar result by Rice and Sorensen (1978), Kfouri and Rice (1977), and Kfouri (1979). We have, for the small scale yielding,

$$R = (c/2)Q^2, \qquad Q \equiv \pi S/2k_0, \qquad (18)$$

where Q is a loading parameter. Assuming Δ and G^Δ to be constant Eqs. (17) and (18) give a relation, $Q = Q(c)$, which represents a stable crack growth. The catastrophic propagation is determined by the condition $dQ/dc = 0$. This idea has been considered by Cherepanov (1968), McClintock (1958), Rice (1968), and Wnuk (1979) among others.

The generalized force acting on the tip of a crack was expressed in terms of the integral (J-integral) of the Maxwell energy-momentum tensor P_{lj} of elasticity taken over a surface. This fact was first pointed out by Eshelby (1951, 1970) and later by Rice (1968).

The three-dimensional expression of the J-integral is defined by

$$J_l = \int_\Gamma Wn_l dS - \int_\Gamma \sigma_{ij} n_j u_{i,l} dS = \int_\Gamma P_{lj} n_j dS \qquad (19)$$

with

$$P_{lj} = W\delta_{lj} - \sigma_{ij} u_{i,l}; \qquad (20)$$

Γ is an arbitrary closed surface (a closed curve for the two-dimensional case) including the crack tip, n is the outward normal vector on dS, and

$$W_{,l} = \sigma_{ij} \varepsilon_{ij,l}, \qquad (21)$$

where $\varepsilon_{ij} = (1/2)(u_{i,j} + u_{j,i})$. It is easy to show that J_l is a path-independent integral. If another path Γ^* is taken, application of Gauss' theorem on integrals leads to $\int_{\Gamma-\Gamma^*} Wn_l \, dS - \int_{\Gamma-\Gamma^*} \sigma_{ij} n_j u_{i,l} \, dS = 0$. Weertman showed that (21) is valid for a growing crack. When (21) can be integrated, we have

$$W = \int_0^{\varepsilon_{ij}} \sigma_{ij} \, d\varepsilon_{ij}. \qquad (22)$$

Denoting V as the domain bounded by Γ, A^+, B^+, A^-, B^-, (19) can be written by the Gauss' theorem as

$$J_l = \int_V W_{,l} \, dD - \int_V \sigma_{ij} u_{i,lj} \, dD + \int_{B^\pm} \sigma_{ij} n_j u_{i,l} \, dS, \qquad (23)$$

where $W_{,l} = \sigma_{ij} u_{i,jl}$, $\sigma_{ij,j} = 0$, and $\sigma_{ij} n_j = 0$ on A. It has been assumed in the derivation of (23) that W on $A^+ + B^+$ is equal to W on $A^- + B^-$. The first and second integrals in (23) cancel each other and $\sigma_{ij} n_j$ on B^+ is $-\sigma_{ij} n_j$ on B^-. Then, (23) becomes

$$J_l = \int_B \sigma_{ij} n_j [u_{i,l}] \, dS, \tag{24}$$

where $[u_{i,l}] = (u_{i,l}$ on $B^+) - (u_{i,l}$ on $B^-)$. From (12) we can also have

$$J_l \delta \xi_l = - \int_B \sigma_{ij} n_j [\delta u_i] \, dS. \tag{25}$$

Expression (6) can, therefore, be written as

$$\mathcal{G}_l \delta \xi_l = \int_D \sigma_{ij} \delta \varepsilon_{ij}^{\mathrm{P}} \, dD + J_l \delta \xi_l. \tag{26}$$

Equation (26) has never been reported except for the elastic case where $\mathcal{G}_l = J_l$. For the perfectly plastic material, $J_1 \to 0$ when B is taken as δA. This contradicts some of Rice's calculations (1968).

If some dislocations move with the crack tip by $\delta \xi_l$, we can write

$$\delta \varepsilon_{ij}^{\mathrm{P}} = -\varepsilon_{jlh} \alpha_{hi} \, \delta \xi_l, \tag{27}$$

where α_{hi} is the dislocation density tensor and ε_{jlh} is the permutation tensor. Then, (26) yields

$$\mathcal{G}_l = - \int_D \sigma_{ij} \varepsilon_{jlh} \alpha_{hi} \, dD + J_l. \tag{28}$$

The integrand is the Peach–Koehler force.

References

Cherepanov, G. P. (1968), "On Quasibrittle Fracture," *J. Appl. Math. Mech.*, **32**, 1050.

Eshelby, J. D. (1951), "The Force on an Elastic Singularity," *Phil. Trans. Roy. Soc.*, **A244**, 87.

Eshelby, J. D. (1970), "Energy Relations and the Energy-Momentum Tensor in Continuum Mechanics," *Inelastic Behavior of Solids*, M. F. Kanninen, W. F. Adler, A. R. Rosenfield and R. I. Jaffee (eds.), Battell Inst. Mater. Sci. Colloquia, McGraw-Hill, New York, 77.

Irwin, G. R. (1958), *Fracture, Handbuch der Physik*, **6**, Springer-Verlag, Berlin, 551.

Kfouri, A. P. (1979), "Crack Separation Energy-rates for the DBCS Model under Biaxial Modes of Loading," *J. Mech. Phys. Solids*, **27**, 135.

Kfouri, A. P., and Rice, J. R. (1977), "Elastic-Plastic Separation Energy Rate for Crack Advance in Finite Growth Steps," *Fracture 1977*, **1**, D. M. R. Taplins (ed.), Univ. of Waterloo Press, Waterloo, Ont., 43.

Kfouri, A. P. and Miller, K. J. "Crack Separation Energy Rates for Inclined Cracks in an Elastic-plastic Material," in these *Proceedings*.

McClintock, F. A. (1958), "Ductile Fracture Instability in Shear," *J. Appl. Mech.*, **25**, 582.

Rice, J. R., (1966), "Plastic Yielding at a Crack Tip," *Proc. 1st Int. Conf. on Fracture at Sendai*, **1**, T. Yokobori, T. Kawasaki and J. L. Swedlow (eds.), 283.

Rice, J. R., (1968), "A Path Independent Integral and the Approximate Analysis of Strain Concentration by Notches and Cracks," *J. Appl. Mech.*, **35**, 379.

Rice, J. R. and Sorensen, E. P., (1976), "Continuity Crack-tip Deformation and Fracture for Plane-strain Crack Growth in Elastic-plastic Solids," *J. Mech. Phys. Solids*, **26**, 163.

Weertman, J., "Fatigue Crack Growth Theory for Ductile Material," in these *Proceedings*.

Wnuk, M. P., (1972), "Accelerating Crack in a Viscoelastic Solid Subject to Subcritical Stress Intensity," *Proc. Int. Conf. on Dynamic Crack Propagation*, G. Sih, (ed.), Lehigh University, 273.

Wnuk, M. P., (1974) "Quasi-static Extension of a Tensile Crack Contained in a Viscoelastic-plastic Solid," *J. Appl. Mech.*, **41**, 243.

Wnuk, M. P., (1979) "Occurrence of Catastrophic Fracture in Fully Yielded Components; Stability Analysis," *Int. J. Fracture*, **15**, 553.

Yokobori, T. and Ichikawa, M. (1966), "The Energy Principle as Applied to the Elastic-plastic Crack Model for Fracture Criterion," *Rep. Res. Inst. Strength and Fracture of Mater.*, **2**, Tohoku University, 21.

Yokobori, T. (1968), "Criteria for Nearly Brittle Fracture," *Int. J. Fracture Mech.*, **4**, 179.

SESSION 3

Chairman: D. McLean
             National Physical Laboratory, Teddington, Middlesex, U.K.

Authors: S. Nemat-Nasser, M. M. Mehrabadi, T. Iwakuma
             J. R. Rice
             F. M. Beremin Research Group

Discussion: J. W. Hutchinson

Reply: F. M. Beremin Research Group

S. Nemat-Nasser, Editor
THREE-DIMENSIONAL CONSTITUTIVE RELATIONS AND DUCTILE FRACTURE
North-Holland Publishing Company (1981) 157–172

ON CERTAIN MACROSCOPIC AND MICROSCOPIC ASPECTS OF PLASTIC FLOW OF DUCTILE MATERIALS

S. NEMAT-NASSER, M. M. MEHRABADI, and T. IWAKUMA

Northwestern University, Evanston, IL, U.S.A.

Microscopic aspects of ductile flow of two-phase alloys are reviewed and the influence of microvoids, microscopic slip, and hydrostatic pressure (tension) on macroscopic plastic flow is discussed. Then, in light of experimental results, a phenomenological plasticity theory with nonassociative flow rule that accounts for pressure sensitivity and plastic volume expansion, as well as possible localization of deformation at the micro-level, is presented. Various associated parameters are discussed and given physical interpretation. As an alternative complementary approach, rate equations for rate independent plastic flow are obtained by the consideration of microslips accompanied by microscopic plastic volumetric changes affected by pressure. Various special cases are discussed and compared with published results. Then, localization in biaxial loading is examined, including pressure and compressibility effects. Finally, some experimental results on localization in axial tension and compression of certain maraging steels are reviewed and used to estimate at the critical state the values of several microscopic parameters that enter the microstructurally based theory.

1. Introduction

Ductility refers to the extent to which a material can deform plastically before fracture. It is dominant in certain metals over a limited temperature range. The existence of nonmetallic inclusions in alloys substantially limits their ductility, because of the formation of microvoids at these inclusions, followed by the growth of these voids and their interconnection which may occur either by internal necking of the matrix that separated two adjacent voids (void coalescence), or by the formation of *void sheets* at smaller precipitates in intense shear bands (localization of plastic distortion) which connect adjacent large voids; for literature review and rather extensive references, see Nemat-Nasser (1977); Goods and Brown (1979). To understand and quantify the phenomenon of ductile flow, the following basic problems must be addressed:

(1) the mechanism of void initiation and void growth;

(2) the mechanism of localization of plastic distortion at the microlevel;

(3) the effect of plastic work accumulation within the matrix, on macroscopic behavior;

(4) the effect of plastic volumetric strain on macroscopic behavior;

(5) macroscopic tension sensitivity.

In addition, and despite the long-held (by the mechanics community) myth of rate-independent plasticity, rate effects are also significant, even at room temperature, as commonly recognized by material scientists, and recently accentuated by Krempl (1979). Here, however, a rate-independent theory is presented in order to bring to focus the essential features enumerated above. Two approaches are considered: (1) a macroscopic phenomenological approach based on a minor but significant generalization of the usual J_2 plasticity theory; and (2) a microscopic approach based on the consideration of slip-lines, plastic localization, and void growth at the microscale. Both approaches permit further generalization to include rate effects. The resulting constitutive relations include: (a) material softening due to initiation and growth of voids; (b) material hardening due to plastic distortion of the matrix; (c) material softening due to the localization of plastic distortion at the microlevel; and (d) pressure (tension) sensitivity.

This paper is organized in the following manner. In the remaining part of the present section, the above-mentioned effects are briefly discussed. In Section 2, a plasticity theory is presented, and various associated parameters are discussed and given physical interpretation; the theory represents an extension of the recent work by Nemat-Nasser and Shokooh (1980). In Section 3, rate equations for rate-independent plastic flow are obtained by consideration of microslips accompanied by microscopic plastic volumetric changes affected by pressure (tension). This part represents a generalization of Asaro's (1979) recent contribution which did not include plastic volume expansion and pressure sensitivity; the general theory is also specialized and results are compared with Asaro's. In Section 4, localization in biaxial loading is examined in light of the theory developed in Section 3, and various special cases are discussed. In particular, experimental results by Anand and Spitzig (1980) on the formation of localized shear bands in maraging steels, are reviewed and used to estimate the critical values of the microscopic constitutive parameters for the special case of the rigid-plastic model of Section 3. Anand and Sptizig have shown that the predictions of the usual J_2 flow theory and those of a deformation theory do not accord with the experimental facts. In view of this, their experimental data are reviewed and are shown to be consistent, satisfying with good accuracy the usual characteristic equation for localization in incrementally linear materials.

1.1.   Void initiation

Recent experiments on various steels have indicated that initially plastic volumetric *expansion* accompanies plastic distortion in all deformation modes, i.e., tension, compression, and torsion. This plastic volume increase appears to be independent of the hydrostatic stress or the value of the principal stresses, and seems to depend only on the total effective plastic distortion. The plastic volume expansion is, however, small, but nevertheless it exists. In fact, experiments by Dyson *et al.* (1976) suggest that the rate of plastic volumetric expansion per unit distortional strain is (initially) constant, its magnitude of course depending on the considered material; see also Spitzig *et al.* (1975, 1976).

At later stages of plastic distortion, void growth becomes a dominant factor. It is reasonable to expect that hydrostatic tension and stress triaxiality can enhance the void growth process.

1.2. *Localization of plastic distortion*

Microheterogeneity in material structure, e.g. grain boundaries, second-phase particles, and voids, promotes plastic distortion by localized deformations. The formation of localized deformations at the microlevel has a macroscopic softening effect which tends to become more significant as microscopic localization tends to magnify during the course of deformation. This is a very significant feature of ductile fracture, which has not yet been fully quantified in terms of macroscopic constitutive relations, although corner theories of plasticity are motivated by this fact; see, e.g., Christoffersen and Hutchinson (1979).

At the scale of tens of microns the existence of voids and inclusions promotes plastic localization. This has been shown for a two-dimensional model by Nemat-Nasser and Taya (1978), where a unit cell consisting of a single void has been subjected on its boundary to the deformation history which has been separately calculated for a necked bar.

1.3. *Pressure sensitivity*

Experiments by Spitzig *et al.* (1975, 1976) show that the yield function is affected by hydrostatic tension or compression, i.e., yield stress increases with increasing hydrostatic pressure. These authors note that, since the pressure sensitivity of yielding exceeds the corresponding plastic volumetric expansion by an order of magnitude, the usual normality rule used in plasticity models for metals does not apply. A consideration of a nonassociative flow rule, however, permits retention of normality with respect to the flow potential, yielding results compatible with experimental facts, as shown in the next section.

2. Macroscopic approach

Based on the above observations the following yield function and flow potential are introduced:

$$f \equiv \bar{\sigma} - F(I, \Delta, \bar{\varepsilon}, \Sigma), \quad \text{yield function};$$
$$g \equiv \bar{\sigma} + G(I, \Delta, \bar{\varepsilon}, \Sigma), \quad \text{flow potential}, \tag{1}$$

where

$$\bar{\sigma}^2 = \frac{1}{2} \sigma'_{ij} \sigma'_{ij}, \qquad I = \sigma_{kk},$$

$$\Delta = \int_0^\theta \frac{\rho_0}{\rho} D^p_{kk} \, d\theta, \qquad \bar{\varepsilon} = \int_0^\theta (2 D^{p'}_{ij} D^{p'}_{ij})^{1/2} \, d\theta; \tag{2}$$

here, a fixed rectangular Cartesian coordinate system is used; σ_{ij} are the corresponding stress components; prime denotes the deviatoric part; ρ_0 and ρ are the reference and current mass densities; D^p_{ij} are the components of the plastic part of the deformation rate tensor; θ is a monotone increasing load parameter; $\bar{\sigma}$ and $\bar{\varepsilon}$ are the effective stress and strain; Δ is the total *plastic* volumetric strain measured per unit reference volume; and Σ is the localization parameter defined on the microscale and characterizes the extent of localization of deformation within the matrix in a typical sample of the material; see Nemat-Nasser and Taya (1978). It is convenient to use the effective plastic strain, $\bar{\varepsilon}$, as the load parameter. Then all superposed dots denote differentiation with respect to $\bar{\varepsilon}$. This is followed in the sequel.

From (1) the plastic part of the strain rate is

$$D^p_{ij} = \dot{\lambda} \frac{\partial g}{\partial \sigma_{ij}} = \dot{\lambda} \left\{ \frac{\sigma'_{ij}}{2\bar{\sigma}} + \frac{\partial G}{\partial I} \delta_{ij} \right\}, \tag{3}$$

which, in view of the consistency relation $\dot{f} = 0$, becomes

$$D^p_{ij} = \frac{1}{H} \left\{ \frac{\sigma'_{ij}}{2\bar{\sigma}} + \frac{\partial G}{\partial I} \delta_{ij} \right\} \left\{ \frac{\sigma'_{kl}}{2\bar{\sigma}} - \frac{\partial F}{\partial I} \delta_{kl} \right\} \mathring{\sigma}_{kl}, \tag{4}$$

where $\mathring{\sigma}_{kl}$ denotes the Jaumann rate of Cauchy stress,

$$\mathring{\sigma}_{kl} = \dot{\sigma}_{kl} - W_{ki} \sigma_{il} - W_{li} \sigma_{ik}; \tag{5}$$

W_{ij} is the spin tensor.

In (4), H is the work-hardening parameter,

$$H = 3 \frac{\rho_0}{\rho} \frac{\partial G}{\partial I} \frac{\partial F}{\partial \Delta} + \frac{\partial F}{\partial \bar{\varepsilon}} + \frac{\partial F}{\partial \Sigma} \frac{\partial \Sigma}{\partial \bar{\varepsilon}}, \tag{6}$$

and $3 \partial G / \partial I$ is the dilatancy factor,

$$3 \frac{\partial G}{\partial I} = \frac{\bar{\sigma}}{\sigma'_{ij} D_{ij}} D^p_{kk} = \frac{D^p_{kk}}{\dot{\bar{\varepsilon}}}, \tag{7}$$

being the rate of plastic volumetric change per unit rate of plastic distortion.

The first term on the right-hand side of (6) is material hardening (softening) due to plastic volumetric changes (void growth). Since $\partial F/\partial \Delta$ is always nonpositive, the sign of this term depends on the sign of D_{kk}^p, which, as pointed out before, at least in the initial stages of deformation, is positive. Thus, the *density-hardening*

$$h_1 = 3(1+\Delta)\frac{\partial G}{\partial I}\frac{\partial F}{\partial \Delta} \tag{8}$$

is negative during the early stages of plastic flow. Moreover, it continues to remain negative during the process of stable void growth. In fact, Eq. (7) shows that $\partial G/\partial I$ has the same sign as D_{kk}^p, which, for void growth, is positive. Hence, h_1 represents material softening due to the geometric effects.

The quantity $\partial F/\partial \bar{\varepsilon}$ in Eq. (6) represents material hardening due to average plastic distortion of the matrix. It is always nonnegative and can be measured at least approximately, on carefully grown very thin specimens which do not include large inclusions.

The last term in the right-hand side of Eq. (6) represents macroscopic material softening due to formation of microscopic intense localized deformations. In the microscale these localized deformations behave in a similar manner as "plastic hinges" in structural frames. Their distribution and geometric pattern are highly affected by the microstructure of the material. Keeping this in mind, it may be assumed that Σ is essentially a function of the distortional plastic strain, i.e.

$$\Sigma = \Sigma(\bar{\varepsilon}). \tag{9}$$

In fact, calculation by Nemat-Nasser and Taya (1978) suggests that

$$\Sigma = \Sigma_0(1 - e^{-k\bar{\varepsilon}}) \tag{10}$$

may be a reasonable approximation, where k and Σ_0 depend on the microstructure, i.e. the material.

It is helpful to pause at this point and review the experimental support for a nonassociative flow rule. In a series of compression-tension tests of several steels, Spitzig *et al.* (1975, 1976) report a pronounced pressure effect on the yield stress and a definite plastic volumetric expansion. If the effective stress, $\bar{\sigma}$, is redefined as

$$\bar{\sigma} = \left(\tfrac{3}{2}\sigma_{ij}'\sigma_{ij}'\right)^{1/2}, \tag{11}$$

then it equals the magnitude of axial stress, σ, imposed on an overall hydrostatic compression (or tension), in a tension-compression test. In this case, $-\partial F/\partial I$ equals the parameter, a, used by Spitzig *et al.* They report that a is essentially independent of strain and therefore, the strength differential (S-D) equals twice this parameter. For an associative flow rule,

$G \equiv -F$ and therefore, $\partial G/\partial I = -\partial F/\partial I$. Based on this assumption, Spitzig et al. calculate the plastic volumetric expansion which turns out to be approximately 15 times larger than the observed values. These authors then conclude that normality (with associative flow rule) may not hold, in general, for materials of this kind.

According to our theory, $\partial G/\partial I \neq -\partial F/\partial I$, in general, because $\partial G/\partial I$ is a kinematical quantity having the same sign as the instantaneous rate of plastic volumetric change, whereas $-\partial F/\partial I$ pertains to the strength of the material and is always positive. Because of the nature of plastic flow, it so happens that for metals both quantities possess the same sign, at least during the early stages of deformation, although under very large pressures negative $\partial G/\partial I$ can be expected (collapse of microvoids). Indeed, for granular materials, $\partial G/\partial I$ changes sign during the normal course of deformation, whereas $-\partial F/\partial I$ remains strictly positive, characterizing the overall frictional resistance.

Based on the experimental data of Spitzig et al., values of $-\partial F/\partial I$ and $\partial G/\partial I$ for several steels are given in Table 1. These values support the use of a nonassociative flow rule.

Since $\partial F/\partial I$ is always negative (F decreases with increasing tension), it is expected that ductile flow should occur at higher stress levels in compression than in tension. However, essentially the same strain patterns are involved in both cases. This has been observed experimentally by Anand and Spitzig (1980), where localized deformations have been seen to occur essentially at the same strain levels (≈ 0.034 axial strain) in tension and compression, but at different stress levels. The patterns of these shear bands are the same in tension and compression, forming an approximately 38° angle with the maximum principal stress axis. Similar results have been reported by other investigators, as discussed, for example, by Asaro (1979) and Goods and Brown (1979). Even for nickel-base superalloys, slip bands are an integral part of the microstructure of plastic flow. Kikuchi and Weertman (1980) have observed that slips emanate from carbides at grain

Table 1

Measured values of pressure sensitivity parameter $-\partial F/\partial I$ and dilatancy parameter $\partial G/\partial I$ for indicated steels; data from Spitzig et al. (1975, 1976)

Steel	$-\dfrac{\partial F}{\partial I}(\%)$	$\dfrac{\partial G}{\partial I}(\%)$
Maraging (unaged)	1.8	<0.03
4310	2.8	0.15
4330	3.0	0.15
Maraging (aged)	3.5	0.23

boundaries that are parallel to the direction of extension in tensile specimens strained about 10% at room temperature, and then annealed for two
hours at 810°C.

The existence of slip bands gives rise to the macroscopic concept of
corner for the yield surface. This implies that further plastic strain rates will
be affected both in their orientation and magnitude by the corresponding
additional stress rates, i.e. the plastic strain rate tensor is noncoaxial with
the stress tensor. All theories which consider a smooth flow potential
depending on stress invariants, lead to coaxiality, and hence cannot accurately account for the effects of the formation of microscopic localized slip
bands.

In the present theory the difficulty may be circumvented by the addition
of a term linear in $\mathring{\sigma}'_{ij}$ to the right-hand side of Eq. (4). If this term is chosen
as

$$A\left\{\mathring{\sigma}'_{ij} - \frac{1}{2\bar{\sigma}^2}\sigma'_{kl}\mathring{\sigma}_{kl}\sigma'_{ij}\right\}, \tag{12}$$

then it does not contribute to the rate of plastic work. Moreover, the
physical meanings of the dilatancy parameter, $\partial G/\partial I$, and the pressure-
sensitivity parameter, $-\partial F/\partial I$, are unchanged. It is interesting to note that
such a term naturally emerges in the double-slip theory of granular materials; Spencer (1964), de Josselin de Jong (1971), and Mehrabadi and Cowin
(1978, 1980). With the addition of this term and upon separation of the
strain rate into the distortional and dilatational parts, it follows that

$$D^{p'}_{ij} = \frac{\sigma'_{ij}}{2H\bar{\sigma}}\left\{\frac{\sigma'_{kl}}{2\bar{\sigma}} - \frac{\partial F}{\partial I}\delta_{kl}\right\}\mathring{\sigma}_{kl} + A\left\{\mathring{\sigma}'_{ij} - \frac{1}{2\bar{\sigma}^2}\sigma'_{kl}\mathring{\sigma}_{kl}\sigma'_{ij}\right\},$$

$$D^p_{kk} = \frac{3}{H}\frac{\partial G}{\partial I}\left\{\frac{\sigma'_{kl}}{2\bar{\sigma}} - \frac{\partial F}{\partial I}\delta_{kl}\right\}\mathring{\sigma}_{kl}. \tag{13}$$

As it stands, it is difficult to interpret the parameter A on a physical basis.
Some interpretations in terms of a secant modulus have been made by
Rudnicki and Rice (1975) and Stören and Rice (1975), using a generalization of linear elasticity in line with deformation plasticity theory. It is,
however, desirable to arrive at modifications of this kind by a physical
microstructural modeling. This is done in the next section for a two-dimensional case, although the basic approach admits generalization to three-
dimensional problems.

3.   Microscopic approach

Plastic deformation of metals involves flow in the form of slip, even
during the early stages of loading. For a single crystal, a system of

double-slip develops symmetrically about the direction of maximum principal stress. For polycrystalline metals also, slip systems are activated in a definite pattern in relation to the maximum stress direction; see, for example, Dyson et al. (1976); Kikuchi and Weertman (1980). As pointed out in the introduction, during the final stages of ductile flow, intense localized deformations form between adjacent voids, which may lead to the generation of void sheets. Therefore, both for single crystals and alloys it is reasonable to consider a microstructural model that consists of slip at active slip systems, accompanied by plastic volumetric change associated with the formation and growth of microvoids. The model then applies even to the later stages of ductile flow, where intense localized deformation bands (formed between adjacent voids) are viewed as individual slip systems. Moreover, the process is, to a large extent, pressure sensitive. In view of these observations, a dilatant, pressure-sensitive set of rate constitutive relations will now be developed by a systematic calculation based on the concept of dilatant, active slip systems having a microscopic pressure-sensitive constitutive response.

3.1. Notation

Quantities pertaining to an individual slip system are denoted by the addition of Greek superscripts which, when repeated, are summed over all instantaneously active slip systems. Since two-dimensional flow (plane strain) is envisaged, Italic subscripts take on values 1 and 2, and the repeated ones are summed. For convenience, dyadic and indicial notations are used, e.g. the stress tensor $\boldsymbol{\sigma}$ with components σ_{ij}, deformation rate and spin tensors, $\boldsymbol{D}$, $\boldsymbol{W}$, with components D_{ij}, W_{ij}, etc. The fourth-order elasticity tensor is denoted by $\boldsymbol{L}$, having components L_{ijkl}. For application, isotropic elasticity is assumed, i.e.

$$L_{ijkl} = G(\delta_{ik}\delta_{jl} + \delta_{il}\delta_{jk}) + \lambda\delta_{ij}\delta_{kl}, \tag{14}$$

where G (shear modulus) and λ are the Lamé coefficients.

3.2. General theory[1]

The total velocity gradient, $\boldsymbol{D} + \boldsymbol{W}$, is obtained by plastic distortion giving rise to $\boldsymbol{D}^{\mathrm{p}} + \boldsymbol{W}^{\mathrm{p}}$, followed by elastic distortion with elastic deformation rate, $\boldsymbol{D}^{\mathrm{e}}$, which is accompanied by a spin denoted by $\boldsymbol{W}^{*}$, so that

$$\boldsymbol{D} = \boldsymbol{D}^{\mathrm{e}} + \boldsymbol{D}^{\mathrm{p}}, \qquad \boldsymbol{W} = \boldsymbol{W}^{*} + \boldsymbol{W}^{\mathrm{p}}, \tag{15}$$

as discussed by Hill (1966), Hill and Rice (1972), Asaro and Rice (1977) and Asaro (1979); (15)$_1$ represents an exact decomposition in the context of continuum plasticity; Nemat-Nasser (1979).

---

[1]Earlier related works are by Taylor (1938), Bishop and Hill (1951), Budiansky and Wu (1962), Mandel (1966), Hill (1967, 1971), Hutchinson (1970), Kocks (1970), Lin (1971), Bui et al. (1972), Zarka (1973), and Havner (1979), where reference to other work can be found.

The plastic parts in (15) are viewed to have stemmed from the plastic slip rates, $\dot{\gamma}^\alpha$, and the plastic dilation rates, $\dot{\theta}^\alpha$, over all active slips. Let $\boldsymbol{n}^\alpha$ be the unit vector normal to the αth slip band, $\boldsymbol{s}^\alpha$ be the unit vector in the direction of the slip, and introduce the following tensors associated with each individual slip system:

$$p_{ij}^\alpha = \tfrac{1}{2}\left(s_i^\alpha n_j^\alpha + s_j^\alpha n_i^\alpha\right) + \tan \nu\, n_i^\alpha n_j^\alpha,$$

$$\omega_{ij}^\alpha = \tfrac{1}{2}\left(s_i^\alpha n_j^\alpha - s_j^\alpha n_i^\alpha\right), \qquad \text{(no sum on } \alpha), \tag{16}$$

where $\tan \nu$ is the dilatancy parameter,

$$\tan \nu = \frac{\mathrm{d}\theta^\alpha}{\mathrm{d}\gamma^\alpha}, \tag{17}$$

assumed, for simplicity, to be the same for all α. Then the plastic constituents in (15) are

$$\boldsymbol{D}^\mathrm{p} = \dot{\gamma}^\alpha \boldsymbol{p}^\alpha, \qquad \boldsymbol{W}^\mathrm{p} = \dot{\gamma}^\alpha \boldsymbol{\omega}^\alpha, \tag{18}$$

where α is summed over all active slip systems.

The elastic parts in (15) give rise to elastic stress rates corotational with the corresponding spin, according to

$$\overset{\triangledown}{\boldsymbol{\sigma}} = \boldsymbol{L} : \boldsymbol{D}^\mathrm{e}, \tag{19}$$

$$\overset{\triangledown}{\boldsymbol{\sigma}} = \dot{\boldsymbol{\sigma}} - \boldsymbol{W}^* \boldsymbol{\sigma} + \boldsymbol{\sigma} \boldsymbol{W}^*. \tag{20}$$

For a typical active slip, α, the shear rate is assumed to be governed by the following constitutive relation:

$$\dot{\tau}^\alpha + \tan \eta\, \dot{\sigma}^\alpha = h^{\alpha\beta} \dot{\gamma}^\beta, \tag{21}$$

where $\tau^\alpha = \sigma_{ij} s_i^\alpha n_j^\alpha$ and $\sigma^\alpha = \sigma_{ij} n_i^\alpha n_j^\alpha$ (α not summed) are the shear and normal stresses transmitted over the αth slip plane, $h^{\alpha\beta}$ is the symmetric work-hardening matrix, and $\tan \eta$ is the pressure-sensitivity parameter. In addition to (21), it is necessary to introduce constitutive assumptions that describe time variations of the unit vectors, $\boldsymbol{n}^\alpha$ and $\boldsymbol{s}^\alpha$. Various possibilities may be entertained, but the final judgment is dictated by the physical modeling associated with actual microstructural behavior. Here, for simplicity, Asaro's (1979) assumption will be used,

$$\dot{n}_i^\alpha = W_{ij}^* n_j^\alpha, \qquad \dot{s}_i^\alpha = W_{ij}^* s_j^\alpha. \tag{22}$$

Then it follows from (21) that

$$\overset{\triangledown}{\sigma}_{ij} q_{ij}^\alpha = h^{\alpha\beta} \dot{\gamma}^\beta, \tag{23}$$

where

$$q_{ij}^\alpha = \tfrac{1}{2}\left(s_i^\alpha n_j^\alpha + s_j^\alpha n_i^\alpha\right) + \tan \eta\, n_i^\alpha n_j^\alpha \qquad \text{(no sum on } \alpha) \tag{24}$$

represents the normal to the yield surface associated with αth slip system.

Let

$$N^{\alpha\beta} = h^{\alpha\beta} + q^\alpha : L : p^\beta, \tag{25}$$

and denote the inverse of matrix $N^{\alpha\beta}$ by $M^{\alpha\beta}$. Then, with the aid of (15)−(24) and in view of (5), it can easily be shown that

$$\overset{\circ}{\sigma} = L : D + M^{\alpha\beta}(\sigma\omega^\alpha - \omega^\alpha\sigma - L : p^\alpha)(q^\beta : L : D). \tag{26}$$

In this equation, the superscripts α and β are summed over all instantaneously active slip systems.

The rate constitutive relations represented by (26) include pressure sensitivity and dilatancy, but nevertheless have a rather simple structure.

3.3. Special cases

To bring the basic structure of rate constitutive Eqs. (26) to light, consider a special case of only two active slip systems, symmetrically oriented with respect to the maximum principal stress, σ_1 at the angle [2] $(\pi/4)+(\phi/2)$; the other principal stress is σ_2. Assume further,

$$h^{11} = h^{22} = h \quad \text{and} \quad h^{12} = h^{21} = h_1, \tag{27}$$

and introduce the following parameters:

$$M = \frac{\sin\eta}{\cos(\phi-\eta)}, \qquad B = \frac{\sin\nu}{\cos(\phi-\nu)},$$
$$2H_1 = (h-h_1)(1-M\sin\phi)(1-B\sin\phi),$$
$$2\mu^* \cos^2\phi = (h+h_1)(1-M\sin\phi)(1-B\sin\phi), \tag{28}$$

where M is the pressure sensitivity parameter, and B the dilatancy parameter proportional to $\partial G/\partial I$ of Section 2, as is seen from

$$B = (D_{11}^p + D_{22}^p)/(D_{11}^p - D_{22}^p). \tag{29}$$

It is now easy to deduce from (26) the following explicit rate equations:

$$\overset{\circ}{\sigma}_{11} = a_1 D_{11} + a_2 D_{22}, \qquad \overset{\circ}{\sigma}_{22} = b_1 D_{11} + b_2 D_{22}, \qquad \overset{\circ}{\sigma}_{12} = 2\mu D_{12}, \tag{30}$$

where

$$a_1 = K[(2G+\lambda)\mu^* + G(G+\lambda)(1-M)(1-B)],$$
$$a_2 = K[\lambda\mu^* + G(G+\lambda)(1-M)(1+B)],$$
$$b_1 = K[\lambda\mu^* + G(G+\lambda)(1+M)(1-B)],$$
$$b_2 = K[(2G+\lambda)\mu^* + G(G+\lambda)(1+M)(1+B)],$$
$$\mu = \frac{G[H_1 + \tau(M-\sin\phi)(1-B\sin\phi)]}{H_1 + G(M-\sin\phi)(B-\sin\phi)},$$
$$K = [\mu^* + G + MB(G+\lambda)]^{-1}, \qquad \tau = \tfrac{1}{2}(\sigma_1-\sigma_2). \tag{31}$$

---

[2] Note that the angle ϕ used here relates to the angle Φ used by Asaro (1979); $\Phi = (\pi/4) \mp (\phi/2)$ for tension and compression, respectively.

Further specialization results if elastic incompressibility is imposed. Care is required when obtaining the limiting form of Eqs. (31), where the quantity $\lambda/2(G+\lambda)$ which equals the Poisson ratio, must approach $1/2$. This results in a new set of parameters, denoted by superposed bar, as follows:

$$\bar{a}_1 = \frac{1}{MB}\left[\mu^* + G(1-M)(1-B)\right],$$

$$\bar{a}_2 = \frac{1}{MB}\left[\mu^* + G(1-M)(1+B)\right],$$

$$\bar{b}_1 = \frac{1}{MB}\left[\mu^* + G(1+M)(1-B)\right],$$

$$\bar{b}_2 = \frac{1}{MB}\left[\mu^* + G(1+M)(1+B)\right], \tag{32}$$

where μ remains unchanged. Observe that the material is still plastically compressible as displayed by the presence of dilatancy parameter B.

In the absence of pressure sensitivity, M vanishes, and then the pressure rate is indeterminate. It follows that

$$\frac{G}{G+\mu^*}(D_{11}-D_{22})-\frac{1}{B}(D_{11}+D_{22})=0. \tag{33}$$

The deviatoric part of the stress rate, $\overset{\circ}{\sigma}'$, is now obtained from (30) and (31) by setting $M=0$.

Finally, when pressure insensitive, incompressible plastic flow is assumed, the indeterminate quantity, $(D_{11}+D_{22})/B$, is eliminated, using (33). The resulting rate equation for the deviatoric part of the stress tensor is summarized by

$$\overset{\circ}{\sigma}' = \frac{G\mu^*}{G+\mu^*}(D_{11}-D_{22})f, \tag{34}$$

where Eq. $(30)_3$ remains the same,

$$f=\begin{pmatrix} 1 & 0 \\ 0 & -1 \end{pmatrix}, \tag{35}$$

and μ^* is obtained from (28) with $M=B=0$, i.e.

$$\mu^* = (h+h_1)/2\cos^2\phi, \tag{36}$$

leading to Asaro's (1979) equations.

4. Localization

Let n with components n_1 and n_2 be the unit vector normal to a localized shear band formed in biaxial plane strain extension (or compression). Then, for materials with constitutive relations (30), the orientation of this band, defined by $c = n_2/n_1$, satisfies,

$$b_2(\mu - \tau)c^4 + [a_1 b_2 - a_2 b_1 - a_2(\mu - \tau) - b_1(\mu + \tau)]c^2 + a_1(\mu + \tau) = 0. \tag{37}$$

Substitution from (31) results in

$$(\mu - \tau)[2(1 - \hat{\nu})\mu^* + G(1 + M)(1 + B)]c^4$$
$$+ 2\{G[2\mu^* - \tau(M - B) - \mu(1 - MB)]$$
$$- 2\hat{\nu}\mu\mu^*\}c^2 + (\mu + \tau)[2(1 - \hat{\nu})\mu^* + G(1 - \hat{\nu})(1 - B)] = 0, \tag{38}$$

where $\hat{\nu}$ is the Poisson ratio.

The rigid plastic limit is obtained when G is taken to infinity, yielding

$$(\mu_R - \tau)(1 + M)(1 + B)c^4 + 2[2\mu^* - \mu_R(1 - MB) - \tau(M - B)]c^2$$
$$+ (\mu_R + \tau)(1 - M)(1 - B) = 0, \tag{39}$$

where μ_R is the limiting value of μ in Eq. $(31)_5$, as $G \to \infty$, i.e.

$$\mu_R = \frac{H_1 + \tau(M - \sin\phi)(1 - B\sin\phi)}{(M - \sin\phi)(B - \sin\phi)}. \tag{40}$$

It should be noted that Eq. (39), with a different μ_R, has been given previously by Mehrabadi and Cowin (1980) in connection with soil plasticity.

Equation (39) reduces to Hill and Hutchinson's (1975) characteristic equation, when M and B are set equal to zero, and μ_R and μ^* are reinterpreted accordingly. This yields

$$(\mu_R - \tau)c^4 + 2(2\mu^* - \mu_R)c^2 + (\mu_R + \tau) = 0, \tag{41}$$

where μ^* is given by (36), and

$$\mu_R = \frac{h - h_1 - 2\tau\sin\phi}{2\sin^2\phi}. \tag{42}$$

The characteristic Eq. (41) is identical with that obtained by Asaro (1979) for the rigid-plastic case. Asaro also considers the more general case of incompressible elastoplastic materials; the corresponding equation is obtained from (38) by setting $\hat{\nu} = \frac{1}{2}$ and $M = B = 0$, arriving at

$$(\hat{\mu} - \tau)c^4 + 2\left(\frac{2G\mu^*}{G + \mu^*} - \hat{\mu}\right)c^2 + (\hat{\mu} + \tau) = 0, \tag{43}$$

where μ^* is given by (36) and $\hat{\mu}$ is obtained from Eq. $(31)_5$,

$$\hat{\mu} = \frac{G(h - h_1 - 2\tau\sin\phi)}{h - h_1 + 2G\sin^2\phi}. \tag{44}$$

4.1. Comparison with experiments

As mentioned before, Anand and Spitzig (1980) report localized shear bands occurring in tension and compression tests of certain maraging steel specimens. The bands are reported to form at angles $\pm(38\pm2)°$ about the maximum principal stress, the maximum principal stress being zero for the compression test. The localization takes place at about 0.034 axial strain in both compression and tension. The corresponding tangent Young modulus, E_t, is reported to be about 391 MPa for tension and 458 MPa for compression. The corresponding secant modulus, E_s, is reported to be 58,656 MPa for tension and 59,046 MPa for compression.

Anand and Spitzig assume incompressibility and a plane strain condition, and apply both the J_2 flow theory and a deformation theory in order to predict their experimental observations. The flow theory predicts shear band orientation of about $\pm45°$ at the critical axial strain of the absolute value of 0.184, whereas the deformation theory predicts shear band orientation of about $\pm42.55°$ (with respect to the maximum stress direction) at the critical strain of the absolute value of 0.085, both for compression and for tension.

The experimental results of Anand and Spitzig are in very good accord with the prediction which can be deduced from the characteristic Eq. (41); this is not reported by the authors who have focused on estimating the critical strain. Indeed, if (41) is rewritten as

$$Ac^4 + 2Bc^2 + C = 0, \tag{45}$$

where

$$A = \mu_R - \tau, \qquad B = 2\mu^* - \mu_R, \qquad C = \mu_R + \tau, \tag{46}$$

then real values for c require $B^2 - AC \geq 0$ or $\tau^2 \geq \mu_R^2 - (2\mu^* - \mu_R)^2$, so that the smallest absolute value of τ is obtained when the critical condition $B^2 = AC$ or

$$\tau^2 = \mu_R^2 - (2\mu^* - \mu_R)^2 \tag{47}$$

holds. The corresponding critical value of c^2 is given by

$$c^2 = \frac{-B}{A} = \sqrt{\frac{C}{A}} = \sqrt{\frac{\mu_R + \tau}{\mu_R - \tau}}, \tag{48}$$

where $2\mu^* < \mu_R$ is assumed, which is almost always the case.

For incompressible materials, $4\mu^* = E_t$ which is measured at the critical state. Therefore, μ_R in (48) can be eliminated in favor of μ^*, with the aid of Eq. (47). Direct substitution results in

$$c^2 = \frac{\tau + 2\mu^*}{\tau - 2\mu^*} = \frac{2\tau + E_t}{2\tau - E_t} \tag{49}$$

which applies at the critical state independently of the particular interpretation of the instantaneous moduli μ^* and μ_R, as long as $2\mu^* < \mu_R$.

To obtain the angle θ which the direction of the shear band makes with the maximum principal stress, it is observed that $\tan^2 \theta = 1/c^2$, so that

$$\tan^2 \theta = \frac{2\tau - E_t}{2\tau + E_t}. \tag{50}$$

Entering the recorded magnitude of the stress of 1,877 MPa in tension, and 2,126 MPa in compression, it is immediately deduced from (50) that $\theta = \pm 39.0°$ in tension and $\pm 38.8°$ in compression, which are in excellent agreement with the observed results.

The recorded experimental data can be used to estimate the values of the microscopic parameters, h, h_1, in terms of ϕ, at the critical state. From (36), (42), and (47) it follows that

$$h = \frac{\tau(\cos 2\theta + \sin \phi)^2}{2\cos 2\theta}, \tag{51}$$

$$h_1 = \tau \cos 2\theta \cos^2 \phi - h, \tag{52}$$

which give the values of h and h_1 at the critical state.

For a single crystal, ϕ is regarded as a crystalline parameter. For a polycrystalline material, on the other hand, ϕ must be viewed as a statistically averaged, microscopically preferred, microlocalization orientation parameter. In this context, it is interesting to observe that when the average microlocalization orientation coincides with the orientation of (macro) shear band, as one would expect that within the framework of the present approximation it should, then the hardening parameter h vanishes, as is seen from Eq. (51).

Acknowledgment

This work has been supported by the National Science Foundation under Grant No. ENG 76-03921 to Northwestern University.

References

Anand, L. and Spitzig, W. A. (1980), "Initiation of Localized Shear Bands in Plane Strain," *J. Mech. Phys. Solids*, **28**, 113.

Asaro, R. J. (1979), "Geometrical Effects in the Inhomogeneous Deformation of Ductile Single Crystals," *Acta Met.*, **27**, 445.

Asaro, R. J. and Rice, J. R. (1977), "Strain Localization in Ductile Single Crystals," *J. Mech. Phys. Solids*, **25**, 309.

Bishop, J. F. W. and Hill, R. (1951), "A Theory of the Plastic Distortion of a Polycrystalline Aggregate Under Combined Stresses," *Phil. Mag.*, **42**, 414.

Budiansky, B. and Wu, T. T. (1962), "Theoretical Prediction of Plastic Strains of Polycrystals," *Proc. 4th U.S. Nat. Congr. Appl. Mech.*, 1175.

Bui, H. D., Zaoui, A. and Zarka, J. (1972), "Sur le Comportement Élastoplastique et Viscoplastique des Monocristaux et Polycristaux Métalliques de Structure Cubique à Faces Centrées," *Foundations of Plasticity*, A. Sawczuk (ed.), Noordhoff Int. Publ., Leyden, 51.

Christoffersen, J. and Hutchinson, J. W. (1979), "A Class of Phenomenological Corner Theories of Plasticity," *J. Mech. Phys. Solids*, **27**, 465.

de Josselin de Jong, G. (1971), "The Double Sliding, Free Rotating Model for Granular Assemblies," *Géotechnique*, **21**, 155.

Dyson, B. F., Loveday, M. S. and Rodgers, M. J. (1976), "Grain Boundary Cavitation Under Various States of Applied Stress," *Proc. Roy. Soc. (London)*, **A349**, 245.

Goods, S. H. and Brown, L. M. (1979), "The Nucleation of Cavities by Plastic Deformation," *Acta Met.*, **27**, 1.

Havner, K. S. (1979), "The Kinematics of Double Slip with Application to Cubic Crystals in the Compression Test," *J. Mech. Phys. Solids*, **27**, 415.

Hill, R. (1966), "Generalized Constitutive Relations for Incremental Deformation of Metal Crystals by Multislip," *J. Mech. Phys. Solids*, **14**, 95.

Hill, R. (1967), "The Essential Structure of Constitutive Laws for Metal Composites and Polycrystals," *J. Mech. Phys. Solids*, **15**, 79.

Hill, R. (1971), "On Macroscopic Measures of Plastic Work and Deformation in Micro-heterogeneous Media," *PMM*, **35**, 31.

Hill, R. and Hutchinson, J. W. (1975), "Bifurcation Phenomena in the Plane Tension Test," *J. Mech. Phys. Solids*, **23**, 239.

Hill, R. and Rice, J. R. (1972), "Constitutive Analysis of Elastic-Plastic Crystals at Arbitrary Strain," *J. Mech. Phys. Solids*, **20**, 401.

Hutchinson, J. W. (1970), "Elastic-Plastic Behavior of Polycrystalline Metals and Composites," *Proc. Roy. Soc. (London)*, **A319**, 247.

Kikuchi, M. and Weertman, J. R. (1980), "Mechanism for Nucleation of Grain Boundary Voids in a Nickel Base Superalloy," *Scripta Met.*, **14**, 797.

Kocks, U. F. (1970), "The Relation Between Polycrystal Deformation and Single-Crystal Deformation," *Met. Trans.*, **1**, 1121.

Krempl, E. (1979), "An Experimental Study of Room-Temperature Rate-Sensitivity, Creep and Relaxation of AISI Type 304 Stainless Steel," *J. Mech. Phys. Solids*, **27**, 363.

Lin, T. H. (1971), "Physical Theory of Plasticity," *Advances in Applied Mechanics*, Vol. **11**, C.-S. Yih (ed.), Academic Press, New York, 255.

Mandel, J. (1966), "Contribution Théorique à l'Étude de l'Écrouissage et des Lois de l'Écoulement Plastique," *Proc. 11th Int. Congr. Appl. Mech.*, Springer-Verlag, Berlin, 502.

Mehrabadi, M. M. and Cowin, S. C. (1978), "Initial Planar Deformation of Dilatant Granular Materials," *J. Mech. Phys. Solids*, **26**, 269.

Mehrabadi, M. M. and Cowin, S. C. (1980), "Pre-failure and Post-failure Soil Plasticity Models," *J. Eng. Mech. Div.*, Trans. ASCE, Special Issue on Mechanics of Heterogeneous Media, **106**, 991.

Nemat-Nasser, S. (1977), "Overview of the Progress in Ductile Fractures," (invited lecture) *Trans. 4th Int. Conf. Struct. Mech. in Reactor Tech.*, San Francisco, CA, August 15–19, Vol. **L**, 1.

Nemat-Nasser, S. (1979), "Decomposition of Strain Measures and their Rates in Finite Deformation Elastoplasticity," *Int. J. Solids Structures*, **15**, 155.

Nemat-Nasser, S. and Shokooh, A. (1980), "On Finite Plastic Flows of Compressible Materials with Internal Friction," *Int. J. Solids Structures*, **16**, 495.

Nemat-Nasser, S. and Taya, M. (1978), "Model Studies of Ductile Fracture," *Continuum Models of Discrete Systems*, J. W. Provan and H. H. E. Leipholz (eds.), Study No. 12, Solid Mech. Div., University of Waterloo Press, Waterloo, Ont., 387.

Rudnicki, J. W. and Rice, J. R. (1975), "Conditions for the Localization of Deformation in Pressure-sensitive Dilatant Materials," *J. Mech. Phys. Solids*, **23**, 371.

Spencer, A. J. M. (1964), "A Theory of the Kinematics of Ideal Soils under Plane Strain Conditions," *J. Mech. Phys. Solids*, **12**, 337.

Spitzig, W. A., Sober, R. J. and Richmond, O. (1975), "Pressure Dependence of Yielding and Associated Volume Expansion in Tempered Martensite," *Acta Met.*, **23**, 885.

Spitzig, W. A., Sober, R. J. and Richmond, O. (1976), "The Effect of Hydrostatic Pressure on the Deformation Behavior of Maraging and HY-80 Steels and its Implications for Plasticity Theory," *Met. Trans.*, **7A**, 1703.

Stören, S. and Rice, J. R. (1975), "Localized Necking in Thin Sheets," *J. Mech. Phys. Solids*, **23**, 421.

Taylor, G. I. (1938), "Plastic Strains in Metals," *J. Inst. Metals*, **62**, 307.

Zarka, J. (1973), "Étude du Comportement des Monocristaux Métalliques. Application à la Traction du Monocristal C.F.C.," *J. Mécanique*, **12**, 275.

S. Nemat-Nasser, Editor
THREE-DIMENSIONAL CONSTITUTIVE RELATIONS AND DUCTILE FRACTURE
North-Holland Publishing Company (1981) 173–184

CREEP CAVITATION OF GRAIN INTERFACES

James R. RICE

Brown University, Providence, RI, U.S.A.

The paper presents an analysis of diffusional cavity growth on grain interfaces. Derivation of the growth rate is reviewed for the simplest case when cavities retain a quasi-equilibrium, spherical caps shape and when the adjoining grains separate in an effectively rigid manner. Modifications of the results are then discussed based on the possibility of nonequilibrium cavity shapes, dislocation creep of the adjoining grains, and constraints which arise when cavitated grain facets are relatively isolated in a polycrystalline aggregate.

1. Introduction

At elevated temperatures, of the order 0.5 T_m and higher, polycrystalline materials generally fracture by the initiation and growth of cavities along grain boundaries. Various analyses suggest that owing to the usual size range of these cavities (say, 0.1 to 1 μm), the dominant growth mechanism is by surface and grain boundary diffusion; see, e.g., the comparison of characteristic times for different matter transport mechanisms by Chuang *et al.* (1979).

This mechanism of cavitation was suggested by Balluffi and Seigle (1957) and modelled quantitatively by Hull and Rimmer (1959). The latter analysis has been corrected and improved in various ways in subsequent work (e.g., Raj and Ashby, 1975). Such studies have been reviewed recently by Chuang *et al.* (1979), who also consider extensions of the Hull–Rimmer model to nonequilibrium cavity shapes, and by Needleman and Rice (1980) who consider the coupling of dislocation creep to the diffusive growth process.

The present paper summarizes results on the diffusive growth process and its various modifications due to nonequilibrium cavity shapes, dislocation creep, and interactions between cavitating grain facets and their surroundings.

2. Analysis of diffusive cavitation

As remarked, the dominant transport mechanisms are usually surface and grain boundary diffusion. Here the basic equations governing the cavitation process are presented based on the assumption that essentially all matter transport is by these mechanisms.

Figure 1a shows a cavitated grain boundary subjected to stress σ. The problem is conventionally simplified by considering (Fig. 1b) the axisymmetric problem of a circular cylindrical bicrystal subject to the same stress σ. The cavity radius is a and the outer radius b is chosen so that $a^2/b^2 = f$, the area fraction of the grain boundary which is cavitated. To simulate interactions between neighboring cavities, the grain boundary flux J_B is set equal to zero at radius $r = b$.

The basic equations describing the diffusion process are those of matter conservation and of the (linearized) kinetic relation between the diffusive flux J and gradient of chemical potential μ along the flow path. Thus,

$$\Omega \nabla \cdot J_S = V_n,$$
$$\Omega J_S = -(D_S \delta_S/kT)\nabla\mu \qquad \text{where} \qquad \mu = -\gamma_S \Omega(K_1 + K_2), \quad (1)$$

on the cavity surfaces, and

$$\Omega \nabla \cdot J_B + \dot{\delta} = 0,$$
$$\Omega J_B = -(D_B \delta_B/kT)\nabla\mu \qquad \text{where} \qquad \mu = -\Omega\sigma_{zz}, \qquad (2)$$

on the grain boundary. Here the fluxes J_S, J_B (of magnitudes J_S, J_B) are

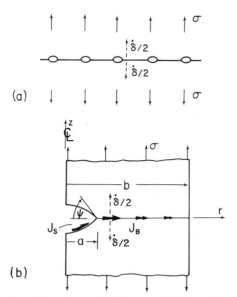

Fig. 1. (a) Cavitated grain boundary. (b) Axi-symmetric bi-crystal configuration used for analysis.

measured in units of atoms crossing unit length per unit time, Ω is atomic volume, ∇ is a surface gradient operator, V_n is the normal recession velocity of the void surface relative to adjoining material, $\dot{\delta}$ is the opening rate across the grain boundary (due to matter addition), $D_S\delta_S$ and $D_B\delta_B$ are diffusion coefficients, kT is the energy measure of temperature, γ_S is the surface energy, K_1 and K_2 are the principal curvatures of the cavity surface, and σ_{zz} is the local normal stress acting along the grain boundary.

In addition, the angle ψ (Fig. 1b) at the cavity tip is assumed to be given by the equilibrium expression,

$$\cos \psi = \gamma_B / 2\gamma_S,$$

where γ_B is the boundary energy, and the potential μ and matter flow are continuous at the cavity tip, so that

$$\gamma_S(K_1 + K_2)_{\text{tip}} = (\sigma_{zz})_{\text{tip}}, \qquad (2J_S)_{\text{tip}} = (J_B)_{\text{tip}}. \tag{3}$$

Completion of the set of governing equations requires a specification of the manner by which the adjoining grains in Fig. 1b deform. Without specifying details of possible constitutive relations, such considerations couple the distribution of stress σ_{zz} on the grain boundary to the distribution of opening displacement $\dot{\delta}$ (precisely, to the distribution of gradients in $\dot{\delta}$). Of course, by overall equilibrium the stress distribution σ_{zz} is constrained to balance the stress σ acting far from the boundary in Fig. 1b.

3. Solution for quasi-equilibrium cavity shape and rigid grain separation

The problem of cavity growth is straightforward to solve in the case for which the following simplifications are made:

(i) Surface diffusion is assumed to be sufficiently rapid compared to boundary diffusion that the cavities retain a quasi-equilibrium, spherical caps shape during growth.

(ii) Grain deformations are assumed to be small enough that the two crystals of Fig. 1b can be taken to move apart rigidly from one another at an opening rate $\dot{\delta}$ that is independent of position along the grain boundary.

When condition (ii) is met, the stress distribution on the grain boundary may be found by integrating Eqs. (2), subject to $J_B = 0$ at $r = b$, which results in

$$\sigma_{zz} = \sigma_0 + (\dot{\delta}/2D)\{b^2 \ln(r/a) - (r^2 - a^2)/2\}, \tag{4}$$

where

$$D = D_B\delta_B\Omega/kT, \tag{5}$$

and where $\sigma_0 \equiv (\sigma_{zz})_{\text{tip}}$, i.e., the value of σ_{zz} at the cavity tip, $r = a$. From this result it follows by overall equilibrium that the applied stress σ is related to σ_0 and $\dot{\delta}$ by

$$\sigma=(1-f)\sigma_0+(b^2\delta/4D)\{\ln(1/f)-(3-f)(1-f)/2\}, \qquad (6)$$

and that the volumetric rate at which matter flows into the grain boundary at the void tip is

$$(2\pi a)(\Omega J_B)_{\text{tip}}=\pi b^2(1-f)\dot{\delta}. \qquad (7)$$

When simplification (i) applies, the curvatures of the void are given by $K_1=K_2=(\sin\psi)/a$, so that the first of Eqs. (3) implies

$$\sigma_0=(2\gamma_S/a)\sin\psi. \qquad (8)$$

Also, the volume of the void may be written as

$$V=(4\pi/3)a^3h, \qquad (9)$$

where (e.g., Chuang et al., 1979)

$$h=h(\psi)=\{1/(1+\cos\psi)-(\cos\psi)/2\}/\sin\psi;$$

$h=1$ when $\psi=90°$ and $h\simeq0.6$ when $\psi=70°$, as is typical. Now, it follows by overall mass conservation for the configuration of Fig. 1b that

$$\int_S (V_{\text{mat}})_n \, dS+2\pi a(\Omega J_B)_{\text{tip}}=\dot{V}=4\pi ha^2\dot{a}, \qquad (10)$$

where S denotes the surface of the void and $(V_{\text{mat}})_n$ is the normal velocity of material points immediately adjacent to the void surface.

If the grains separate rigidly at rate $\dot{\delta}$, the integral in Eq. (10) is easily shown to equal $\pi a^2\dot{\delta}$ so that, using Eq. (7), one obtains

$$\pi a^2\dot{\delta}+2\pi a(\Omega J_B)_{\text{tip}}=\pi b^2\dot{\delta}=\dot{V}=4\pi ha^2\dot{a}. \qquad (11)$$

Hence $\dot{\delta}=4hf\dot{a}$, and using this in Eq. (6) one obtains

$$\dot{a}=(D/ha^2)\{\sigma-2(1-f)(\gamma_S/a)\sin\psi\}$$
$$\times\{\ln(1/f)-(3-f)(1-f)/2\}^{-1}. \qquad (12)$$

This last equation expresses the cavity growth rate when simplifications (i), pertaining to the quasi-equilibrium cavity shape, and (ii), pertaining to rigid grains, apply. Note, however, that Eqs. (8) to (10) apply whenever the quasi-equilibrium shape can be assumed, whether or not the grains deform, whereas Eqs. (4) to (7) apply whenever the grains can be assumed to separate rigidly, whether or not the cavity has the quasi-equilibrium shape.

Equation (12) was given by Needleman and Rice (1980), who observe that it corrects an undue factor of $(1-f)$ in previous presentations of the result (e.g., Raj and Ashby, 1975, Chuang et al., 1979); the factor arose from neglect of the surface integral in Eq. (10). They observe also that the derivation tacitly assumes that the surface tension (in the sense of surface stress, and as distinct from surface energy) is zero. The inclusion of surface tension T_s adds a term $(-2fT_s/a)$ sin ψ to the bracketed term in Eq. (12) that contains σ. This is generally a small effect.

If the growth rate law is written symbolically as $\dot{a}=1/F(a)$, then the time to rupture, t_r, associated with cavity growth from some initial size, a_i, to coalescence, $a=b$, is given by

$$t_r = \int_{a_i}^{b} F(a)\,\mathrm{d}a. \tag{13}$$

This is straightforward to compute when the applied stress σ is constant throughout the growth process and is much larger than the "sintering stress" $(2\gamma_S/a_i)$ sin ψ at the onset of growth. In that case the expression for $\dot{a}$ of Eq. (12) leads to

$$t_r = \tfrac{16}{315}(hb/D\sigma)\Big\{1 - \tfrac{105}{16}f_i^{3/2}\ln(1/f_i)$$

$$- \tfrac{1}{8}f_i^{3/2}\Big(63f_i^{1/2} - \tfrac{175}{4} - \tfrac{45}{4}f_i\Big)\Big\}, \tag{14}$$

where $f_i = a_i^2/b^2$. This rupture time is inversely proportional to stress and to the grain boundary diffusion coefficient $D_B\delta_B$. This estimate of the rupture time neglects details of cavity nucleation (see Raj and Ashby, 1975), but the result does not depend strongly on the initial value of a_i. Indeed, the bracketed term involving f_i can be replaced by unity for f_i less than $1/10$ or so.

The validity of simplifications (i) and (ii) is now examined.

4. Nonequilibrium cavity shape

For this section simplification (ii) is retained, but not (i). That is, the possibility is considered that surface diffusion may not be fast enough to allow retention of the quasi-equilibrium, spherical caps shape.

The problem in this form was addressed by Chuang *et al.* (1979), who point out that according to an appropriate solution of Eqs. (1), the relaxation time for a doubly periodic surface disturbance with half-wavelength equated to cavity diameter, $2a$, is approximately $a^4/24B$, where

$$B = D_S\delta_S\Omega\gamma_S/kT. \tag{15}$$

A characteristic time associated with exposure of new cavity surface to the surface diffusion process is $a/\dot{a}$, and this identifies the dimensionless parameter

$$a^3\dot{a}/24B \tag{16}$$

with the following interpretation: When the parameter is small compared to unity the quasi-equilibrium analysis of the last section applies, whereas when it is much greater than unity the cavity can be assumed to grow in a narrow "crack-like" mode as discussed by Chuang and Rice (1973) and Chuang et al. (1979).

This crack-like mode of cavity growth emerges in the analysis of Chuang et al. (1979) as a limiting singular perturbation form of the solutions to Eqs. (1) as the parameter $(B/a^3\dot{a})\to 0$ and all lengths near the cavity tip are scaled by the length parameter $(B/\dot{a})^{1/3}$. In this case one obtains the asymptotic results

$$(K_1+K_2)_{\text{tip}}\to 2\,\sin(\psi/2)(\dot{a}/B)^{1/3},$$

$$(\Omega J_S)_{\text{tip}}\to 2\,\sin(\psi/2)(B\dot{a}^2)^{1/3}. \tag{17}$$

These expressions may be used in Eqs. (3) to calculate $(\sigma_{zz})_{\text{tip}}$ [$\equiv\sigma_0$ of Eqs. (4) and (6)] and $(\Omega J_B)_{\text{tip}}$. The latter is then used in Eq. (7) to calculate δ. The result is that the term σ_0 and δ in Eq. (6) are known in terms of the cavity growth rate and, when the expression is inverted, one obtains (Chuang et al., 1979)

$$\dot{a}=\frac{27}{64}\frac{B}{b^3\Delta}\frac{\left\{(1+Q\Sigma\Delta)^{1/2}-1\right\}^3}{(1-f)^3Q^3}. \tag{18}$$

Here,

$$Q=\left\{3f^{1/3}/2(1-f)^3\right\}\left\{\ln(1/f)-(3-f)(1-f)/2\right\},$$

$$\Sigma=4\sigma b/3\gamma_S\,\sin(\psi/2), \tag{19}$$

and

$$\Delta=D_S\delta_S/D_B\delta_B. \tag{20}$$

The function Q is not strongly variable; it lies between approximately 0.5 and 0.65 for $0.01<f<1$, although $Q\to 0$ as $f\to 0$.

By comparisons of predictions of Eqs. (12) and (18) against a family of self-similar solutions for cavity growth, Chuang et al. (1979) conclude that Eq. (12) is valid for $a^3\dot{a}/B$ less than approximately 10, whereas Eq. (18) is valid for $a^3\dot{a}/B$ greater than approximately 30. A tolerable approximation, valid over the entire range of $\dot{a}$, was found to be that of taking $\dot{a}$ as the maximum of the expressions of Eqs. (12) and (18).

Chuang et al. (1979) present various inequalities for deciding which of the two limiting growth models more closely applies to a given case. A somewhat different but related procedure is followed here. Setting $\psi=70°$, as is representative, one finds from (12) that a growth rate satisfying $a^3\dot{a}/24B=1$

is predicted when

$$\sigma = (2\gamma_S/a)\{0.94(1-f)+7.2\Delta[\ln(1/f)-(3-f)(1-f)/2]\}$$

and from (18) when

$$\sigma = (2\gamma_S/a)\{1.65(1-f)$$

$$+8.32\Delta[\ln(1/f)-(3-f)(1-f)/2]/(1-f)\}.$$

These results are not very different from one another, and their average, namely

$$\sigma^* = (2\gamma_S/a)\{1.3(1-f)+[(7.8-3.6f)/(1-f)]$$

$$\times[\ln(1/f)-(3-f)(1-f)/2]\Delta\}$$

$$\sim(2\gamma_S/a)\{1.3+7.8\ln(1/4.5f)\Delta\} \qquad \text{for small } f,$$

may be interpreted as an approximate dividing stress level in that when σ is less than σ^* growth occurs in the quasi-equilibrium mode, and when σ is greater than σ^* growth occurs in the crack-like mode.

Estimated values of Δ in tabulations by Chuang et al. (1979) cover a considerable range, from approximately 0.1 to 100 and sometimes higher, depending on material and temperature. Since $2\gamma_S/a$ is essentially the sintering stress limit, and is usually much lower than applied stress levels, it is evident that growth will frequently be in the crack-like mode when Δ is of the order unity. On the other hand, the more typical cases of large Δ will generally correspond to growth in the quasi-equilibrium mode.

5. Deformability of the adjoining grains

Now simplification (i) is retained, i.e. it is assumed that Δ is large enough to retain the quasi-equilibrium cavity shape, but simplification (ii) is relaxed in that deformability of the adjoining grains in Fig. 1 is considered. Various estimates of the effect of elastic deformability of the grains suggest that at least for cases of sustained loading and for cavities that are not spaced at great distances from one another, elastic effects can be disregarded. (Different conclusions may apply for cyclic loading and transient effects.)

Plastic dislocation creep of the adjoining grains provides a more significant mechanism, however, at least at sufficiently high stress levels. The mechanism was analyzed in a preliminary manner by Beere and Speight (1978) and later by Edward and Ashby (1979) and Rice (1979). A definitive analysis based on the formulation by Rice has been given recently by Needleman and Rice (1980), and is summarized briefly here.

The chief effect of dislocation creep is to shorten the diffusive path length. For example, because nonrigid grains can open nonuniformly across their boundary, it is not necessary that matter be transported over a distance comparable to the inter-void spacing. Instead it is found that matter can be accommodated by nonuniform openings which take place over a distance of order L ahead of the void (at least when $a+L$ is less than b), where L is a stress and temperature dependent parameter defined by

$$L=(D\sigma/\dot{\varepsilon}_{cr})^{1/3}. \qquad (21)$$

Here, $\dot{\varepsilon}_{cr}$ is the creep rate of the adjoining grains due to stress σ.

The analysis assumes that the grains undergo nonlinear power-law creep with deformation rates $D_{\alpha\beta}$ given by

$$D_{\alpha\beta}=\tfrac{3}{2}\Lambda^{-n}\{3\sigma'_{\gamma\delta}\sigma'_{\gamma\delta}/2\}^{(n-1)/2}\sigma'_{\gamma\delta}, \qquad (22)$$

where $\sigma'_{\gamma\delta}$ is the deviatoric stress and Λ and n are constants. The outer radius, $r=b$, in Fig. 1b is subjected to zero shear tractions and to a radial velocity $v_r=-\dot{\varepsilon}_{cr}b/2$, where $\dot{\varepsilon}_{cr}=(\sigma/\Lambda)^n$. The following functional F is defined for all axisymmetric velocity fields v_α, with associated deformation rates $D_{\alpha\beta}$ and grain boundary fluxes $j(=\Omega J_B/2)$, where

$$D_{\alpha\beta}=\text{symmetric part of }\partial v_\alpha/\partial x_\beta,$$

and

$$\nabla\cdot j+v_z=0 \text{ on the grain boundary,}$$

with the understanding that j vanishes at the outer radius and that v_α meets the velocity boundary conditions on $r=b$:

$$F=\int_{\text{Vol.}}\left\{\frac{n}{1+n}\Lambda\left(\tfrac{2}{3}D_{\alpha\beta}D_{\alpha\beta}\right)^{(1+n)/n}\right\}\,d(\text{Vol.})$$
$$+\int_a^b(j^2/D)2\pi r\,dr-\int_0^b\sigma(v_z)_{\text{top}}2\pi r\,dr+2\pi a\sigma_0 j_0. \qquad (23)$$

Here, "Vol." denotes the volume of the upper crystal in Fig. 1b, $(v_z)_{\text{top}}$ is the vertical velocity of the upper loaded surface in Fig. 1b, σ_0 is the stress at the void tip as given by Eq. (8), and j_0 is the value of j at $r=a$. The functional F is minimized by the true velocity field, and this principle was used by Needleman and Rice (1980) to construct a finite-element procedure for determining the velocity field. By solving the deformation rate problem in this way and hence evaluating the left side of Eq. (10) to determine the cavity growth rate, they found that results could be put in either of the (equivalent) forms.

$$\dot{a}=(D\sigma/a^2)\times(\text{a function of }a/L, f \text{ and } \sigma_0/\sigma)$$

or

$$\dot{a} = a\dot{\varepsilon}_{cr} \times (\text{another function of } a/L, f \text{ and } \sigma_0/\sigma).$$

The former form enables comparisons with the result given in Eq. (12) for rigid grain behavior. The case of greatest interest is that for which σ is much greater than the sintering limit σ_0. Results of Needleman and Rice (1980) for this case are summarized in Table 1, where the ratio of the predicted value of $\dot{a}$ for creeping grains to that calculated on the basis of Eq. (12) for rigid grains is shown.

Evidently, large differences from the rigid-grains analysis, Eq. (12), result when $a + L$ is significantly less than b. In fact, based on an observation of Dr. I. W. Chen of M.I.T. (privately communicated by Prof. A. S. Argon), a reasonable approximation to the results of Needleman and Rice (1980), which predicts the cavity growth rate to within an error of about 30% for all a/L less than 10 or so, is to predict $\dot{a}$ from Eq. (12) but to use for f the maximum of $(a/b)^2$ and $\{a/(a+L)\}^2$.

Needleman and Rice (1980) tabulate values of L for various pure metals. They observe, following Rice (1979), that when power-law creep occurs with an activation energy equal to that for bulk (lattice) diffusion, L can be expressed in the form

$$L = L_0 \exp(\kappa T_m/T) \left(10^{-3}\mu/\sigma\right)^{(n-1)/3}, \tag{24}$$

where L_0 and κ are constants and μ is the shear modulus. For fcc metals at $T = 0.5T_m$ and $\sigma = 10^{-3}\mu$, L ranges from 2 to 8 μm (although lower for Al and higher for Ag); for bcc metals under the same conditions L ranges from 0.25 to 0.35 μm. A lower stress of $\sigma = 10^{-4}\mu$ at the same temperature increase these numbers by about a factor of 20. Hence, at $0.5T_m$ the assistance of diffusive rupture by plastic creep flow can be significant at $\sigma = 10^{-3}\mu$, especially for bcc metals, but will generally be insignificant,

Table 1
Ratio of cavity growth rate $\dot{a}$ between creeping grains to that for rigid grain response (eq. (12)); parameter L defined by Eq. (21); from Needleman and Rice (1980) for stress σ much larger than the sintering threshold

a/L \ a/b	1/10	1/5	1/3	2/3
0.10	1.04	1.01	1.00	1.00
0.32	1.72	1.18	1.04	1.00
1.0	6.8	3.95	2.14	1.09
3.2	66	38	19	3.46
10.0	1330	746	385	6.0

certainly for fcc metals, at $\sigma = 10^{-4}\mu$. The parameter κ ranges from 2.4 to 3.9 for fcc metals and from 1.7 to 2.2 for bcc metals. Hence, increases in temperature from, say, $0.5T_m$ to $0.8T_m$ decrease L by a factor between approximately 0.05 and 0.28, depending on the material, for fixed σ/μ.

6. Discussion

The paper has presented an analysis of diffusional cavity growth based on the standard assumptions of quasi-equilibrium cavity shapes and effectively rigid grain separations, and has indicated the range of validity of each of these assumptions. Results of fuller analyses, based on nonequilibrium, crack-like cavity shapes and on dislocation creep of the adjoining grains have been summarized.

In applications of these results it is important to properly choose the stress σ in relation to the applied stress and to the geometry of the cavitating grain boundary facets in a polycrystalline aggregate. Dyson (1976) has emphasized the constraints on diffusive cavity growth that can occur if the cavitated grain facets are well separated from one another. In such cases, the requirement that the opening rate δ across such cavitated facets be compatible with the deformation rate of the surroundings serves to reduce σ from the value (say, σ_∞) which would act across a similar grain facet in an uncavitated polycrystal. A recent analysis of such constraint effects has been presented by Rice (1981), who considers an isolated, cavitated facet of diameter d and observes that approximately the opening rate is given by

$$\dot{\delta} = \alpha\{(\sigma_\infty - \sigma)/\sigma_\infty\}\dot{E}_\infty d, \tag{25}$$

where $\dot{E}_\infty$ is the strain rate of a similarly loaded but uncavitated polycrystal. Here, the coefficient α is estimated as equal approximately to 1 for power-law creep with nonsliding grain boundaries and to 2 for freely sliding boundaries, at least if the cavitated facets are far-separated and effectively noninteracting; α would be extremely large in the opposite limit when all boundaries approximately perpendicular to the tensile direction are cavitated, especially with freely sliding boundaries. Assuming that σ is within the range for which Eq. (12) is valid and calculating a related expression for $\dot{\delta}$ from Eq. (11), Rice (1981) shows that the constraint effects, namely, that the expressions for $\dot{\delta}$ based on Eqs. (11), (12) and (25) should agree, reduces σ from σ_∞ and leads to a growth rate

$$\dot{a} = (D/ha^2)\{\sigma_\infty - 2(1-f)(\gamma_S/a)\sin\psi\}/\{(4L_\infty^3/ab^2d)$$
$$+ \ln(1/f) - (3-f)(1-f)/2\}, \tag{26}$$

where

$$L_\infty = (D\sigma_\infty / \dot{E}_\infty)^{1/3}. \tag{27}$$

(Rice (1981) used the notation L, but L_∞ is used here to avoid confusion with the "local" L of Eq. (21).)

The associated rupture time, calculated as in Eq. (13), is found to be given by

$$t_r = (t_r)_1 + (t_r)_2, \tag{28}$$

where $(t_r)_1$ is given by Eq. (14) with σ replaced by σ_∞ in that expression and where

$$(t_r)_2 = (4hb/3\alpha\dot{E}_\infty d)(1 - f_i^{3/2}). \tag{29}$$

For small f_i the ratio of $(t_r)_2$ to $(t_r)_1$ is $(105/4\alpha)(L_\infty^3/b^2 d)$. Conditions might often be encountered when this ratio is large (e.g., low σ_∞ and well separated facets), and in such cases the analysis suggests a Monkman–Grant (1956) type of correlation of t_r with $\dot{E}_\infty$.

Acknowledgment

This paper was prepared under support of the DARPA Materials Research Council, funded through the University of Michigan.

References

Balluffi, R. W. and Seigle, L. L. (1957), "Growth of Voids in Metals During Diffusion and Creep," *Acta Met.*, **5**, 449.

Beere, W. and Speight, M. V. (1978), "Creep Cavitation by Vacancy Diffusion in a Plastically Deforming Solid," *Metal Sci.*, **12**, 172.

Chuang, T-j. and Rice, J. R. (1973), "The Shape of Inter-Granular Cracks Growing by Surface Diffusion," *Acta Met.*, **21**, 1625.

Chuang, T-j., Kagawa, K. I., Rice, J. R., and Sills, L. (1979), "Nonequilibrium Models for Diffusive Cavitation of Grain Interfaces," *Acta Met.*, **27**, 265.

Dyson, B. F. (1976), "Constraints on Diffusional Cavity Growth Rates," *Metal Sci.*, **10**, 349.

Edward, G. H. and Ashby, M. F. (1979), "Intergranular Fracture During Power-law Creep," *Acta Met.*, **27**, 1505.

Hull, D. and Rimmer, D. E. (1959), "The growth of Grain Boundary Voids under Stress," *Phil. Mag.*, **4**, 673.

Monkman, F. W. and Grant, N. J. (1956), "An Empirical Relation between Rupture Life and Minimum Creep Rate," *Proc. ASTM*, **56**, 593.

Needleman, A. and Rice, J. R. (1980), "Plastic Creep Flow Effects in Diffusive Cavitation of Grain Boundaries," *Acta Met.*, **28**, 1315.

Raj, R. and Ashby, M. F. (1975), "Intergranular Fracture at Elevated Temperature," *Acta Met.*, **23**, 653.

Rice, J. R. (1979), "Plastic Creep Flow Processes in Fracture at Elevated Temperature," *Time-Dependent Fracture of Materials at Elevated Temperature*, S. M. Wolf (ed.), U. S. Dept. of Energy Report CONF790236 UC-25, 130.

Rice, J. R. (1981), "Constraints on the Cavitation of Isolated Grain Boundary Facets in Creeping Polycrystals," *Acta Met.*, **29**, to appear.

S. Nemat-Nasser, Editor
THREE-DIMENSIONAL CONSTITUTIVE RELATIONS AND DUCTILE FRACTURE
North-Holland Publishing Company (1981) 185–205

EXPERIMENTAL AND NUMERICAL STUDY OF THE DIFFERENT STAGES IN DUCTILE RUPTURE: APPLICATION TO CRACK INITIATION AND STABLE CRACK GROWTH

F. M. BEREMIN*

Analyses of the stress and strain fields inside axisymmetric notched tensile specimens have been carried out using finite element computations. The results are used to determine experimental criteria which characterize the three stages of ductile rupture, i.e. the formation of cavities from inclusions, cavity growth, and void coalescence. The experiments are carried out on a A508 pressure vessel steel which contains MnS inclusions. A model for crack initiation and stable crack growth which is based on the attainment of a critical void growth rate is discussed in the light of metallographic observations near the crack tip.

1. Introduction

Many investigations have been devoted recently to the study of ductile tearing which takes place after large plastic deformations in low and medium strength materials for which the methodology of linear fracture mechanics cannot be applied. The main objective of these investigations is to characterize crack initiation and stable crack growth, and to interpret the results in terms of criteria for crack growth. In both aspects two types of approach are still under development. The first one assumes that elastic-plastic fracture can be described in terms of "global" parameters like J integral (e.g. Begley and Landes, 1972) or crack tip opening displacement (COD approach) (e.g. Wells, 1963). The second type of approach is more mechanistic and more "local." It is essentially based on computational fracture models which use damage functions (e.g. EPRI Report, 1978;

*F. M. Beremin is a research group including: Y. d'Escatha[1], P. Ledermann[1]; J. C. Devaux[2]; F. Mudry[3], A. Pineau[3] and J. C. Lautridou[3].

[1]Bureau de Contrôle de la Construction Nucléaire, 3, rue De Vosges, F-21000 Dijon, France.
[2]Division des Fabrications de Framatome, B.P. 13, F-71380 Saint-Marcel, France.
[3]Centre des Matériaux de l'Ecole des Mines de Paris, B.P. 87, F-91003 Evry Cédex, France. Equipe associée au CNRS no. 767.

d'Escatha and Devaux, 1979). The computational method involves the node release technique in finite element calculations whilst the damage functions are experimentally established. In some cases this function models the physical microfracture processes taking place at the tip of a crack which is subjected to any kind of loading.

The aim of the paper is to describe recent studies on the ductile rupture of a medium strength steel, which deal with the second type of approach. In this respect the work is an extension of recently published studies (d'Escatha and Devaux, 1979; F. M. Beremin, 1979). The material investigated is a C–Mn steel, A.508 steel, used for the fabrication of P.W. nuclear reactors. The composition of the heats which are employed and their conventional properties are given in Appendix A.

This paper is divided into two parts. Firstly, we deal with the determination of criteria which can be used to describe the three stages of ductile rupture. These criteria are established using various geometries of notched specimens which are calculated by finite element analysis. For each of these stages which can be considered to model ductile rupture, i.e. the formation of cavities from MnS inclusions, the subsequent growth of those cavities and their final coalescence, various specimens' geometries are used in order to determine the influence of the stress and strain state. All the experiments are performed on axisymmetric specimens in order to avoid difficulties

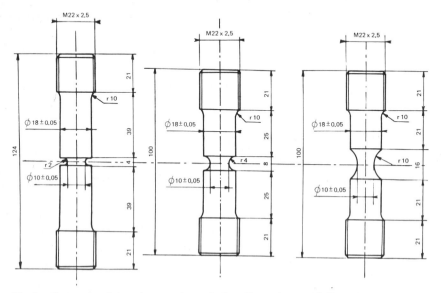

Fig. 1.   Geometries of the axisymmetric notched tensile specimens. All dimensions are given in mm.

arising from 3-D computations, on the one hand, and from the necessary assumptions of plane stress or plane strain conditions for plate specimens, on the other hand. The dimensions of the specimens are shown in Fig. 1. These circumferentially notched specimens were chosen in order that rupture processes involve a rather important volume of material which is subjected to quasi-homogeneous stress and strain conditions. This facilitates metallographic observation of the damage which takes place in the bulk of the material.

The second part of the paper deals with the study of crack initiation and stable crack growth of precracked specimens. It is shown how the results obtained in the first part can tentatively be used to model the fracture processes which take place at the tip of a stationary or a propagating crack. In a previous study a simplified damage function had already been used in order to test the numerical possibilities of the model (d'Escatha and Devaux, 1979). In this paper the results of metallographic observations are emphasized.

2. Investigation of the three elementary stages involved in the ductile rupture of bulk specimens

Many investigations have shown that ductile rupture involves three successive damage processes which are the nucleation of cavities from inclusions, cavity growth and void coalescence. The results related to these three stages are described successively.

2.1. Cavity nucleation

A recent paper by Goods and Brown (1979) has reviewed a number of theoretical and experimental results related to the first stage of ductile rupture. From these results it can be concluded that, in the case of rather big, widely spaced inclusions, a continuum mechanics approach can be used to estimate the stresses and the strains inside the inclusions and in the neighboring matrix when the material is plastically deformed. It is generally assumed that the initiation of cavities from inclusions takes place when either the inclusion or the matrix-inclusion interface is subjected to a critical normal stress. Several types of calculations can be applied in order to determine the local stresses and strains as a function of the macroscopic stresses and strains applied to the material. For this purpose the Eshelby (1961) theory of inclusions and inhomogeneities has been used by Tanaka *et al.* (1971). This theory which is based on an elastic analysis is strictly valid only when the applied strains are not too important. In the case where cavity initiation takes place after large plastic deformation Argon and Im

(1975a) have given an expression which fits their experimental results. This expression can be written as,

$$\sigma_m + k\sigma_{eq} = \sigma_c, \tag{1}$$

where σ_m is the hydrostatic stress, σ_{eq} is the equivalent von Mises stress, σ_c is a critical stress and k is a geometric factor ($k = 1$ for spherical particles).

In the materials investigated in this study the size of the inclusions is such that it can be reasonably assumed that the continuum mechanics approach can be used. The conditions prevailing for the formation of cavities from MnS inclusions were studied using axisymmetric notched specimens.

These specimens have been deformed at various temperatures ($-196°C$, $25°C$, $100°C$). The specimens were cut either in the longitudinal or in the transverse direction of a nozzle shell of a P.W.R. reactor in order to investigate the effect of the shape of MnS particles. The specimens were calculated with an elastic-plastic finite element program up to an overall deformation $\bar{\varepsilon}$ as large as 10%. The overall strain is defined as $\bar{\varepsilon} = 2 \ln \phi_0/\phi$, where ϕ_0 is the initial diameter and ϕ is the actual diameter. The specimen geometry corresponding to $\rho = 2$ mm was calculated up to $\bar{\varepsilon} = 16\%$. This has been possible by taking into account the geometric modifications of the mesh. The details of the finite element calculations and most of the results are described elsewhere (Beremin, 1980b). Figure 2 shows a very close agreement between the calculated and the measured loading curves $\bar{\sigma}$ vs $\bar{\varepsilon}$, where $\bar{\sigma}$ is the mean tensile stress $\bar{\sigma} = 4P/\pi\phi^2$ and P is the load. This

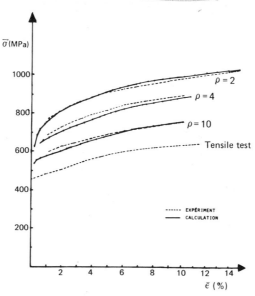

Fig. 2. Comparison of the loading curves ($\bar{\sigma}$ vs $\bar{\varepsilon}$) calculated and obtained experimentally; ρ indicates the notch radius.

gives confidence in the values of stresses and strains which are calculated at every point inside the specimens.

The experimental procedure which has been used to determine the damage of the inclusions has been described elsewhere (Beremin, 1980a). The specimens were given various amounts of overall strain. Then they were longitudinally sectioned and carefully polished in order to observe damaged inclusions using a technique similar to that employed by Argon *et al.* (1975b). In the longitudinal direction it was observed that most of the inclusions were broken whilst in the transverse direction damage corresponded to the decohesion of the interface between MnS particles and the matrix. The sections of the specimens allowed the determination of a limiting curve for a given specimen geometry and a given applied overall strain $\bar{\varepsilon}$. Inside this curve all the inclusions are damaged whilst outside this frontier all the inclusions are not yet damaged. An example of such limiting curves is shown in Fig. 3. The knowledge of the strains and stresses all along this frontier permits us to derive a criterion for inclusion decohesion. This method leads to the results which are given in Fig. 4 for two types of material. Steel A was studied at 100°C both in the longitudinal and in the transverse direction. Steel B was investigated at -196°C, 25°C, and 100°C

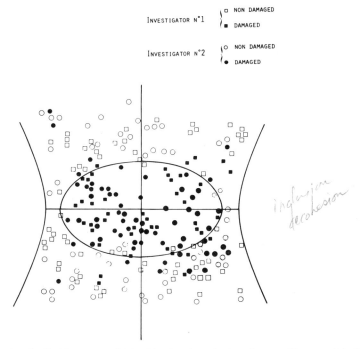

Fig. 3. Example of a limiting curve obtained by two investigators in a $\rho = 10$ mm notched specimen which was given an overall deformation $\bar{\varepsilon} = 0.17$.

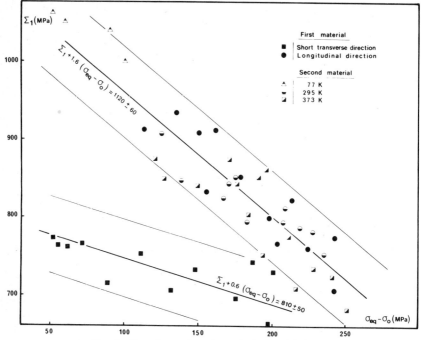

Fig. 4. Experimental results on inclusion decohesion.

in the longitudinal direction only. These results indicate that cavity initiation can be described using the following relationship:

$$\Sigma_1 + k(\sigma_{eq} - \sigma_0) = \sigma_c, \tag{2}$$

where Σ_1 is the maximum principal stress, σ_{eq} is the von Mises equivalent stress, σ_0 is the yield strength of the material, σ_c is a critical stress and k is a geometrical parameter which is a function of the shape of the inclusions ($k = 1.6$ for the longitudinal direction and $k = 0.6$ for the transverse direction). This expression is very similar to that already given by Argon *et al.* (1975b). It has been shown (Beremin, 1980a) that it can be derived from an extension of Eshelby's theory by Berveiller and Zaoui (1979).

It is worth noting that the strain necessary to nucleate cavities, ε_d, is small as compared to the strain at failure since nucleation represents less than 15% of the ductility even in a conventional tensile test. Moreover the comparison of the results obtained for the two directions shows that the lower ductility of the transverse direction cannot be explained only in terms of the difference in nucleation strain.

2.2. Cavity growth

Several theoretical analyses have been proposed in order to model the growth of voids in a material which is plastically deformed under certain stress conditions. Amongst them, McClintock's model (1968) and that proposed by Rice and Tracey (1969) are the most widely used. They are based on a certain number of assumptions. In particular they assume that no interaction takes place between two neighboring voids. Very few experimental studies have been performed in order to test the applicability of these theoretical models. However, it is worth mentioning McClintock's (1968) experiments carried out on plasticine and those performed by Perra and Finnie (1977) on coarse grain copper. In this last case, it was found that the theoretical relation proposed by McClintock reproduced the tendency of cavity growth but seemed to underestimate its size.

In our study, only a few experiments have been made to investigate the process of cavity growth. For this purpose a special technique was developed. Some specimens were deformed at 100°C very close to the final fracture strain, $\bar{\varepsilon}_f$. Then, after machining a very sharp notch in the minimum section of the axisymmetric specimens, these specimens were broken at -196°C. A close examination of the fracture surfaces similar to those shown in Fig. 5 allowed an estimation of the growth of voids which originated from the inclusions left in the bottom of the holes. The micrographs shown in Fig. 5 illustrate the fact that the cavity growth is more important in the center of the specimens as compared to the periphery, as expected.

The cavity growth was computed using the relationship derived by Rice and Tracey (1969) for a single spherical void in an infinite rigid plastic material, i.e.,

$$dR/R_0 = 0.28 \, d\varepsilon_{eq}^\infty \exp\left(\tfrac{3}{2}\sigma_m^\infty / \sigma_{eq}^\infty\right), \tag{3}$$

where R_0 is the initial cavity radius which is equal to the inclusion size, R is its actual mean radius, σ_m^∞ is the hydrostatic stress at infinity and $d\varepsilon_{eq}^\infty$ is the equivalent von Mises strain increment at infinity. The values of σ_m^∞ and σ_{eq}^∞ at every point in the minimum section were obtained from the results of finite element calculations. In the original expression proposed by Rice and Tracey, we have taken for the yield stress, the equivalent von Mises stress (σ_{eq}^∞) in order to allow for the effect of work hardening. This differential equation was integrated from the strain corresponding to cavity initiation up to that corresponding to the actual value at a given point in the section. Although the results are insufficient to conclude definitely, it can be stated that the values of R/R_0 calculated from the Rice and Tracey expression are not drastically different from those which are observed using the experimental procedure employed in this study.

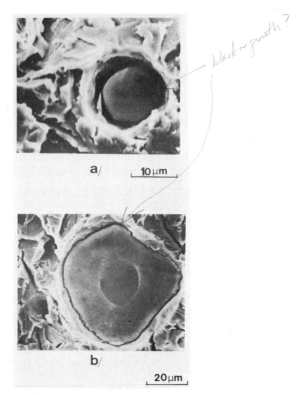

black ~ growth ?

Fig. 5. Scanning electron micrograph of the fracture surface of a $\rho=4$ mm-specimen which has been given 18.2% deformation at 100°C and then broken at -196°C. (a) Center of the specimen; (b) Close to the specimen surface.

2.3. Cavity coalescence

The last stage of ductile rupture is poorly understood, partly because it is generally accepted that several phenomena can intervene to interrupt the process of void growth. Our observation of the specimens which were broken at 100°C indicates that the size of the large dimples measured on the fracture surfaces are much larger than the size of the voids calculated with the method described previously. Moreover, a close examination of sections of specimens loaded just before fracture indicates that in our case the coalescence phenomenon takes place in a very narrow band linking two neighboring voids which originate from the MnS inclusions (Fig. 6).

These observations suggest that the last process of ductile rupture can be tentatively characterized in terms of a critical growth ratio $(R/R_0)_c$, where R_c is the mean radius of the cavities at failure. It has been shown previously that the strain at failure, ε_f, can be unambiguously defined using a diametral

(local ?)

? (rim & data from a)

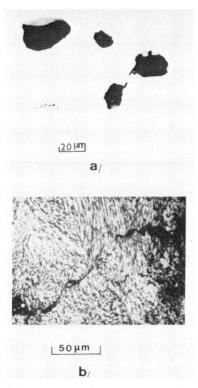

Fig. 6.   Optical micrographs illustrating void coalescence. (a) Illustration of the coalescence beginning between two cavities; (b) Micrograph showing a narrow band linking two bigger cavities (Nital Etching).

extensometer. Interrupted tests have equally shown that, in all the geometries employed in this study, ductile rupture occurred in the center of the specimens (Beremin, 1980b; Rousselier, 1979). These observations enable us to calculate the value of $(R/R_0)_c$ by integration of the Rice and Tracey relationship from ε_d to ε_f. These values for ε_d and ε_f are taken in the element located in the center of the specimens. These calculations give the variation of $(R/R_0)_c$ as a function of stress triaxiality (σ_m/σ_{eq}) using the results obtained with the various geometries. The results are given in Fig. 7. These calculations required an extrapolation of the numerical results for $\bar{\varepsilon} > 10\%$. We assumed that the stress (resp., the strains) distribution divided by $\bar{\sigma}$ (resp., $\bar{\varepsilon}$) was constant. This property was verified in the range of $\bar{\varepsilon}$ between 6% and 10%.

In Fig. 7 it can be noted that the variation of $(R/R_0)_c$ as a function of σ_m/σ_{eq} is rather low. This is one reason why the use of $(R/R_0)_c$ criterion is, at least for the material that we have considered, more appropriate than a

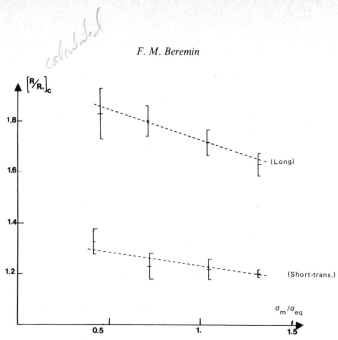

Fig. 7. Results of critical hole growth evaluated from the rupture of axisymmetric notched tensile specimens. Steel A tested at 100°C in two directions.

more global ductility criterion which simply describes the variation of ε_f as a function of σ_m/σ_{eq}. In our steels, it was shown previously that the extrapolation of ε_f vs σ_m/σ_{eq} to larger values of stress triaxiality was very inaccurate (Beremin, 1980b). The results given in Fig. 7 clearly show that $(R/R_0)_c$ tends to decrease as the stress triaxiality in increased.

The results derived from the above analysis call for a final remark. In the materials investigated in this study, the distance between MnS inclusions is about 10 times their average radius. At failure the mean distance between the voids is of the same order of magnitude since the final cavity size is smaller than twice its initial size in all cases. Thus there is a catastrophic local instability which involves the material located between the adjacent voids. The explanation of this local instability requires further investigation. At least two directions can be considered. The first one is based solely on continuum mechanics. Strain localization could occur in a narrow band because the material, becoming porous, exhibits different plastic behavior (Gurson, 1977; Yamamoto, 1978; Rousselier, 1979). The second one is more mechanistic. It can be thought that the carbides themselves nucleate cavities all around the large voids formed from MnS inclusions because of the strain concentration near the holes. The process of carbide decohesion "weakens" the material located between the large inclusions. Both viewpoints could explain the tendency of the variation of $(R/R_0)_c$ as a function of the stress triaxiality.

3. Application to the study of crack initiation and stable crack growth

In this section an attempt is made to illustrate how the results presented in the previous part can be applied to the study of crack initiation and stable crack growth. A large part of this section relies upon experimental work performed recently by Lautridou (1980). This author has investigated three materials which have mechanical properties very close to those of the steels used in this study. One of these materials is Steel A. This permits us to use directly the results of the previous part in order to assess the crack initiation properties and the stable crack growth behavior of this material. In his study Lautridou has used two types of compact tension (CT) specimens ($B=30$ and 50 mm) which were loaded at 100°C in order to determine the $J-\Delta a$ resistance curves. Both the longitudinal and the transverse directions have been investigated. Moreover a sectioning technique has been used in order to determine directly the crack opening displacement (COD) just before crack initiation takes place.

3.1. Crack initiation

The main difficulty encountered in the study of crack initiation is associated with the fact that there are very steep stress and strain gradients at the crack tip. Several approaches can be considered to deal with this problem.

The first approach introduces a process zone over which average or minimum values for the stresses and the strains are used more or less explicitly. This approach has been used by several authors, particularly McClintock (1971). This process zone must have dimensions in connection with the "metallurgical" lengths—such as the mean distance between inclusions—which are involved in the process of ductile rupture. Several criteria like a critical COD (Knott and Green, 1976), a critical ductility (Mackenzie *et al.*, 1977; Ritchie *et al.*, 1979) or a critical void growth at a critical distance (Rice and Johnson, 1970; McMeeking, 1976) have been used to assess the process of crack initiation. Some of the difficulties associated with the application of these criteria are considered later.

A second approach attempts to model more closely the different stages involved in the process of ductile rupture. An example of such an approach is that already presented by Rice and Johnson (1970). These authors have calculated the growth rate of voids which are assumed to lie at a given distance in front of the crack tip. They assumed that crack initiation takes place for a critical value of the void growth. This value is necessarily chosen more or less arbitrarily unless experimental results similar to those given in the previous part of this study are available. This second type of approach requires a very accurate knowledge of the stress and strain profiles at the crack tip in conjunction with a differential equation which describes the "kinetics" of the damage. One of the main interests of this approach is that it can be directly used with a statistical analysis of the inclusion distribution, which is certainly a very important aspect of ductile rupture.

3.1.1. Approach using the concept of a process zone

As stated previously one example of this approach is that which compares the COD at crack initiation and the distance between inclusions. This comparison has been widely used (Knott and Green, 1976; Rice and Jonhson, 1970). However the use of this concept gives rise to several difficulties. Most often the distance to the nearest neighbor is used. This distance is easily defined for an inclusion distribution which obeys a Poisson's law which is not necessarily the case. A second difficulty is associated with the measurement of the COD which most often involves the determination of the position of a rotation center.

In Table 1 the results obtained by Lautridou are given. Two different distances to the nearest neighbor are used. The first one, Δ_3, corresponds to a distance in the volume. It is calculated as $\Delta_3 = 0.554/\sqrt[3]{N_v}$, where N_v is the number of inclusions per unit volume. The second one Δ_2 corresponds to a distance in a plane. It is given by $\Delta_2 = 0.50/\sqrt[2]{N_A}$, where N_A is a function of the plane of observation. In Table 1 Δ_2 is computed from the number of inclusions on a plane normal to the crack line. These results indicate that the COD at crack initiation is of the same order of magnitude as Δ_2. Pawals and Gurland (1976) in a spheroidized carbon steel have also found a better correlation for the critical COD with Δ_2 than with Δ_3.

In spite of this correlation it is not clear why the size of the process zone should be different in the two directions. Perhaps it would seem more appropriate to consider that the size of the process zone is the same in the two directions but that the ductility at fracture is different. This approach in terms of a fracture strain has been recently applied by Ritchie et al. (1979)

Table 1
Comparison of measured COD and distance to the nearest neighbor in a plane normal to the crack front (Δ_2) and in the volume (Δ_3)

Steel	Direction of loading	COD (μm)	Δ_2 (μm)	Δ_3 (μm)
A	Longitudinal	210	228	58
	Short-transverse	125	145	58
C	Longitudinal	180	149	38
	Short-transverse	115	107	38
D	Longitudinal	200	157	43

to the prediction of the upper shelf toughness of nuclear pressure vessels steels. In one grade of steel these authors have found that crack initiation is consistent with the attainment of a critical fracture strain (i.e. "the ductility at a highly triaxial stress state") over a distance of roughly one inclusion spacing from the crack tip. In another steel, they have found that the characteristic distance was larger and was of the order of 6 to 7 times the inclusion spacing.

From this discussion, it can be concluded that the choice of the size of the process zone for ductile rupture cannot be easily made and that the characteristic distances are most often chosen to best fit the experimental data.

3.1.2. Approach using a critical cavity growth factor combined with a statistical analysis

The distributions of stress and strain associated with stationary cracks in various elastic-plastic materials are available from analytical and numerical solutions (Rice and Johnson, 1970; Beremin, 1979). These solutions have emphasized the importance of the effect of crack blunting. Models for fracture have already been discussed in the light of the results derived from these studies including a criterion based on the growth of voids. The results presented in this section lend themselves to this type of approach. This is attempted in the case of Steel A for which experimental results related to crack initiation and to the processes of cavity initiation and cavity growth are available.

The application of the criterion derived for cavity nucleation shows that a very small strain is required to form voids at the crack tip. This is because of the highly triaxial stress field which exists in front of a crack. Therefore, in the interpretation of our experiments it is quite reasonable to neglect this first step of the process of ductile rupture and to take into account only the process of cavity growth and void coalescence. This can be done using the results of Rice and Johnson (1970) or those of McMeeking (1976) for the distribution of stress and strain. In spite of the fact that it has clearly been shown that the experiments performed by Lautridou do not correspond to plane strain state (Devaux *et al.*, 1978) we will use the results available for this stress state.

To take into account the distribution of the inclusions which are located at the crack tip, we compute the probability that an inclusion is located between r and $r + dr$ in an angle of span $\pi/2$ in front of the crack (Fig. 8). The calculations are presented in Appendix B. This can be used to calculate the mean value of void growth $\langle R/R_0 \rangle$ by integrating up to the critical COD which was measured by Lautridou (Table 1). The results lead to a mean value of 1.6 in the longitudinal direction and 1.25 in the transverse

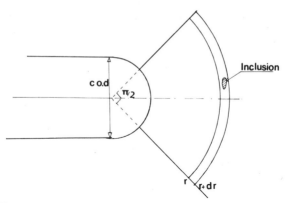

Fig. 8. Diagram of the statistical modeling of crack initiation.

direction. These two values are very close to those which were measured in axisymmetric notched specimens. However the extrapolation of the results obtained in bulk specimens to higher values for the stress triaxiality which are close to those existing at a crack tip should lead to lower values (see Fig. 7). It is felt that this difference can partly be attributed to the fact that in the analysis of the experiments on bulk specimens the size effect has not been taken into account. The incorporation of the size effect to the analysis of the results on bulk specimens should lead to a lower slope of the $(R/R_0)_c$ vs (σ_m/σ_{eq}) curve since the amount of material involved in the process of ductile rupture decreases with the notch radius of the specimens.

3.2. Crack propagation

The numerical results presented by d'Escatha and Devaux (1979) have shown that the main features of stable crack propagation can be obtained using for the conditions of node release a criterion based on a constant value for $(R/R_0)_c$. In these calculations a value of $(R/R_0)_c = 1.30$ was used. This value is reasonable as compared to those which were obtained in bulk specimens. Thus it is felt that the results presented in this section can be used to study crack propagation.

It is well know that the main result which emerges from experimental investigations on crack propagation is that the "resistance curve"—such as the J–Δa curve—has a positive slope. The fact that the utilization of the concept of $(R/R_0)_c$ can explain the main features of this curve suggests that the distribution of stress and strain ahead of a propagating crack is different from the same distribution associated with a stationary crack. The observation of the geometry of a crack just before initiation and after propagation can give some insight into this difference. With this in mind metallographic observations have been made to study the crack profile, and a recrystallization technique has been used to evaluate the strain distribution.

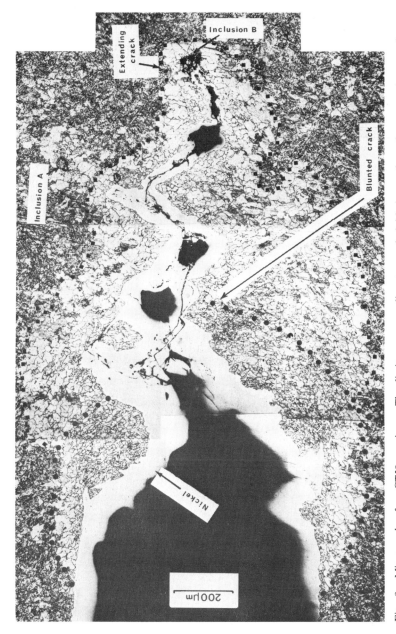

Fig. 9. Micrograph of a CT50 specimen. The limits corresponding to crack initiation and crack propagation are drawn approximately. They were obtained by a recrystallization technique (675°C, 4 h), and they correspond to $\varepsilon \simeq 20\%$. Note the highly strained regions around two inclusions (A and B).

F. M. Beremin

Figures 9 and 10 show an example of the results which can be obtained with this procedure. In these figures it is observed that the main difference between a stationary crack and a propagating one is associated with the absence of a significant crack blunting effect in the latter case. Moreover, it has been shown more quantitatively by Lautridou (1980) that the strain distribution derived from the measurement of recrystallized grain size in front of a blunted crack is consistent with the analytical results given by Rice and Johnson (1970) for fully plastic situations. An example of the strain distribution associated with a blunted crack is given in Fig. 10. A close examination of these micrographs clearly show that, when the same type of grain size measurements are made at the tip of a propagating crack, smaller strains are found as compared to those corresponding to a blunted stationary crack. This difference is still clearer in the case corresponding to the specimens cut in the transverse direction. In this case, it was almost impossible to observe recrystallized grains in front of a propagating crack, which means that the associated strain is lower than 20–25%. In light of these observations the fact that the existence of a constant value of $(R/R_0)_c$ might be used to model both crack initiation and stable crack growth

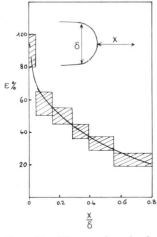

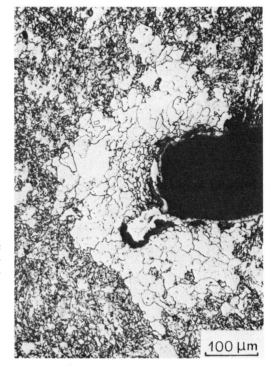

Fig. 10. Micrograph of the blunted crack and strain distribution evaluated with the recrystallization technique.

100 µm

implies that the stress distributions are different in the two situations. This difference is associated with the geometry of the crack since crack blunting strongly affects the stress triaxiality in front of the crack.

It is worth comparing these observations to the results of a certain number of theoretical studies devoted to the analysis of the deformation at an extending crack tip (Slepyan, 1974; Rice and Sorensen, 1978; Amazigo and Hutchinson, 1977). These studies have shown that in small scale yielding the strain singularity at the tip of a propagating crack is smaller than that corresponding to a stationary crack. Although in our materials the determination of the strain distribution in front of a growing crack cannot be made very accurately, our observations emphasize equally the difference in the deformation field between a stationary and an extending crack.

4. Conclusions

4.1. Various stages during the ductile rupture of bulk specimens of A 508 steel have been characterized.

4.1.1. Cavity formation from elongated MnS inclusion occurs differently in the longitudinal and the transverse direction. In both cases the criterion for the initiation of cavities is a critical local stress.

4.1.2. The application of the Rice and Tracey relation for cavity growth describes the general trend for cavity growth.

4.1.3. Void coalescence can be described in terms of a critical growth factor $(R/R_0)_c$. This factor is weakly dependent on stress triaxiality. Size effect should be taken into account to interpret more closely the variation of $(R/R_0)_c$ as a funtion of stress triaxiality.

4.2. Metallographic observations can differentiate the distribution of strain between stationary and extending cracks. Plastic strains corresponding to stationary blunted cracks just before crack initiation are much larger than those associated with growing sharp cracks.

4.3. In addition to the statistical analysis of the inclusion distribution near the tip of a crack, these observations are used to discuss models for crack initiation and stable crack propagation. It is concluded that a model using a critical void growth $(R/R_0)_c$ in conjunction with the calculation of the stress and strain fields by the finite element method can lead to useful results.

Appendix A

Four different A508 steels have been used. They were taken from different forged components.

Steel A and B are parts of a nozzle shell of a P.W.R. nuclear vessel. These two steels are very similar. Steel C is part of a tubular plate whilst Steel D is a special low sulphur steel taken from a rolled plate.

The chemical compositions are given in Table A1.

Table A1
Chemical compositions

Steel	C	S	P	Si	Mn	Ni	Cr	Mo	Cu
A	0.15	0.010	0.009	0.325	1.47	0.714	0.157	0.516	0.082
B	0.144	0.009	0.011	0.185	1.225	0.700	0.260	0.510	0.060
C	0.128	0.013	0.008	0.235	1.20	0.715	0.185	0.505	0.06
D	0.155	0.005	0.09	0.275	1.395	0.61	0.09	0.500	0.13

The conventional mechanical properties at 100°C are given in Table A2.

Table A2
Tensile properties

Steel	R_E (MPa)	R_T (MPa)	Σ (%)
A	466	590	75
B	421	580	76
C	290	490	71.6
D	432	615	74

An analysis of the inclusion distribution has been made on three orthogonal faces on Steels A, C and D in order to evaluate the mean length of particles and the number of inclusions per unit area (Table A3). L is long direction, T is long-transverse direction, ST is short-transverse direction. The number of inclusions per unit volume (N_v) was evaluated assuming elliptical shaped inclusions. The results are given in Table A4.

Table A3
Mean length of particles and number of inclusions per unit area

Steel	Face	Length (μm)	Number of inclusions (mm^{-2})
A	L–T	14.1	4.8
	T–ST	10	11.8
	L–ST	14.2	10.8
C	L–T	8.8	11.3
	T–ST	7.5	22
	L–ST	7.5	18.5
D	L–T	7.5	10.2
	T–ST	6.3	15.7
	L–ST	11	14.6

Table A4
Number of inclusions per
unit volume, N_v

Steel	N_v (mm^{-3})
A	870
C	3200
D	2100

Appendix B

Let N_A be the number of inclusions per unit area in the plane normal to the crack front (Fig. 8). Since the number of inclusions in a square of 1 mm by 1 mm follows almost a Poisson law, the probability that no inclusion lies between 0 and r and that at least one inclusion exists between r and $r+dr$ is, in the initial configuration (COD$=0$),

$$P(r)\,dr = \exp\left(-N_A \times \frac{\pi r^2}{4}\right) \times N_A \times \frac{\pi r}{2}\,dr.$$

Since the plastic deformation in plane strain induces no change of area, the probability is the same when a finite COD exists.

The calculations of Rice and Johnson (1970) or McMeeking (1976) give the variation of the growth rate of an assumed hole located at a distance r of the original crack tip. The curves are given for different angles θ of the crack-tip-inclusion line to the plane of the crack. The variation is rather weak. Therefore, as a first approximation, a mean curve between $\theta=0$ and $\theta=\pi/4$ was used and was assumed to be independent on θ, i.e.,

$$\frac{R}{R_0}(r,\text{COD}) = f(r/\text{COD}).$$

Therefore the mean growth rate of the first inclusion in front of the crack is, for a given COD,

$$\left\langle \frac{R}{R_0}(\text{COD}) \right\rangle = \int_0^\infty f(r/\text{COD}) \exp\left(-N_A \times \frac{\pi r^2}{4}\right) \times N_A \times \frac{\pi r\,dr}{2},$$

where $\langle(R/R_0)_c\rangle$ is computed for the critical COD. The same N_A has been used for the longitudinal and the short transverse directions since the probability to find an inclusion at a given distance of the crack line is the same in both cases: $N_A = (N_{A1}\cdot N_{A2}\cdot N_{A3})^{1/3}$ where N_{A1}, N_{A2}, N_{A3} are measured on perpendicular faces (Lautridou, 1980).

204 F. M. Beremin

Acknowledgments

This study has been partly supported by *Délégation Générale à la Recherche Scientifique et Technique* under contracts no. 76-7-1206, no. 78-7-2691 and no. 78-7-2692. Financial support from FRAMATOME for numerical calculations is also greatly acknowledged.

References

Amazigo, J. C. and Hutchinson, J. W. (1977), "Crack-tip in Steady Crack Growth with Linear Strain-hardening," *J. Mech. Phys. Solids*, **25**, 81.

Argon, A. S. and Im, J. (1975a), "Separation of Second Phases Particles in Spheroidized 1045 Steel, Cu–0.6% Cr Alloy, and Maraging Steel in Plastic Strain," *Met. Trans.*, **6A**, 839.

Argon, A. S., Im, J. and Safoglu, R. (1975b), "Cavity Formation from Inclusions in Ductile Fracture," *Met. Trans.*, **6A**, 825.

Argon, A. S. (1976), "Formation of Cavities from Nondeformable Second Phase Particles in Low Temperature Ductile Fracture," *Trans. ASME, J. Eng. Mat. Techn.*, **98**, 60.

Begley, J. A. and Landes, J. D. (1972), "The *J* Integral as a Fracture Criterion," *ASTM STP 514*, p.1.

Beremin, F. M. (1979), "Study of Growing Cracks Using Damage Functions: Application to Warm Prestress Effect," *CSNI Specialist Meeting on Plastic Tearing Instability*, Saint-Louis (Missouri), 25–27 September 1979.

Beremin, F. M. (1980a), "Cavity Formation from Inclusions in Ductile Fracture of A508 Steel," *Met. Trans.*, to appear.

Beremin, F. M. (1980b), "Calculs Elasto-Plastiques par la Méthode des Éléments Finis d'Éprouvettes Axisymétriques Entaillées Circulairement," *J. Mécanique Appl.*, **4**, 307.

Berveiller, M. and Zaoui, A. (1979), "An Extension of the Selfconsistent Scheme to Plastically-flowing Polycrystals," *J. Mech. Phys. Solids*, **26**, 325.

d'Escatha, Y. and Devaux, J. C. (1979), "Numerical Study of Initiation, Stable Crack Growth, and Maximum Load with a Ductile Fracture Criterion Based on the Growth of Holes," *ASTM STP 668*, p. 229.

Devaux, J. C., Lautridou, J. C., Mudry, F., and Pineau, A. (1978), "Experimental and Numerical Studies Undertaken for the Definition of Damage Functions for Numerical Modeling of Ductile Tearing and Cleavage," *5th SMIRT Conference*, Berlin, 13–17 August 1979.

Eshelby, J. D. (1961), "Elastic Inclusions and Inhomogeneities," *Prog. Solid Mech.*, **11**, I. N. Sneddon and R. Hill, (eds.), North-Holland Publishing Company, Amsterdam.

EPRI, Report (1978), "EPRI Ductile Fracture Research Review Document," Report No. NP-701-SR, EPRI, 3412 Hillview Avenue, Palo Alto, CA.

Perra, M. and Finnie, I. (1977), "Void Growth and Localization of Shear in Plane Strain Tension," *ICF4 Conference*, University of Waterloo Press, Waterloo, Ont., **2**, 413.

Goods, S. J. and Brown, L. M. (1979), "The Nucleation of Cavities by Plastic Deformation," *Acta Met.*, **27**, 1.

Gurson, A. L. (1977), "Continuum Theory of Ductile Rupture by Void Nucleation and Growth: Part I—Yield Criteria and Flow Rules for Porous Ductile Media," *J. Eng. Mat. Tech.*, **44**, 2.

Knott, J. F. and Green, G. (1976), "The Initiation and Propagation of Ductile Fracture in Low Strength Steels," *J. Eng. Mat. Techn.*, **98**, 37.

Lautridou, J. C. (1980), "Etude de la Déchirure Ductile d'Aciers á Faible Résistance—Influence de la Teneur Inclusionnaire," Thése de Docteur Ingénieur à l'École des Mines de Paris, 60, Boulevard Saint-Michel, 75006 Paris.

McClintock, F. A. (1968), "Ductile Rupture by the Growth of Holes," *J. Appl. Mech.*, **35**, 36.

McClintock, F. A. (1971), "Plasticity Aspects of Fracture," *Fracture III*, H. Liebowitz (ed.), Academic Press, New York, 47.

Mackenzie, A. C., Hancock, J. W. and Brown, D. K. (1977), "On the Influence of State of Stress on Ductile Failure Initiation in High Strength Steels," *Eng. Fract. Mech.*, **9**, 167.

McMeeking, R. M. (1976), "Finite Deformation Analysis of Crack Tip Opening in Elastic-Plastic Materials and Implications for Fracture Initiation," Technical Report C00(3084/44) Division of Engineering, Brown University, Providence, R. I.

Pawals, S. P. and Gurland, J. (1976), "Observations on the Effect of Cementite Particles on the Fracture Toughness of Spheroidized Carbon Steels," Technical Report E(11-1) 3084/44 Division of Engineering, Brown University, Providence, R.I.

Rice, J. R. and Tracey, D. M. (1969), "On the Ductile Enlargment of Voids in Triaxial Stress Fields," *J. Mech. Phys. Solids*, **17**, 201.

Rice, J. R. and Johnson, M. A. (1970), "The Role of Large Crack Tip Geometry Changes in Plane Strain Fracture," *Inelastic Behavior of Solids*, M. F. Kanninen et al. (eds.), McGraw-Hill, New York, 641.

Rice, J. R. and Sorensen, E. P. (1978), "Continuing Crack-tip Deformation and Fracture for Plane Strain Crack Growth in Elastic-Plastic Solids," *J. Mech. Phys. Solids*, **26**, 163.

Ritchie, R. O., Server, W. L. and Wullaert, R. A. (1979), "Critical Fracture Stress and Fracture Strain Models for the Prediction of Lower and Upper Shelf Toughness in Nuclear Pressure Vessels Steels," *Met. Trans.*, **10A**, 1557.

Rousselier, G. (1979), "Contribution à l'Étude de la Rupture des Métaux dans le Domaine de l'Élastoplasticité," Thèse d'Etat, Ecole Polytechnique, 91120 Palaiseau.

Slepyan, L. I. (1974), "Growing Crack During Plane Deformation of an Elastic-Plastic Body," *Izv. AN SSR., Mekhanika Tverdogo Tela*, **9**, 57.

Tanaka, R., Mori, T. and Nakaruma, T. (1971), "Decohesion at the Interface of a Spherical, Fiber or Disc Inclusion," *Trans. ISIJ*, **11**, 383.

Wells, A. A. (1963), "Application of Fracture Mechanics at and Beyond General Yielding," *Brit. Welding J.*, **42**, 563.

Yamamoto, H. (1978), "Conditions for Shear Localization in the Ductile Fracture of Void Containing Metals," *Int. J. Fract.*, **14**, 347.

S. Nemat-Nasser, Editor
THREE-DIMENSIONAL CONSTITUTIVE RELATIONS AND DUCTILE FRACTURE
North-Holland Publishing Company (1981) 207-208

DISCUSSION ON SESSION 3

J. W. HUTCHINSON

Harvard University, Cambridge, MA, U.S.A.

In the treatment of Nemat-Nasser *et al.*, on the influence of void growth on plastic constitutive behavior, the voids are assumed to exist at the start of straining and nucleation of voids from inclusions with straining is not considered. In certain metals the nucleation process appears to be as important as the void growth process itself. For example, in the steels studied by the Beremin[1] Group the voids appear to undergo only about a factor of two growth in volume over their starting volume at nucleation when the material at the crack tip fails.

Nemat-Nasser *et al.* use the same bifurcation analysis as Anand and Spitzig to predict the angle to the principal stress direction at which shear bands first formed in the tests conducted by the second group of authors. Nemat-Nasser *et al.* use a formula involving the stress and tangent modulus and they use the experimental values for these quantities in obtaining their prediction. Anand and Spitzig used an alternative expression involving the tangent modulus and the instantaneous shear modulus. The shear modulus was not measured directly but was derived from two different plasticity theories and was evaluated at the experimentally determined critical stress. The difference between the two predictions most likely reflects the inadequacy of the plasticity theories.

Most of the work of Rice discussed here has been concerned with high temperature void growth under uniaxial stress states. The role of stress state on void growth should be of some importance. When growth is dominated by power-law creep, hydrostatic tension significantly influences the growth-rate. Hydrostatic tension can develop under high constraint conditions such as at the tip of a plane strain crack. It also seems likely that hydrostatic tension might develop locally at a grain boundary or junction due to grain boundary sliding even though the overall stress state is uniaxial. Whether the alteration of the local stress state under which the voids grow is likely to lead to significantly different predictions is not clear.

The fracture criterion adopted by the Beremin Group assumes material failure when a void has experienced a certain fractional increase of volume.

---

[1] F. M. Beremin is a research group including: Y. d'Escatha, J. C. Devaux, P. Ledermann, F. Mudry and A. Pineau.

It does not bring in the void spacing in any manner. The critical fractional increase is obtained experimentally for their steels. The Group has had success in applying the criterion to a variety of deformation situations. It seems unlikely that this criterion can be a fundamental one in that the spacing between voids (or for example the void volume fraction) plays no role and there may be some difficulty in applying the criterion to a wider range of materials.

S. Nemat-Nasser, Editor
THREE-DIMENSIONAL CONSTITUTIVE RELATIONS AND DUCTILE FRACTURE
North-Holland Publishing Company (1981) 209

REPLY BY F. M. BEREMIN*

The void spacing was not introduced in our presentation which concerned essentially one single material. In another study, J.C. Lautridou (1980, see Reference in our paper) obtained results for $(R/R_0)_c$ in various materials. He showed the important effect of the inclusion distribution on this parameter. Hence we completely agree with Dr. Hutchinson's comment.

More generally, the criterion could be rewritten in the following form:

$$R/\lambda = k \qquad \text{instead of} \qquad R/R_0 = (R/R_0)_c = k(\lambda/R_0),$$

where λ is a distance between inclusions, R_0 and R are the initial and the actual radii of the cavities, and k should be a constant. The ratio λ/R_0 is closely related to the inclusion distribution. It must be statistically defined since the statistically nearest inclusions will first coalesce. In order to assess λ, it is necessary to introduce a local volume fraction of voids which must be size dependent. In our experiments, when the notch radius of the specimen is reduced, (λ/R_0) should increase and, hence $(R/R_0)_c$, if k is a constant. Figure 7 showed exactly the reverse trend. Therefore, this indicates that k is not actually a constant but decreases as stress triaxiality increases.

---

*F.M. Beremin is a research group including: Y. d'Escatha, J. C. Devaux, P. Ledermann, F. Mudry, A. Pineau and J. C. Lautridou.

SESSION 4

Chairman: Y. Yamada
             University of Tokyo, Tokyo, Japan

Authors: H. P. Stüwe
             F. M. Burdekin

Discussion: R. Roche

S. Nemat-Nasser, Editor
THREE-DIMENSIONAL CONSTITUTIVE RELATIONS AND DUCTILE FRACTURE
North-Holland Publishing Company (1981) 213–221

THE PLASTIC WORK SPENT IN DUCTILE FRACTURE

Hein Peter STÜWE

Erich-Schmid-Institut für Festkörperphysik, Österreichische Akademie der Wissenschaften, Leoben, Austria

The plastic work used in forming the visible relief of a ductile fracture surface is estimated from the stress–strain curve and the shape of the relief. The result is compared to measured K_{Ic} values. Since fair agreement is reached it is concluded that very little strain energy is spent in other parts of the specimen when the specimen is big enough to yield valid K_{Ic} values.

1. Estimate of the plastic work spent to form the fracture surface

The scanning electron micrograph of a typical ductile fracture surface is shown in Fig. 1. One sees dimples separated by sharp-edged ridges. This relief is produced by severe plastic deformation.

The specific plastic work done on any volume element dV of this "landscape" is given by,

$$W = \int_0^\phi \sigma \, d\phi, \tag{1}$$

where ϕ is the local logarithmic strain and σ the flow stress. The total work used to form the landscape is thus given by,

$$A = \oint W \, dV, \tag{2}$$

where the integral has to be extended over the total volume. Eq. (1) can be evaluated numerically when the stress–strain curve of the material is known. It is easier, however, to write,

$$W = \bar{\sigma} \phi, \tag{3}$$

where $\bar{\sigma}$ is a suitable average stress. Where large strains are involved—as they are in this paper—$\bar{\sigma}$ is at least equal to the maximum engineering stress and probably about 50% higher. An estimate for $\bar{\sigma}$ shall be given in the Appendix of this paper.

Equation (2) can be simplified by assuming,

$$\phi = \ln \frac{F_0}{F}, \tag{4}$$

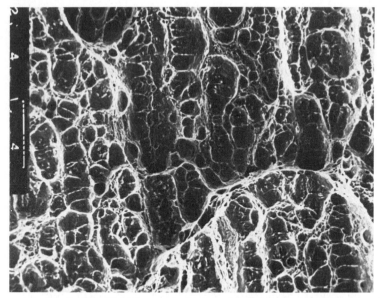

Fig. 1.   Scanning micrograph of a typical ductile fracture surface.

where F is the metallic part of a cross section at height h through the fracture surface (Fig. 2a) and F_0 is the original cross section of the specimen (Fig. 2b). This assumption leads to what is called the "elementary theory" in plastomechanics, Lippmann and Mahrenholtz (1967) (plane cross sections remain plane). While certainly not correct in detail, this approximation is known to give results in close agreement with experiment.

F is a function of h or $x = h/h_0$ (Fig. 2a), and runs from zero at $h = x = 0$ at the highest "summits" of the fracture surface to $F = F_1$ at $h = h_0$ or $x = 1$ at the bottom of the deepest "valleys". This function contains the geometry of the fracture surface.

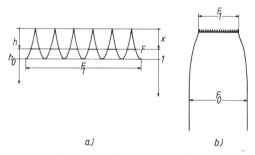

Fig. 2.   Definition of coordinates in describing the fracture surface.

With these approximations the specific work to form a fracture surface on the original cross section F_0 becomes,

$$\gamma = \frac{A}{F_0} \approx \bar{\sigma} h_0 S, \tag{5}$$

where

$$S = \int_0^\infty \frac{F}{F_0} \ln \frac{F_0}{F} \, dx; \tag{6}$$

or, with $z = F/F_0$,

$$S = S_1 + S_2 = \int_0^1 z \ln \frac{1}{z} \, dx + \int_1^\infty z \ln \frac{1}{z} \, dx. \tag{7}$$

2. Numerical evaluation of S for a rigid bulk specimen

If we assume that the plastic work is used exclusively to form the fracture surface and that the bulk of the specimen ($x > 1$!) does not deform at all, then $F_1 = F_0$ and

$$S = S_1 = \int_0^1 z \ln \frac{1}{z} \, dx, \tag{8}$$

where z runs from zero to one.

If the elementary feature of the fracture surface is the wall around a dimple of parabolic cross section, (Stüwe, 1980), then

$$z = x \quad \text{and} \quad S = 0.25. \tag{9}$$

While $z = x$ does seem a rather reasonable assumption, the result for S is quite insensitive to deviations from linearity between z and x.

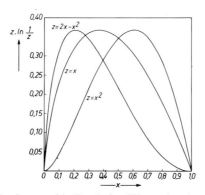

Fig. 3.   Integrand in Eq. (8) for different functions $z(x)$.

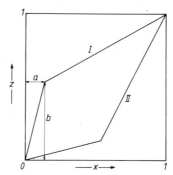

Fig. 4.   Topography of fracture surfaces with dimples of different size.

The reason is that any integrand in Eq. (8) will form a curve running from zero for $x=0$ to zero at $x=1$ passing through a maximum of 0.368 at $z=1/e$ (which means at $\phi=1$!). The integral over any curve passing through these points should be quite insensitive to details. This should be evident from the three examples shown in Fig. 3 and shall be further discussed for the two structures shown schematically in Fig. 4.

Figure 4a symbolizes a fracture surface which is made up in part by large dimples and in part by small ones.

For $0 \leqslant x \leqslant a$, z increases rapidly and from then on slowly with x: this corresponds to curve I in Fig. 5 ($a < b$).

Figure 4b symbolizes a population of small dimples at the bottom of large dimples. In that case, z increases slowly for $0 \leqslant x \leqslant a$ and rapidly from then on like curve II in Fig. 5 ($a > b$).

The two sections of the curve are described by,

$$z = \frac{b}{a}x \quad \text{for} \quad 0 \leqslant x \leqslant a,$$

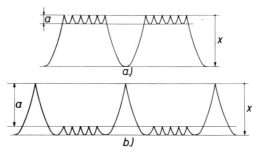

Fig. 5.   Function $z(x)$ for topographies as in Fig. 4: curve I corresponds to Fig. 4a, curve II to Fig. 4b.

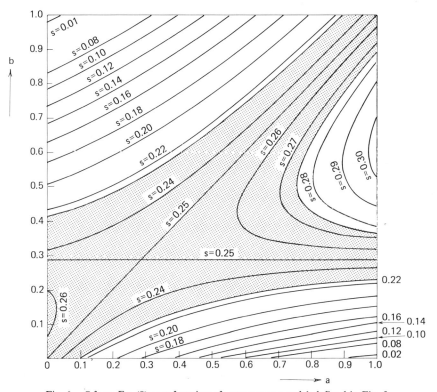

Fig. 6. S from Eq. (8) as a function of parameters a and b defined in Fig. 5.

and

$$z = \frac{1-b}{1-a}x + \frac{b-a}{1-a} \qquad \text{for} \qquad a \leqslant x \leqslant 1. \tag{10}$$

The result for S (Eq. (8)) is then a function of the parameters a and b. This function can be found analytically and is shown in Fig. 6. Obviously the special result given in Eq. (9) is a good approximation for a wide variety of combinations of the parameters. The region where S deviates from 0.25 by less than 10% is shaded in Fig. 6a. One sees that the approximation,

$$S \approx 1/4 \tag{11}$$

is quite good except for very "special" geometrics of the fracture surface so that Eq. (5) simplifies to

$$\gamma \approx \frac{\bar{\sigma} h_0}{4}. \tag{12}$$

3. Comparison with experiment

The fracture toughness K_{Ic} can be interpreted as,

$$K_{Ic} = \sqrt{2\gamma_B E}, \tag{13}$$

where γ_B is the energy necessary to form a unit of fracture surface. If we assume that this energy is, indeed, spent entirely in the visible relief of the fracture surface (i.e. that Eq. (8) is a reasonable approximation and that $S_2 \ll S_1$ in Eq. (7)) then,

$$K_{Ic} \approx \sqrt{\frac{\bar{\sigma} h_0 E}{2}}. \tag{14}$$

The last section was meant to show that details of geometry do not strongly affect the evaluation of Eq. (8). Therefore, the numerical values obtained from Eq. (14) are mostly affected by the choice of values for $\bar{\sigma}$ and h_0.

In a previous paper, (Stüwe, 1980), the author suggested using the maximum engineering stress σ_B as an appropriate average flow stress $\bar{\sigma}$ and the diameter of the large dimples in the fracture surface as an estimate for h_0.

In that way, reasonable agreement between computed and measured values for K_{Ic} could be found in a number of cases. A few examples are shown in Table 1.

Current work at the Erich-Schmid-Institute shows that both assumptions must be corrected somewhat. h_0 seems to be smaller than the diameter of the largest dimples (Kolednik and Stüwe, 1981). On the other hand the average stress $\bar{\sigma}$ that should be used in Eqs. (3) and (14) is larger than σ_B (see Appendix). Since neither correction is very large and since they tend to cancel each other, quantitative results such as those in Table 1 are hardly affected.

Table 1
Computed and measured values of K_{Ic}

Material	(N^σ/mm^2)	$h_0(\mu)$	$K_{Ic}(MN/m^{3/2})$ Eq. (14)	$K_{Ic}(MN/m^{3/2})$ measured	$r(mm)$
X2NiCoMo 18.8.5	1808	50	100	95	0.42
X2NiCoMo 18.9.5	1982	40	81	81	0.14
X2NiCoMo 18.9.5	1800	30	76	85	0.35
StE47	690	20	38.5	(125)	(5.2)

We conclude that the fracture toughness can be computed—at least in principle—from the energy spent in forming the relief of the fracture surface which in turn is determined by the stress–strain curve of the material and the geometry of the relief.

This implies, however, that in such cases it is permitted to neglect the plastic work done in other parts of the specimen, i.e. to neglect the second term in Eq. (7). We shall now examine the conditions under which such an approximation can be useful.

4. Evaluation of S for a specimen where the plastic zone is large

We assume that the function $z(x)$ is given by,

$$z = cx \quad \text{for} \quad 0 \le x \le 1,$$

$$z = \frac{1-c}{d-1}x + \frac{cd-1}{d-1} \quad \text{for} \quad 1 \le x \le d. \tag{15}$$

This function is shown in Fig. 7: at $x = h/h_0 = 1$, z has not yet risen to one but to a smaller value $c = F_1/F_0 = 1 - \delta$.

From there on the curve rises much more slowly until it finally reaches the value 1 at d, which is something like the "radius of the plastic zone" usually ascribed to a ductile fracture. Examples computed in the usual way are given in Table 1. They show that d may be much larger than 1. On the other hand we shall assume that

$$\delta \ll 1, \tag{16}$$

i.e., that the fracture is not preceded by extensive necking. Equation (7) now reads,

$$S = \int_0^d z \ln \frac{1}{z} \, dx = \frac{1}{c} \int_0^c z \ln \frac{1}{z} \, dz + \frac{d-1}{1-c} \int_c^1 z \ln \frac{1}{z} \, dz. \tag{17}$$

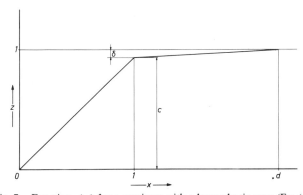

Fig. 7. Function $z(x)$ for a specimen with a large plastic zone (Eq. (15)).

Equation (16) allows us to approximate the solution as,

$$S = S_1 + S_2 \approx \frac{1}{4} + \frac{d\delta}{2} = \frac{1}{4}(1 + 2 d\delta).\qquad(18)$$

This shows that the estimate for K_{Ic} given in Eq. (14) is reasonable only if,

$$2 d\delta \ll 1.\qquad(19)$$

We may reverse the argument: Since there are examples where Eq. (14) gives reasonable estimates for K_{Ic}, Eq. (19) puts a limit on the amount of deformation that can be expected in the plastic zone outside the relief of the fracture surface. The author is inclined to believe that Eq. (19) is also the underlying condition for the *measurement* of a valid K_{Ic}-value.

It should be stressed that there are numerous examples where Eq. (19) is not fulfilled. Wherever fracture is preceded by extensive necking of the specimen one finds,

$$S_2 \gg S_1,\qquad(20)$$

which means that the conventional concepts of fracture mechanics do not apply.

Appendix. Estimate for $\bar{\sigma}$ in Eq. (3)

Equation (5) should be written more exactly as,

$$\gamma = \int_0^{h_0} \int_0^{\phi} \sigma(\phi)\, d\phi\, \frac{F}{F_0}\, dh.\qquad(21)$$

If we use the standard approximation for the stress–strain curve,

$$\sigma = \sigma_0 \phi^n,\qquad(22)$$

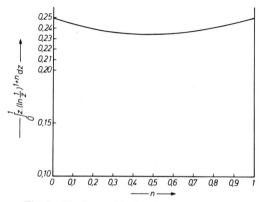

Fig. 8.   The integral in Eq. (23) as a function of n.

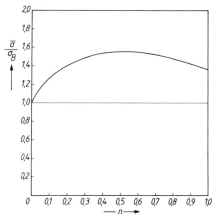

Fig. 9. Estimate for $\bar{\sigma}$ in Eq. (3) as a function of n (see Appendix).

this transforms to

$$\gamma = \frac{h_0 \sigma_0}{1+n} \int_0^1 z \left(\ln \frac{1}{z} \right)^{1+n} dz. \tag{23}$$

Figure 8 shows that in the relevant region $0 < n < 1$ the integral is very insensitive to n and always close to 0.25. Comparison with Eq. (12) gives thus,

$$\bar{\sigma} \approx \frac{\sigma_0}{1+n}, \tag{24}$$

or, with

$$\sigma_B = \sigma_0 n^n e^{-n}, \tag{25}$$

$$\frac{\bar{\sigma}}{\sigma_B} \approx \frac{e^n}{(1+n)n^n}. \tag{26}$$

This function is shown in Fig. 9. While the numerical values are, of course, only as good as the approximation Eq. (22) permits, it is certainly true that σ_B is a lower limit to what should be used for $\bar{\sigma}$ and $\bar{\sigma} \approx 1.5 \, \sigma_B$ is a probable value.

References

Kolednik, O. and Stüwe, H.P. (1981), "Zur quantitativen Beschreibung einer Verformungsbruchfläche," *Z. Metallk.*, to be published.

Lippmann, H. and Mahrenholtz, O. (1967), *Plastomechanik der Umformung metallischer Werkstoffe*, Springer-Verlag, Berlin.

Stüwe, H.P. (1980), "The Work Necessary to Form a Ductile Fracture Surface," *Eng. Fract. Mech.*, 13, 231.

S. Nemat-Nasser, Editor
THREE-DIMENSIONAL CONSTITUTIVE RELATIONS AND DUCTILE FRACTURE
North-Holland Publishing Company (1981) 223–242

STABILITY OF TEARING FRACTURE IN STRUCTURAL STEELS

F. Michael BURDEKIN

University of Manchester Institute of Science and Technology, Manchester, U.K.

The various factors contributing to the experimentally observed R-curve behavior in structural steels are considered. It is noted that both initiation toughness and tearing modulus can be dependent on specimen type, size and thickness.

Further analysis on the strip yield model with crack extension is reported, emphasizing the need for careful definition of the crack length and exact position for which COD or the J-contour integral are determined. The strip yield model would only predict a rising R-curve behavior provided increasing COD and plastic zone size occurred at the moving crack tip.

Exploratory experimental work on 0.2 mm thick impact PVC, which shows plastic zones by a stress-whitening effect, demonstrated that under plane stress conditions the crack advances with a sharp tip and constant flank angle. Tests on center cracked and double edge cracked tension strips show identical R-curve behavior in terms of COD at the original crack tip against crack extension, and the plastic zones grow larger as the crack tip advances. These tests also give results compatible with the COD design curve, provided the actual crack length is used once crack extension occurs.

The effect of crack curvature on the apparent R-curve behavior in thick material is discussed.

It is concluded that in structural steels, the effects of triaxiality on initiation toughness and tearing modulus must be taken into account. Critical values of toughness taken at 95% maximum load in full thickness standard bend or compact tension specimens will generally give conservative predictions of allowable defect sizes in service, when used in conjunction with the COD design curve.

1. Introduction

One of the difficulties affecting all of the fracture mechanics treatments which seek to extend beyond the plane strain linear-elastic regime, is the occurrence of ductile tearing as the first stage of crack extension from a pre-existing crack. This tearing may often commence before the attainment of maximum load in a fracture toughness test. The problem then arises as to whether the critical value of the fracture toughness parameter being measured should be taken as the value at onset of ductile tearing, or as a greater value such as that just before maximum load in the test.

It is found in practice that the adoption of a critical value of crack opening displacement (δ_i) or the J-contour integral (J_i) at initiation of

223

ductile tearing is very restrictive on material selection requirements for critical applications such as nuclear pressure vessels or offshore structures, particularly for welded joints in stress concentration regions. To advance beyond the position of ductile tearing initiation as the critical stage in a fracture toughness test requires an understanding of the factors controlling both the onset and the stability of tearing fracture.

In structural steels the situation may be further complicated by the occurrence of cleavage fracture in thicker sections, at low temperatures, or at high strain rates. The strain rate effect may lead to an initial ductile tear developing into an unstable cleavage fracture. Cleavage fractures in steel invariably appear to be unstable.

It is intended in this paper to explore and discuss various factors which affect ductile tearing stability, and to examine the implications on fracture toughness testing, and application of fracture mechanics to service problems.

2. Resistance curves and tearing stability

The concept of material "resistance curves" is now well known, and a "tentative recommended practice for R-curve determination" has been published by ASTM. Many research investigations have also been carried out using the R-curve approach to initial ductile tearing, and a useful summary of the present position has been given recently by Turner (1979).

In their original form, R-curves for materials with only limited plasticity before fracture (close to LEFM conditions) were expressed as a plot of crack extension force, G, (or stress intensity factor, K) against increment of crack extension. Various shapes of resistance curve have been postulated to explain difference phenomena, typical examples being shown in Fig. 1. The point of instability of fracture is predicted from R-curves by the tangency of the particular driving force curve for the structural component with the resistance curve, i.e.,

$$\frac{dG}{da} \geqslant \frac{dR}{da}, \qquad G = R. \tag{1}$$

With the extension to LEFM methods by the crack opening displacement and J-contour integral concepts, it was natural that the R-curve approach should develop for elastic plastic fracture mechanics by plotting COD or J against increment of crack extension, Δa. It has been found experimentally for materials on which much of this work has been carried out (particularly A533B steel), that the R-curve in terms of J against Δa, rises linearly at first. This observation has been used by Paris et al. (1979), as part of an analysis of the stability of such tearing fractures. These authors suggested that unstable fracture could occur if the elastic contraction remote from a crack

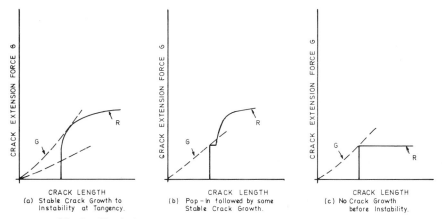

Fig. 1. Classical *R*-curve concepts for stable and unstable crack extension.

between fixed pints in a structure was greater than the opening displacement across the crack, when the crack grew by a small increment. Using the relationship $J = m\sigma_Y\delta$, between *J*-contour integral and COD, Paris et al. derived a term which they called tearing modulus, T_{mat}, defined by,

$$T_{mat} = \frac{E}{\sigma_Y^2}\frac{\mathrm{d}J}{\mathrm{d}a}, \tag{2}$$

where $\mathrm{d}J/\mathrm{d}a$ is the slope of the *R*-curve for the material in terms of *J* against Δa. Instability is predicted when a corresponding term T_{app}, representing severity of conditions in the structure, is greater than T_{mat}.

Similar results were obtained in a different analysis by Turner (1978), who introduced the concept of energy release rate in the presence of plasticity, *I*. Turner makes clear his interpretation of *J* for real elastic plastic materials as defining work absorption rate with incremental crack extension (J_r). He also uses a geometric factor η to relate J_r to total work done on a cracked specimen divided by area of ligament. In his instability analysis, Turner also finds the parameter T_{mat}, as defined by Paris et al. (1979), to be of prime importance, although other terms may also be important in some cases.

These studies are all based upon the assumption that the *R*-curve is a material property, fixed for a given material. It is the author's view that whilst *R*-curves arc a convenient representation of resistance to fracture for a particular configuration, they are not a material parameter as such, but are dependent on other factors. In the same way, in the elastic plastic range, neither COD, δ_c, nor *J*-contour integral, J_c, are true material properties independent of other factors, although of course they are extremely useful parameters.

More detailed analyses of local behavior at a crack tip under elastic-plastic conditions have been carried out by Rice (1976) to seek an explanation for the *R*-curve behavior. The classical explanation for increasing resistance to fracture with incremental crack extension has always been the formation of plane stress shear lips, requiring greater fracture work at the surfaces, than the plane strain region at mid-thickness. Rice linked the *R*-curve development to crack tip plasticity, since the extent of plane stress and plane strain zones in the thickness depends on the ratio of plastic zone size to thickness. More recently Rice (1978) has examined the mechanism of crack extension with a small Prandtl plastic field moving along with the advancing crack tip, leaving a constant crack opening angle behind. The constant flank angle had previously been suggested by Green and Knott (1975).

Rice's analysis led to a relationship between crack opening $2v$ at a distance r from the advancing tip as follows:

$$2v = r\left\{ \frac{\alpha}{\sigma_Y}\frac{\mathrm{d}J}{\mathrm{d}a} + \frac{\beta\sigma_Y}{E}\left(1 + \log_e\left(\frac{z}{r}\right)\right)\right\}, \tag{3}$$

where α, β are dimensionless coefficients.

In ductile materials the first term is large compared to the second provided r is not extremely small. This then reduces to the constant flank angle behind the advancing crack tip.

2.1. *Experimental observations on thick steels*

Although there have been a number of investigations into the use of critical values of J at fracture as a plane strain fracture toughness parameter, J_{1c}, with lower thickness and size requirements than for LEFM, specimen geometry effects are reported by others. For instance work at the Welding Institute, at Cranfield Institute of Technology, and at Glasgow University has shown different values of J_{1c} can be obtained on the same material with different specimen geometries.

The early work on crack opening displacement as a fracture criterion, showed a dependence of critical values on specimen type and geometry, although the major effect was one of thickness (Burdekin and Stone, 1966).

An example of the geometric effect of thickness is given in the results in Fig. 2, obtained recently in the author's research program. Apart from the effect of thickness on initiation toughness values for the particular mild steel being tested, the transition in fracture mode from shear to cleavage above 40 mm thickness is most striking. Fracture surfaces are shown in Fig. 2 for 20–80 mm thick bend specimens.

It is generally considered that these effects on initiation toughness values are due to variations in triaxiality of stress ahead of the crack tip in the

Fig. 2. Fracture surfaces of notched bend specimens of mild steel in thicknesses 20–80 mm, tested at +20°C, showing transition from shear to cleavage above 40 mm thickness.

plastic zone. Whilst the ratio of plastic zone size to thickness is a major factor affecting triaxiality, the type of specimen, specimen geometry and crack geometry also have an effect. Detailed analyses for all cases are not available, and are urgently needed to define triaxiality factors. The experimental observations at the Welding Institute, however, have shown much greater values of J_{1c} or δ_c for center cracked tension test plates, than for standard full thickness bend tests, and lower values for double edge notched tension plates where yielding occurs. At Cranfield a consistent variation of J_{1c} and δ_c with notch depth ratio in bend specimens has been shown. At Glasgow University it has been found that a wide variation in initiation toughness values with triaxiality exists in different specimens.

A similar effect of different specimen types, and of triaxiality has been found for the slope of the J, Δa-R-curves by each of the three research groups above. An example taken from the work of Garwood (1980) at the Welding Institute is shown in Fig. 3. This particular figure shows the different initiation values, initial slope and different overall shape for resistance curves on 110 mm thick A533B bend specimens. Center cracked tension tests in the same investigation gave higher resistance curves than the bend tests.

The Cranfield work on resistance curves also showed that as the notch depth ratio changed so the slope of the resistance curve also changed. It is

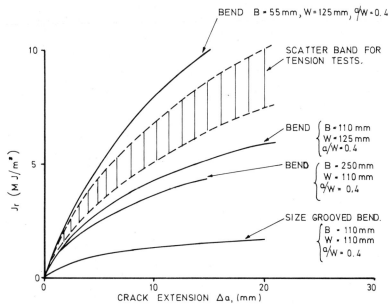

Fig. 3.   R-curves for different geometry specimens in A533B (Garwood, 1980).

understood that Glasgow University work has shown variation of the slope of the resistance curve with triaxiality in different types of test specimen on the same material.

3. Crack growth resistance in thin sheets

3.1. The strip yield model with crack growth

The strip yield model, originally introduced to analyze plastic zones by Dugdale gives a simplified analysis of crack tip conditions which has been used as a guide in some elastic-plastic fracture mechanics analyses. The present author used the model in early COD analyses, (Burdekin and Stone, 1966) although it was later found that experimental results departed from the theory, and the COD design curve was based upon the trend of experimental observations of COD against strain. The COD design curve has been the subject of recent debate with alternative proposals put forward at the International Institute of Welding, and alternative analyses now put forward in terms of *J*. These matters are being taken further elsewhere; suffice it to say that the COD design curve method has given valuable service over about twelve years, and whilst developments and improvements are necessary the basic concept remains valid.

The strip yield analysis based on the anti-plane strain version by Bilby, et al. (1963) and subsequent developments is the basis of interpolation between elastic and collapse behavior in the CEGB two-criteria method (Dowling and Townley, 1975).

Although the strip yield model is simplistic for real materials, and applies strictly only to center cracked tension geometry under plane stress, it is instructive to take the analysis one stage further to examine what happens with an increment of real crack extension. Details of the analysis are given in Appendix A.

The analysis is based upon determining the work done in the plastic zone ahead of the crack, when the crack extends an incremental distance, by integrating the product of force and displacement throughout the plastic zone (i.e. from the real crack tip to the notional elastic crack tip). It would be expected that the work absorption rate per unit crack extension should be closely related to *J*, as determined experimentally as a work absorption rate. It would also be expected that the integration procedure should be closely related to the contour integral definition of *J*. Previous discussions on these points have taken place between Rice and Chell (see Chell and Heald, 1975; Rice, 1975).

If one considers a simplified case, in which the moving crack travels with a constant shape and size of plastic zone, Eq. (7A) from the Appendix A shows that,

$$\frac{\mathrm{d}U}{\mathrm{d}a} = \sigma_Y \delta. \tag{4}$$

If these restraints are not applied, however, the analysis appears to predict that there are additional terms in the expression for $\mathrm{d}U/\mathrm{d}a$ as shown in Eq. (9A). Similar conclusions were reached by Cherepanov (1979). The full analysis considers the effect of variations in applied stress, σ, as the crack length changes, and can be used to assess stability of the system, when $\mathrm{d}U/\mathrm{d}a$ is regarded as a material toughness property.

To see the reasons for the additional terms of Eq. (9A), it is helpful to consider their physical implications. The strip yield model formulation is essentially based on equilibrium aspects of load bearing capacity on the crack plane, without crack tip singularity effects. The displacement at the real crack tip is a function of crack length and applied stress. An increment in crack length, still maintaining the basic formulation, will lead to an increase in opening displacement at the new crack tip and an increase in plastic zone size. If these conditions occurred, without a change in applied stress, the implication is that the rate of work absorption in the plastic zone would continue to increase as the crack extended, i.e. a rising R-curve effect. Thus to achieve a rising R-curve under plane stress conditions, the model predicts that the plastic zone size should continue to increase, and the COD value at the new crack tip should increase as the crack extends. If these two parameters do not increase, but are essentially material properties at least for given testing conditions then the R-curve effect should not exist for two dimensional plane stress conditions.

3.2. Experimental observations on thin plastic strips

To examine the shape of the extending crack, and the spread of yield zones with crack growth, some investigations have been commenced using 0.2 mm thick impact PVC strips. This material shows up yield zones by a "stress-whitening" effect, although the strains at which the whitening develops are of the order of 1%, and the whitening effect does not show the full extent of zones at initial nonlinearity.

At the present stage, exploratory tests only have been carried out on two center cracked tension samples, and two double edge cracked samples, but the technique is reported as being simple and of interest, without being as tedious as the etch grid pattern method reported by Bergkvist and Anderson (1972). Although the stress whitening technique is not as accurate as grid pattern methods it does provide useful observations.

The tests reported here were on strips of 41 mm width, and 60 mm gauge length, tested in a displacement controlled universal testing machine, with glass plates clamped loosely each side of the test-piece to prevent buckling at the crack surfaces. The specimens were photographed at intervals during the tests, to enable crack and plastic zone profiles to be analyzed later. Examples showing the development of plastic zones and subsequent tearing for each case are shown in Figs. 4–7.

In view of recent discussions on the COD design curve, it is of interest first to examine the behavior of these specimens in terms of the relationship between COD, δ and overall strain. The results in this form are shown in Fig. 8. It can be seen that up to the start of crack extension the results fall below (on the safe side of) the design curve. The dashed lines in Fig. 8 are continuations after the start of tearing crack extension, with points based on the actual crack length, but on COD at the original crack tip. A cautionary note should be sounded over application of the design curve when tearing occurs, without using the corrected crack length. The effect is, of course, greater for tearing crack extensions from short cracks, where either high strains or poor material are likely to be involved. Other considerations of the design curve relationships are being made in separate publications.

The photographs of the specimens show clearly the initial extension of yield zones from the center cracked samples on the crack plane, as assumed in the strip yield model. The yield zones then develop into slip line bands at general yield for the short crack case, but remain confined to the ligament for the long crack case. In the case of the edge cracked samples, the yield zones again spread along the crack plane, and heavy deformation is confined to this band, although there is some evidence of light yielding in fan-shaped zones above and below the ligament between cracks. For both types of sample, the extension of the crack takes place into material heavily deformed by plastic deformation.

For each of the samples, initial crack extension occurs at a crack tip opening displacement value of approximately 0.2 mm. In all cases, the plastic zone spreads further and at a faster rate than the extension of the crack. The load deflection records show that crack extension commenced at or just before maximum load, and the majority of the crack extension shown has occurred with increasing strain but falling load. It would be expected that failure would have occurred in an unstable manner either under load controlled conditions, or if the specimens had been long enough to provide sufficient energy release capability under the controlled displacement conditions which actually applied.

Figure 9 shows the results of plotting COD at the original crack tip against crack extension, i.e. COD *R*-curves. The results for the two center cracked tension samples are virtually identical up to a crack extension of 1 mm, at which stage the longer crack specimen failed with unstable propagation, whereas the short crack specimen developed general yield slip lines and stable crack growth to 3 mm under falling load. The double edge

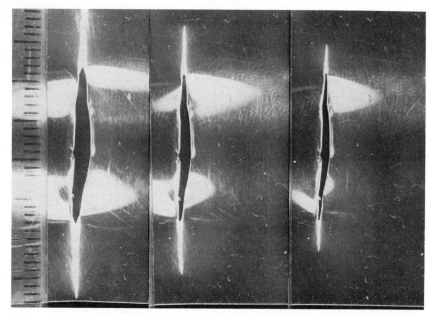

Fig. 4.   Development of plastic zones and crack extension in 0.2 mm thick impact PVC—center cracked tension $2a/W=0.5$.

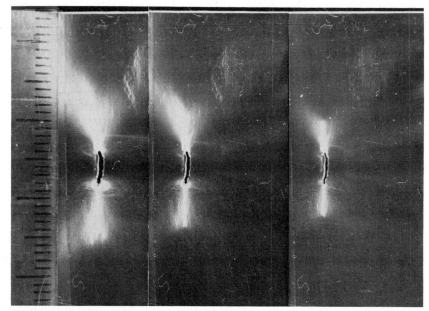

Fig. 5.   Development of plastic zones and crack extension in 0.2 mm thick impact PVC—center cracked tension $2a/W=0.15$.

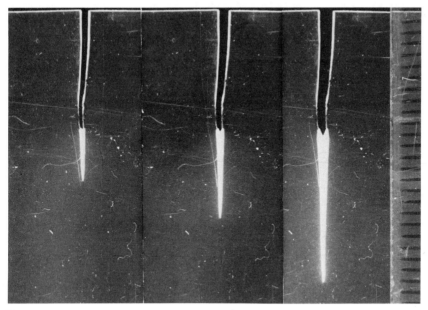

Fig. 6. Development of plastic zones and crack extension in 0.2 mm thick impact PVC—double edge cracked tension $2a/W=0.5$.

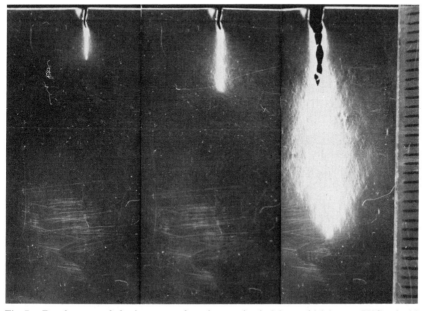

Fig. 7. Development of plastic zones and crack extension in 0.2 mm thick impact PVC—double edge cracked tension $2a/W=0.11$.

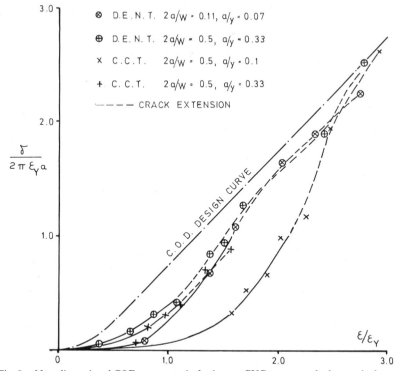

Fig. 8.   Non-dimensional COD versus strain for impact PVC, center and edge cracked tension.

cracked specimen results in Fig. 9 also follow the identical linear graph up to 1 mm crack growth, as the center crack tension specimens, for both crack length to specimen width ratios, at least within the accuracy of these tests. It should be noted, of course, that these samples are effectively under plane stress conditions, and that no triaxiality effects are present to affect the results.

The lengths of plastic zones at different stages are plotted in Fig. 10 for the different samples, against crack opening displacement at the original crack tip. These results again show a common linear relationship between plastic zone size and COD for the two center cracked samples, and for the two edge cracked samples, at least up to the stage where boundary effects and general yield intervened.

It should be noted from the photographs of the developing cracks, Figs. 4–7, that once the crack has started to extend its tip appears to remain sharp, contrary to the suggestions of others of a travelling COD value at the moving tip. The results of Fig. 9 imply the same constant flank angle behind the crack in the early stages of crack extension for both center crack and

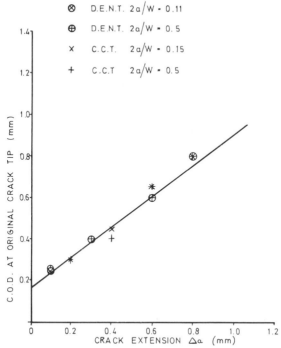

Fig. 9. *R*-curves for 0.2 mm thick impact PVC.

double edge cracked specimens, under plane stress conditions, although the flank angle diminishes with extensive propagation. The results of Fig. 10 confirm that the plastic zone size does extend faster than the crack extension in the early stages. It is likely that the intervention of general yielding has affected the extent of tearing before attainment of maximum load, and further tests of this type will be carried out on wider strips to enable the relationships to be studied further.

The results indicate that although initiation of crack extension occurs at a constant value of COD, δ, at the original crack tip for the particular material and testing conditions, the concept of crack opening displacement at the tip of a moving crack has relatively little meaning. Furthermore, the relationship between crack length and plastic zone size does not conform to the strip yield model analysis once crack extension takes place by ductile tearing, although in the very thin strips examined here the basic strip yield analysis will be upset by thinning in the heavily deformed plastic zones.

Nevertheless the results are consistent with some increase in resistance to crack extension developing at first, under plane stress conditions, due to the plastic zone having to extend further when the crack grows with a constant

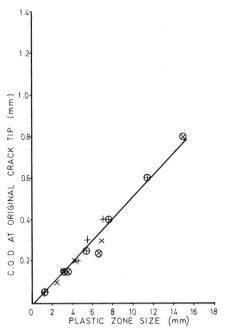

Fig. 10.   Growth of plastic zones for 0.2 mm thick impact PVC.

flank angle and a sharp tip. They are also consistent with Rice's analysis of crack displacements mentioned previously, provided J is defined as referring to the original crack tip. In any of these analyses, the exact position of definition for J or COD will make a considerable difference to the analysis of results.

4.   Crack growth resistance in thick plates

As mentioned previously, the classical explanation for the R-curve behavior is the development of shear lips at the surface, having higher resistance to crack extension than the plane strain regions at mid-thickness. It appears from the work described above, that the plane stress regions themselves may well show an increased resistance with crack extension due to further growth of plastic zones to allow the crack to extend with constant flank angle, and a similar though lesser effect may also occur under plane strain conditions with inherently smaller plastic zone.

In thick samples, with straight crack fronts, fracture initiation generally occurs at mid-thickness where conditions are nearer to plane strain. In general, however, crack extension in stable or unstable fracture under monotonic loading does not maintain a straight crack front. For example the classical chevron markings on fracture surfaces of unstable cleavage fractures are orthogonal to the curved crack front of the propagating fracture, and under tension conditions, when such fractures arrest they invariably do so with a curved crack profile. This curved crack front behavior under tension loading occurs, whether shear lips are present or not.

The author has observed in his own work, that when stable tearing fracture occurs in steel bend specimens, a curved crack profile develops, whereas if cleavage fracture occurs and then arrests, the arrest profile is straight. A further program of work on effects of crack curvature is currently in progress.

It is instructive to refer briefly to results of boundary integral analyses carried out recently by Lee at Imperial College, sponsored by the Welding Institute. The results are for linear-elastic conditions, and for compact tension specimens with a/W ratios in the range 0.45 to 0.55. Lee's (1979) results show that the mid-thickness stress intensity factor increases very slightly but then decreases as the crack front extends into a curved shape, fixed at the surfaces. The stress intensity factor at the surfaces is initially below that at mid-thickness for a straight crack front, but the surface value increases with crack curvature in the compact tension specimen until when the crack front radius is three times the thickness, a uniform K value is found along the whole crack front. This suggests that the K values calculated on average crack length for a curved crack front in fact over-estimate the true K value, and that crack curvature effects make a significant contribution to the apparent R-curve behavior. The effects of crack curvature will undoubtedly be different for different specimens, structural and crack geometries, and further differences are anticipated in elastic-plastic conditions as compared to LEFM. As pointed out by Neale (1978) the crack curvature effect is a much more logical explanation of the pop-in metastable phenomenon than the stepped resistance curve postulated by some as in Fig. 1b.

The R-curve results of Garwood (1980) on different specimen geometries, show different initiation values and different initial slopes, i.e. different tearing moduli defined as dJ/da referred to the original crack length. Firstly there is a clear effect of absolute thickness, with minimum values obtained in three point bend specimens with heavy side grooves. This last case still gives an R-curve with a positive slope. In the author's view this positive slope, where shear lip effects have been eliminated, must be due to a combination of the plastic zone size effect and the crack curvature effect, with the latter probably dominating.

It is highly significant that the effect of increasing thickness with side grooves providing the extreme case, is to lower both the initiation toughness value, and the tearing modulus. Also the results for increasing a/W in bend specimens show a decrease for both properties. Furthermore center cracked tension specimens show higher results for both properties than bend or compact tension tests. The work at Glasgow University has shown a strong dependence of initiation values of toughness under elastic-plastic conditions, on triaxiality of stress. Simple theoretical analyses suggest that the tearing modulus dJ/da should be related to the initiation toughness value J_{1c}. The experimental results referred to above all support the view that both the initiation toughness and the tearing modulus under elastic-plastic conditions are strongly affected by triaxiality of stress, and satisfactory correlations between laboratory tests and service performance can only be expected if these and the other factors discussed above are taken into account.

5.   Conclusions

(1) The R-curve phenomenon is probably a result of three effects in combination as follows:

(i) the need for increased growth of plasticity to allow crack extension to progress with constant flank angle and sharp tip through a wake of yielded material,

(ii) variations in severity of crack tip conditions with crack curvature,

(iii) the classical effect of high toughness surface layers (plane stress) combined with lower toughness layers at midthickness (plane strain).

(2) R-curves are not a unique material property, but are a convenient representation of the resistance to crack growth of a particular material under particular circumstances.

(3) Different R-curves may be obtained under elastic-plastic conditions for different types of specimen. Both the initiation toughness value, and the tearing modulus appear to depend on triaxiality of stress, and the two properties are probably related for a given material.

(4) R-curve behavior is observed with very thin materials under plane stress conditions. Impact PVC is a material which shows up yield zones, by stress whitening, and convenient for investigating the R-curve phenomenon and crack tip deformation behavior under plane stress conditions.

(5) Tests on center cracked tension specimens and double edge cracked tension specimens of 0.2 mm thickness impact PVC, under plane stress conditions have shown an identical R-curve for short and long crack length to specimen width ratios and both specimen types. These R-curves have been expressed in terms of COD at the original crack tip against crack extension. After a critical COD up to initiation, the cracks extended with a sharp tip and constant flank angle up to 1 mm growth.

(6) In view of the various factors affecting initiation toughness and tearing modulus, it is recommended that these properties should be determined either on full thickness specimens, or on specimens of sufficient thickness to satisfy normal plane strain criteria if less than full thickness. Bend specimens of 2×1 full thickness dimensions with $a/W = 0.5$ will give triaxiality and constraint conditions more severe than most practical service conditions, and may be regarded as giving minimum levels for initiation toughness and tearing modulus. Such specimens may be excessively conservative compared to service situations with tension loading and part thickness cracks with a curved crack front.

(7) The COD design curve gives safe predictions of the relationship between overall strain, crack tip COD and defect size up to the stage of slow stable tearing. After tearing commences, particularly from short initial cracks, it is necessary to use the actual (extended) crack length in COD design curve relationships. The higher triaxiality of the bend and compact tension tests compared to most practical situations means that critical values of toughness taken at 95% maximum load in full thickness bend or compact tension tests, will generally give conservative predictions of allowable defect sizes in service, when used in conjunction with the COD design curve.

Appendix A. Crack extension in the strip yield model

The strip yield model can be formulated by the Westergaard type stress function,

$$Z = \frac{2\sigma_Y}{\pi} \left[\cot^{-1} \frac{k}{z} \left(\frac{z^2 - a_1^2}{1 - k^2} \right)^{1/2} \right], \tag{1A}$$

where $2a$ is the length of real crack, $2a_1$ is the length of notional elastic crack containing internal restraining forces σ_Y from a to a_1 simulating the plastic zone, and $k = a/a_1$.

From this it follows that,

$$\int Z \, dz = \bar{Z} = \frac{2\sigma_Y}{\pi} \{ z\theta_1 - a\theta_2 \}, \tag{2A}$$

where

$$\theta_1 = \left[\cot^{-1} \frac{k}{z} \left(\frac{z^2 - a_1^2}{1 - k^2} \right)^{1/2} \right], \qquad \theta_2 = \cot^{-1} \left(\frac{z^2 - a_1^2}{a_1^2 - a^2} \right)^{1/2}.$$

Displacements between the notional elastic crack faces are given by the following expression, for plane stress:

$$2v = \frac{4}{E} \operatorname{Im} \bar{Z}, \tag{3A}$$

Displacements at $x=a$, $y=0$, corresponding to the real crack tip are,

$$\delta = 2u = \frac{8\sigma_Y a}{\pi E} \log \sec\left(\frac{\pi\sigma}{2\sigma_Y}\right). \tag{4A}$$

Displacements in the region $a < x < a_1$, $y=0$ are,

$$2v = \frac{8\sigma_Y}{\pi E}\left\{ a \tanh^{-1}\left(\frac{a_1^2 - x^2}{a_1^2 - a^2}\right)^{1/2} - x \tanh^{-1}\left[\frac{k}{x}\left(\frac{a_1^2 - x^2}{1 - k^2}\right)^{1/2}\right]\right\}. \tag{5A}$$

Work done in plastic zone at one end of the crack, due to increment of crack length da is

$$dU = 2\sigma_Y \int_a^{a_1}\left\{\frac{\partial v}{\partial a} da + \frac{\partial v}{\partial \sigma}\frac{\partial \sigma}{\partial a} da\right\} dx. \tag{6A}$$

Note that if crack extension occurs with translation of the crack tip with a plastic zone of fixed size, and constant displacement pattern within the plastic zone part of the notional elastic crack, then the following simplification occurs by changing coordinate systems:

$$\frac{\partial v}{\partial a} = \frac{\partial v}{\partial x}, \qquad \frac{\partial v}{\partial \sigma} = 0.$$

Equation (6A) then reduces to,

$$\frac{dU}{da} = 2\sigma_Y \int_a^{a_1}\frac{\partial v}{\partial x} dx = \sigma_Y \delta, \tag{7A}$$

where δ is the crack opening displacement at the tip of the real crack, $x=a$. For the general case of Eq. (6A),

$$\frac{dU}{da} = \frac{4\sigma_Y}{E}\int_a^{a_1}\left\{\frac{\partial}{\partial a}(\operatorname{Im}\bar{Z}) + \frac{\partial}{\partial\sigma}(\operatorname{Im}\bar{Z})\frac{\partial\sigma}{\partial a}\right\} dx.$$

Now

$$\frac{\partial}{\partial a}(\operatorname{Im}\bar{Z}) = \frac{2\sigma_Y}{\pi}\left\{\tanh^{-1}\left(\frac{a_1^2 - x^2}{a_1^2 - a^2}\right)^{1/2}\right\},$$

and

$$
\int_a^{a_1} \frac{\partial}{\partial a}(\text{Im } \bar{Z})dx = \left[\frac{2\sigma_Y}{\pi} \left\{ x \tanh^{-1}\left(\frac{a_1^2 - x^2}{a_1^2 - a^2} \right)^{1/2} \right. \right.
$$

$$
- a \tanh^{-1}\left[\frac{k}{x}\left(\frac{a_1^2 - x^2}{1 - k^2} \right)^{1/2} \right]
$$

$$
\left. \left. - a_1(1 - k^2)^{1/2} \cos^{-1}\left(\frac{x}{a_1} \right) \right\} \right]_a^{a_1}
$$

$$
= \frac{2\sigma_Y}{\pi} a \left\{ \log\left(\frac{1}{k} \right) + \frac{(1 - k^2)^{1/2}}{k} \cos^{-1}(k) \right\}.
$$

$$(8A)$$

Hence

$$
\frac{dU}{da} = \frac{8\sigma_Y^2 a}{\pi E} \left\{ \log \sec\left(\frac{\pi\sigma}{2\sigma_Y} \right) + \left(\frac{\pi\sigma}{2\sigma_Y} \right) \tan\left(\frac{\pi\sigma}{2\sigma_Y} \right) \right\}
$$

$$
+ \frac{\partial\sigma}{\partial a} \frac{4\sigma_Y}{E} \int_a^{a_1} \frac{\partial}{\partial\sigma}(\text{Im } \bar{Z})dx,
$$

and therefore,

$$
\frac{dU}{da} = \sigma_Y \delta + \frac{4\sigma_Y \sigma a}{E} \tan\left(\frac{\pi\sigma}{2\sigma_Y} \right) + \frac{\partial\sigma}{\partial a} \frac{4\sigma_Y}{E} \int_a^{a_1} \frac{\partial}{\partial\sigma}(\text{Im } \bar{Z})dx.
$$

$$(9A)$$

For stability,

$$
\partial\sigma/\partial a > 0, \quad \text{and thus} \quad \frac{dU}{da} > \sigma_Y \delta + \frac{4\sigma_Y \sigma a}{E} \tan\left(\frac{\pi\sigma}{2\sigma_Y} \right).
$$

References

ASTM (1973), "Fracture Toughness Evaluation by *R*-curve Methods," ASTM STP 527, ASTM Philadelphia.

ASTM (1978), "Tentative Recommended Practice for *R*-curve Determination," E561-76T, ASTM Philadelphia.

Bergkvist, H. and Anderson, H. (1972), "Plastic Deformation at a Stably Growing Crack Tip," *Int. J. Fract. Mech.*, **8**, 2.

Bilby, B. A., Cottrell, A. H. and Swinden, K. H. (1963), "The Spread of Plastic Yielding from a Notch," *Proc. Roy. Soc.*, A272.

Burdekin, F. M. and Stone, D. E. W. (1966), "The Crack Opening Displacement Approach to Fracture in Yielding Materials," *J. Strain Analysis*, **1**, 2.

Burdekin, F. M. and Dawes, M. G., (1971), "The Practical Use of Linear Elastic and Yielding Fracture Mechanics with Particular Reference to Pressure Vessels," Conf. on Practical Application of Fracture Mechanics to Pressure Vessels, *I. Mech. E.*, London.

Chell, G. G. and Heald, P. T. (1975), "The Path Dependence of the *J*-Contour Integral," *Int. J. Fract.*, **11**, 349.

Cherepanov, G. P. (1979), *Mechanics of Brittle Fracture*, Section 5–6, McGraw-Hill Inc., New York.

Dawes, M. G. (1979), "Elastic-Plastic Fracture Toughness Based on the COD and *J*-Contour Integral Concepts," *ASTM STP668*, ASTM, Philadelphia, Pa.

De Castro, P. M. (1979) Ph.D. Thesis, Cranfield Inst. of Technology, Cranfield.

Dowling, A. R. and Townley, C. H. A. (1975) "Effect of Defects on Structural Failure: A Two Criteria Approach," *Int. J. Pressure Vessels and Piping*, **3**.

Dugdale, D. A. (1960), "Yielding of Steel Sheets Containing Slits," *J. Mech. Phys. Solids*, **8**.

Garwood, S. J. (1980), "Specimen Thickness Considerations For Structural Predictions in A533B Class I Steel," *Welding Institute Report* 7301 06/80/203.2, Welding Institute, Cambridge.

Green, G. and Knott, J. F. (1975), "On Effects of Thickness on Ductile Crack Growth in Mild Steel," *J. Mech. Phys. Solids*, **23**, 167.

Hancock, J. W. and Cowling, M.J. (1977), "The Initiation of Cleavage by Ductile Tearing," *Fracture* 1977, 4th Int. Conf. on Fracture, University of Waterloo, Waterloo, Ont.

Harrison, R. P., Loosemore, K. and Milne, I. (1976), Assessment of the Integrity of Structures Containing Defects, *CEGB Report* R/H/R6.

IIW Colloquium on Significance of Defects (1979), Bratislava.

Lee, K. H. (1979) The Effect of Crack Front Curvature on Stress Intensity Factor in Compact Tension Specimens Using the Boundary Integral Equation Method, *Welding Institute sponsored Project*, Report 7302.07/79/222.3, Part of Ph.D. Thesis to be submitted to Imperial College of Science and Technology, London.

Neale, B. K. (1978), *Proc. First Int. Conf. Numerical Methods in Fract. Mech.*, Swansea, 218.

Paris, P. C., Tada, H., Zahoor, A. and Ernst, H. (1979), "A Treatment of the Subject of Tearing Instability," *Elasto-Plastic Fracture*, ASTM STP668, ASTM Philadelphia.

Rice, J. R. (1975), Discussion on the path independence of the J-contour integral by Chell and Heald, *Int. J. Fract.*, **11**, 352.

Rice, J. R. (1976), *The Mechanics of Fracture*, F. Erdogan (ed.), ASME, New York.

Rice, J. R. (1978), *Numerical Methods in Fracture Mechanics*, A. R. Luxmore and M.J. Owen (eds.), University College Swansea.

Turner, C. E. (1978), Description of stable and unstable crack growth in the elastic-plastic regime in terms of J_r resistance curves, 11*th Symp. on Fracture Mechanics*, ASTM STP677, ASTM Philadelphia.

Turner, C. E. (1979), *Developments in Fracture Mechanics I*, G. G. Chell (ed.), Chapter 4, Applied Science Publishers, London.

S. Nemat-Nasser, Editor
THREE-DIMENSIONAL CONSTITUTIVE RELATIONS AND DUCTILE FRACTURE
North-Holland Publishing Company (1981) 243–246

DISCUSSION ON SESSION 4

R. ROCHE

C.E.A., Gif-sur-Yvette, France

1. The three beautiful lectures given this afternoon cover a very large field in fracture mechanics. Each uses a different approach. It seems that Prof. Burdekin's lecture is more related to global criteria with a special emphasis on practical applications, Prof. Francois' lecture explores connections between global criteria and material properties, and that of Prof. Stüwe is a deep and precise study of the local behavior of material on the surface of the propagating crack.

Reading these works reminds me of the thinking of Blaise Pascal, the French philosopher, concerning man between two infinities. Pascal pointed out that "infinity in smallness is less easy to see." Perhaps it would be useful to give some attention to the smallest scale that can be usefully considered in fracture mechanisms. I suppose the authors can give their opinion about this point.

2. Before speaking about micromechanisms, I would like to devote a few moments to the human aspects of this problem. From this point of view it must be observed that global criteria are needed for practical application in design. As emphasized by Prof. Burdekin "all methods need to be able to define material fracture toughness properties from laboratory tests, and to predict behavior of cracked structural components." Such requirements are almost identical to those related to the current analysis of uncracked structures. In current practice, used material properties are limited to tensile properties. Everyone knows that the tensile test curve is only a rough indication of the necessary material properties, but designers must work with it. This brings us to an important question: *"What is the definition of a valuable material property?"* As far as fracture mechanics is concerned, I am afraid that the answer is not easy and can be subjective. On this point Prof. Francois has provided remarkable results on the difference of toughness properties in LT and TL directions. The anisotropy of the material is an important factor both in practical applications and in the studies of micromechanisms. Prof. Stüwe and Prof. Burdekin are certainly providing us with valuable indications about anisotropy.

3. I now reach a point which I think is of very great interest: it is the *definition of crack extension*. When stable propagation occurs, the current

243

practice of analysis concentrates on the R curves. In these curves a global parameter like J or COD is plotted as a function of a *geometrical factor called crack extension*. At first the definition seems obvious, but today's lectures do not support such an opinion. I intend to return to this definition in a few moments, but attention must be paid to the fact that global parameters like J or COD can be affected by specimen geometry, crack shape, and type of loading. This effect is generally attributed to triaxiality, but it is important to pay more attention to the *definition of the used parameter* (J or COD) and to the way its value is extracted from experimental test results. We must be grateful to Prof. Burdekin for insisting on the definition of J and to mention the work of Prof. Turner.

4. If you allow me, I would also like to deal with the definition of J because different definitions can lead to misinterpretations. Avoiding any assumption on material behavior, attention can be given to the *principle of virtual work* applied to continuous media.

Stress working variation = Virtual work of external forces.

This is the conventional form of this principle. But what are the virtual displacements? As you know, any displacement is not sufficient because equations of compatibility must be satisfied. They are not satisfied when fracture is considered, therefore the conventional form of the principle of virtual work cannot be applied in fracture mechanics.

It is possible to generalize this principle by introducing *material displacement*, δx_i, which is the displacement (or flow) of material properties (including holes and cracks) through the body. With regard to material displacement, material forces j_i are also introduced in the manner proposed by Casal. As a matter of fact, material rotation $\delta\Omega_i$ and material couple l_i are also needed. With these definitions a *general form of the principle of virtual work* can be given. Besides the virtual work of conventional external forces, the virtual work of material forces, j_i and l_i must be considered,

$$-\int (j_i\, \delta x_i + l_i\, \delta\Omega_i)\, dV,$$

hence we have,

Strain working variation = Virtual work of external forces
$\qquad\qquad\qquad\qquad\qquad - $ Virtual work of material forces.

It is this form that is convenient in fracture mechanics. Now, in practical application, the main problem is whether the field of material displacement can be defined by a limited number of parameters, δa_k (generalized material displacements). If such a description is possible, material forces are defined by the dual variables J_k of a_k (generalized forces), such that

$$\delta W = X_i\, \delta u_i - J_k\, \delta a_k,$$

W = stress working; X_i = conventional external forces, (generalized); u_i = spacial displacement.

This expression is a *more general definition of J than the conventional ones, but it gives a set of scalars* J_k and not only one (or three or six). Moreover, *crack extension is given by a set of scalars* a_k. The current practice is to reduce this set to only one, but evidence is given by Prof. Burdekin and Prof. Francois that this is rarely justified. I think the question *"How crack extension must be defined"* is of great interest.

5. From the preceding consideration, it can be seen that *the use of J is always the result of simplification* of the real problem. Moreover, experimental determination of the J value is not straight forward but involves various assumptions (e.g. strain working being only a function of deflexion and crack growth—effect of crack length on load deflexion curve is known etc.). Therefore, comparison between results obtained from specimens with different geometries is difficult, and a critical evaluation of the different experimental procedures is needed in order to avoid misinterpretations. Perhaps the authors can give us their opinion on the determination of global parameter values (J or COS) in different structures.

6. Nevertheless, it seems that there is a general agreement on the strong effect of triaxiality on fracture parameters. Unfortunately, to go further, *triaxiality must be strictly defined*, and it seems that no adequate definition has been given when a sample or a structure is taken as a whole. May the authors give us their ideas on that subject? It must be pointed out that the triaxiality concept in fracture mechanics is very particular. In conventional structural analysis, geometrical similitude can be applied. This is not the case when fracture is concerned. There is a strong thickness effect, perhaps it would be better to speak of a *"size effect"* because it is well-known that structures or samples made out of the same material, presenting the same shape but of different sizes, can exhibit strong differences in fracture behavior, the smallest being more stable than the largest. These facts imply that *one given length must be one of the toughness material properties* and that this characteristic length is probably needed to define triaxiality. It can be noted that usual toughness properties lead to a characteristic length like

$$COD, \qquad \frac{EJ}{\sigma_y^2} \quad or \quad \frac{(KI)^2}{\sigma_y}.$$

The relation between characteristic length as a global parameter and micromechanisms is a good subject for discussion. A very clear interpretation is given by Prof. Stüwe in Eq. (14) where h_0 is a local characteristic length. On the contrary Prof. Francois shows us the complexity of such an analysis and points out that experimental evidence is scarce and contradictory. There is a question to be asked of these two authors, concerning the effect of the size of dimples on the fracture behavior when crack propagation occurs. According to Rousseau, there is a noticeable increase of the value of the tearing modulus T when the material is submitted to external pressure, but this increase is not linked to a modification of the size of dimples. What explanation of this effect is proposed?

Concerning the subject of characteristic lengths of the material, the *"traveling COD"* needs some attention. It seems that Prof. Burdekin and Prof. Francois are not sure that this criterion is a good one and prefer *crack opening angle* (flank angle). This latter one is not a length, and a discussion would be useful on that subject. It can be noted that no experimental results are given about the way the crack moves: is it step by step as suggested in some studies?

7. The effect of *triaxiality* or size effect is probably one of the most important phenomena connected with *micromechanisms*. Before fracture occurs, and then during fracture itself, a *redistribution of strain and stress* occurs as the load increases. This redistribution is really not a part of fracture mechanism itself. Such a redistribution occurs in uncracked structures and must be taken into account in structural analysis. When global criteria are considered, they more or less include this redistribution, but it is difficult to know in what condition redistribution is considered. As strain and stress redistribution is strongly dependent on the shape of the structure, it could be thought that shape effect is rather a matter of correct strain analysis than fracture process. Prof. Stüwe made a clear distinction between the two phenomena, it seems that the fracture micro-process is really distinct from general plastic behavior of samples. Perhaps he can add some comments about it: Are the micromechanisms affected by the general behavior of structure? Is the dimple size unaffected by surface proximity or by stable crack propagation. Prof Burdekin gives a fine analysis of the effect of plastic deformation. Moreover, he points out that the fracture process can change the geometry in curving the crack front. There is some "feedback" from the fracture process towards the strain redistribution process. It would be worthwhile if the authors could give their opinions on the amount of computational work which would be needed for fracture analysis if fracture mechanisms were well-known.

It is possible that *mixing between fracture process and strain redistribution* can lead to some confusion. Authors do not seem to agree on the meaning of dJ/da or T as a material characteristic. One writes that the R curves are not unique material properties, the other that the slope dJ/da (or the T parameter) is a characteristic of the material (and of the direction of propagation). It can be suspected that the discrepancy is only artificial, and due to difference in definitions, but some clarification would be fruitful.

8. I am afraid my questions are a reflection of my ignorance but I hope the authors will excuse it, as I hope that the relevant questions coming from the floor will produce a fruitful discussion.

SESSION 5

Chairman: R. Noel
                E.D.F.–Septen, Paris la Defense, France

Authors: B. Marandet, G. Sanz
                J. Christoffersen
                Th. Lehmann

Discussion: B. Storåkers

Reply: Th. Lehmann

S. Nemat-Nasser, Editor
THREE-DIMENSIONAL CONSTITUTIVE RELATIONS AND DUCTILE FRACTURE
North-Holland Publishing Company (1981) 249–273

A QUANTITATIVE DESCRIPTION OF
FRACTURE TOUGHNESS UNDER PLANE STRESS
CONDITIONS BY THE R-CURVE METHOD

B. MARANDET and G. SANZ

Institut de Recherches de la Sidérurgie Française (IRSID), 78105 St-Germain-en-Laye, France

The crack growth resistance curve, or the R-curve is a plot of the stress intensity factor, K_R, versus the absolute crack extension, Δa. It appears well established today that this curve is a characteristic of the fracture toughness of material for a given thickness and temperature. The test methods described here make it possible to determine the R-curve of high-strength steel or alloy sheet materials under plane stress conditions.

The critical parameters of the instability of a cracked thinwall structure (K_c, Δa_c) are the coordinates of the point of contact between the R-curve of the material and the curve representing the variation in the stress intensity factor, K_I, as a function of crack length, a. Several practical application examples (specimens, pressure vessels) support the use of this concept for predicting instability conditions under ductile crack growth.

1. Introduction

Most welded metallic structures of large dimensions with relatively thin walls (hulls, pressure vessels, pipelines, etc.) are subjected to a service environment such that, in practice, plane stress conditions prevail. Ductile crack growth of Δa_c, of greater or lesser extent depending on the temperature, thickness, geometry or loading mode, can then take place in a stable manner before suddenly becoming unstable at a maximum load P_c corresponding to a critical value K_c of the stress intensity factor.

Tests conducted in the laboratory using specimens of different dimensions show that the critical value K_c measured under plane stress conditions or at least under non plane strain conditions is not like K_{Ic} a property of the fracture toughness of the material at given strain rate and temperature. Many investigators consequently consider today that the calculation of the critical instability conditions P_c and Δa_c of a structure must be based not upon a single critical value measured at failure (K_c) or at the moment of crack initiation (J_{Ic}, $(COD)_c$) but rather upon a set of values which describe quantitatively the ability of a material to resist stable crack propagation from initiation up to fracture. The critical parameters J_{Ic}, $(COD)_c$ or K_c are then simply regarded as particular points of a crack resistance curve referred to as the R-curve.

Developing an idea which was set forth in 1954, Irwin suggested a few years later (1960) the possibility of representing, by means of a parabolic curve, the variation in the crack growth resistance of a material as a function of relative stable crack extension. He moreover defines the critical conditions of instability G_c (or K_c), P_c and Δa_c in any cracked structure subjected to an increasing load as the coordinates of the point at which the crack driving force curve, G, becomes tangent to the crack resistance curve, R, of the material. At the same time, Kraft, Sullivan, and Boyle (1961) assumed that, for a given material, thickness, temperature and loading rate, there is a unique relationship between the amount of stable crack growth and the corresponding stress intensity factor. Experiments conducted afterward, on both laboratory specimens and real geometries, appear to corroborate this assumption.

2. Crack stability or instability criteria under an imposed force

Let us consider an infinite plate of unit thickness with a central notch of length $2a$ and subjected to a nominal stress σ perpendicular to the notch plane (Fig. 1a). The load is applied, for example, by hanging a weight from one end of the plate.

2.1. Plane strain conditions—definition of symbols

Let us assume firstly that the material considered is very brittle and that the plane strain conditions are satisfied over the entire thickness of the plate. According to the Griffith criterion, the initial crack can propagate as soon as the system composed of the crack body and of the external forces is just capable of delivering the energy necessary for an incremental increase da in crack length. The condition required for propagation is thus,

$$\frac{\mathrm{d}}{\mathrm{d}a}(F-U)=\frac{\mathrm{d}W}{\mathrm{d}a}, \tag{1}$$

where U is the elastic energy contained in the body, F if the work performed by external forces, and W is the energy consumed by crack propagation.

In the case of fixed grips, the work of the external forces is zero. It is then easily demonstrated that crack propagation will result only in a decrease in the elastic energy of the crack body.

By definition, $G=(\mathrm{d}/\mathrm{d}a)(F-U)$ is the elastic energy release rate, or the crack extension force, and $R=\mathrm{d}W/\mathrm{d}a$ is the energy dissipation rate for decohesion, or the crack growth resistance force.

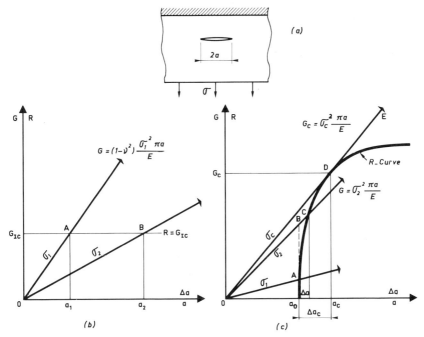

Fig. 1. Crack stability or instability criteria: (a) Theoretical model of the infinite plate with a center crack; (b) Graphic representation of the energy criterion in plane strain conditions; (c) Graphic representation of the energy criterion in plane stress conditions (Irwin, 1960).

Under plane strain conditions, the resistance $R = dW/da$ remains roughly constant during propagation. It is in fact shown experimentally that specimens containing cracks of various sizes fail at the same value G_{Ic} of G. This critical value is written as:

$$G_{Ic} = (1 - \nu^2) \frac{K_{Ic}^2}{E}. \tag{2}$$

This fracture criterion can also be depicted graphically as indicated in Fig. 1b. Crack growth resistance R and the crack extension force G are plotted as functions of notch length, a. In the case examined here, R is independent of the crack size; it is thus represented by a straight horizontal line such as $R = G_{Ic}$. For the considered geometry, the stress intensity factor, K, is given by the relationship,

$$K = \sigma \sqrt{\pi a}. \tag{3}$$

The crack extension force is thus,

$$G = (1 - \nu^2) \frac{\sigma^2 \pi a}{E}. \tag{4}$$

It is seen in Fig. 1b that to produce the growth of a crack having an initial length a_1, the applied stress must be raised to a value of σ_1 such that the condition $G_{Ic} = R$ is satisfied (point A). The propagation of a longer crack $a_2 > a_1$ would occur under a stress $\sigma_2 < \sigma_1$ (point B).

2.2. Plane stress conditions—description of the R-curve

Let us now consider the case of a material which is sufficiently ductile so that plane stress conditions are predominant in the considered thickness. It is known by experience that a stable crack can begin under an applied stress σ_i, but propagation stops soon after and the stress must be raised again in order for the crack to continue its propagation. The gradual raising of the applied stress finally leads to unstable tearing of the remaining ligament.

Crack initiation and stable propagation take place as long as the crack extension force G is equal to the crack growth resistance force R. This phenomenon can be represented graphically, as in Fig. 1c. For the considered model, the variation in G as a function of crack length, a, is a straight line with the equation,

$$G = \frac{\sigma^2 \pi a}{E}.$$ (5)

If an initial crack of length a_0 is loaded to a stress σ_1, the available crack extension force G is given at the point A but it is insufficient to cause crack growth. Propagation can, however, begin under a higher load. Let us imagine that it is possible to increase the stress to σ_2 while preventing progression during loading (point B) and let us release the crack. Under an imposed load, the crack extension force G will thus increase along the line defined by Eq. (5). The crack finally stops at point C after an extension Δa. By thus raising the applied stress step by step, one obtains a series of equilibrium conditions which describe a curve called the R-curve. It is, however, seen in Fig. 1c that if the stress is raised to σ_c, the crack extension force G will follow the path DE. As the energy made available by the system is always greater than that necessary for propagation, the crack will become unstable at the contact point D between the curve G–a and the curve R–a. Crack instability conditions under an imposed load are thus,

$$\frac{\partial G}{\partial a} = \frac{\partial R}{\partial a},$$ (6)

$$G = R.$$ (7)

Hence, the unstable breaking of the ligament will take place at a critical value G_c of G and after an amount of stable crack growth $\Delta a_c = a_c - a_0$; Fig. 1c.

It is current practice to express fracture toughness in terms of the stress intensity factor K rather than in terms of the energy G (Fig. 2). Let us point

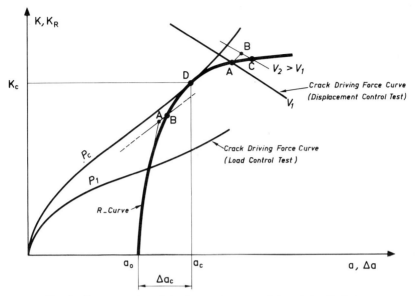

Fig. 2. Representation of the *R*-curve in terms of stress intensity factor.

out that, in the case of plane stress conditions, we have,

$$EG = K^2. \tag{8}$$

By analogy, we write,

$$ER = K_R^2. \tag{9}$$

The instability conditions (6) and (7) hence become,

$$\frac{\partial K}{\partial a} = \frac{\partial K_R}{\partial a}, \tag{10}$$

$$K = K_R. \tag{11}$$

These conditions are satisfied at the point where the curve *K–a* becomes tangent to the curve K_R–Δa (point D in Fig. 2). Unstable tearing thus takes place theoretically for a critical load P_c corresponding to a particular value K_c of the stress intensity factor, the subcritical crack growth being $\Delta a_c = a_c - a_0$. In the plane defined by the "stress intensity factor"—"crack length *a*" axes, the *R*-curve is in the locus of the points which share the stability and instability ranges of a crack for a given material, thickness, temperature and loading rate. For the same material, it would be possible to consider investigating the influence of one of these parameters on the shape of the *R*-curve and to present the envelope of the curves obtained in a three-dimensional system.

In practice, a cracked body can be subjected to an imposed load or an imposed displacement.

—Under an imposed load, the curve representative of the crack driving force (K versus a) has a positive slope (Fig. 2). We saw that it becomes tangent to the R-curve at a point having the coordinates K_c, Δa_c where crack growth suddenly becomes unstable (point D). It is thus observed by simple graphic construction that, for a material having a certain thickness, the coordinates of the instability point depend on both the initial crack length a_0 and on the shape of the curve of K vs a, and hence on the considered geometry.

—Under an imposed displacement, the curve representative of the crack driving force (K vs a) has a negative slope (Fig. 2). Contrary to the preceding case, it never becomes tangent to the R-curve, so that the instability condition (10) cannot be satisfied. From the experimental viewpoint, the test method with a "controlled displacement" is therefore preferable because it makes it possible to record the R-curve in its entirety.

3. Experimental determination of the R-curve

The ASTM draft standard E561-76T entitled "Tentative Recommended Practice for R-Curve Determination" (1978) is the fruit of experience acquired during the past 10 years or so by a small group of American researchers among whom should be mentioned in particular Heyer (1973), Novak (1976) and McCabe (1972–1979). The test methods we have developed are based to a great extent on this ASTM draft standard. They are suited preferably to the study of high-strength steel or alloy sheet materials under confined plasticity conditions.

3.1. Test methods

The Crack Line Wedge Loading (CLWL) specimen is designed for the performance of constant displacement tests with the gradual insertion of a wedge in the bore made in the mechanical notch (Fig. 3). Its outer dimensions (height H, width W) are in a ratio of $H/W = 0.6$, the product being studied in the original thickness B.

The mechanical notch is extended by a fatigue crack under the conditions called for by ASTM E399 up to a length a such that $a/W \sim 0.350$. This very sharp crack constitutes the initiation of a stable ductile tear whose slow progression is followed from deformations measured between points beyond ($V_1 = 0.1576\ W$) and on this side ($V_2 = 0.303\ W$) of the loading axis by means of two special clip gages.

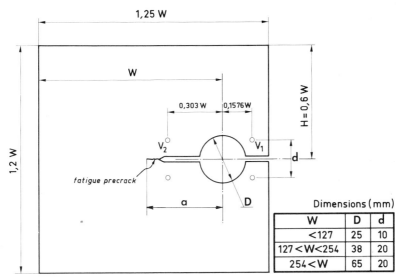

Fig. 3.   Crack-line wedge-loaded specimen ($H/W=0.6$).

During a test with an imposed displacement, the force applied to the specimen decreases as the crack length increases. The mechanical loading system must thus be as stiff as possible in order to minimize variations in displacement capable of resulting from a certain relaxation of the stressed parts. Though very satisfactory in this respect, the wedge technique turns out to be difficult to apply and has the drawback of constantly tying up a tensile testing machine. We were thus led to develop an independent testing device whose design is original as concerns the loading principle (Fig. 4).

The notch is spread at the bore thanks to a quasi-elliptical cam actuated by an electric motor via a reducer (1 : 720). The forces are transmitted to the specimen through two semicircular pieces (Fig. 4a) which slide freely in a transverse guide fixed on the lower supporting plate. The purpose of this guide is to neutralize the torque and to impose a force perpendicular to the axis of the notch. The specimen is held with a minimum clearance between two ball pads fixed on rigid plates in order to avoid any lateral deformation capable of producing buckling (Fig. 4b).

The rotation of the cam is controlled automatically by a DEC PDP-11/05 minicomputer which scans and analyzes the data delivered by each clip gage. The operating principle of the control system is the following:

The cam moves by a fraction of a turn, thereby opening the notch. The crack thus begins to propagate. The electric signal delivered by the clip gage V_2 is periodically compared. The variation in this signal with respect to time indicates crack propagation. As soon as stability is reached, the electric

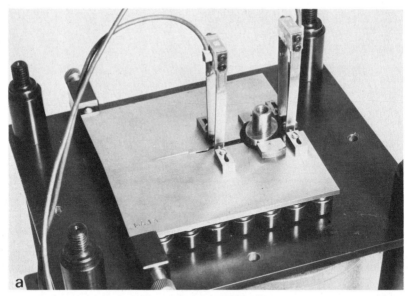

Fig. 4. Apparatus used at IRSID for the experimental determination of R-curves under controlled displacement: (a) The crack opening is kept constant during the test by means of a cam controlled by a motor via a reducer (the upper holding plate has been removed); (b) The specimen is held between two ball pads.

signals delivered by the two clip gages V_1 and V_2 are stored by the computer which delivers the starting command to the stepping motor which actuates the cam. The cam again moves by a fraction of a turn and the crack begins to propagate again. The test is interrupted when the crack length becomes such that $a/W = 0.7$.

3.2. Calculation of the R-curve

The coordinates K_R and Δa of the R-curve are calculated point by point using the elastic small-scale plastic deformation method.

The "effective" crack length, a_{eff}, is determined from a double compliance function established beforehand either by elastic calculation or by simple calibration (Fig. 5),

$$\frac{2V_1}{2V_2} = f_1\left(\frac{a}{W}\right),$$ (12)

where $2V_1$ and $2V_2$ are the measured displacements on the specimen at points V_1 and V_2, each time the crack stops.

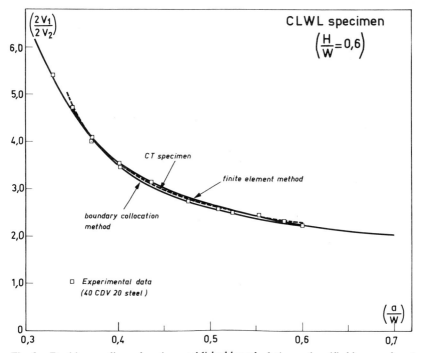

Fig. 5. Double compliance function established by calculation and verified by experiment.

The measurements carried out by means of fatigue-precracked CLWL specimens ($0.350 < a/W < 0.700$) loaded under elastic conditions have confirmed the calculations conducted independently by Newman (1976) and Boissenot (1976) using different methods. It is noted in Fig. 5 that the characteristic functions of the CLWL and CT specimens are altogether comparable.

The load P applied to the specimen is determined from a single compliance function obtained as previously either by elastic calculation or by calibration (Fig. 6),

$$EB\frac{2V_1}{P} = f_2\left(\frac{a}{W}\right), \tag{13}$$

where

$$f_2\left(\frac{a}{W}\right) = 101.9 - 948.9\left(\frac{a}{W}\right) + 3961.5\left(\frac{a}{W}\right)^2 - 6065\left(\frac{a}{W}\right)^3 + 4054\left(\frac{a}{W}\right)^4; \tag{14}$$

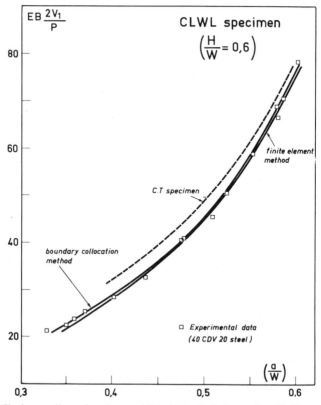

Fig. 6.   Single compliance function established by calculation and verified by experiment.

and where $E=$ Young's modulus (206,000 MPa for steels); $B=$ Specimen thickness; $2V_1=$ displacement measured at V_1; and $a=a_{eff}=$ effective crack length.

The measurements carried out by means of fatigue-precracked CLWL specimens $(0.350 < a/W < 0.700)$ loaded under elastic conditions have confirmed the calculations carried out by two different methods. Figure 6 shows in particular that the function characteristic of the CLWL specimen is slightly different from that relative to the CT specimen.

Finally, the stress intensity factor K_R is calculated using the following formula according to Srawley (1976):

$$K_R = \frac{P}{B\sqrt{W}} f_3\left(\frac{a}{W}\right),$$ (15)

where

$$f_3\left(\frac{a}{W}\right) = \left[2+\left(\frac{a}{W}\right)\right]\left[0.886+4.64\left(\frac{a}{W}\right)-13.32\left(\frac{a}{W}\right)^2\right.$$
$$\left. +14.72\left(\frac{a}{W}\right)^3-5.6\left(\frac{a}{W}\right)^4\right],$$
$$\times\left[1-\left(\frac{a}{W}\right)\right]^{-3/2},$$ (16)

and $a=a_{eff}=$ effective crack length.

The calculation and automatic plotting of the coordinates K_R and $\Delta a_{eff} = (a_{eff} - a_0)$ are carried out by a DEC PDP-11/10 computer. The calculation principle is indicated in Fig. 7.

The R-curve obtained by this method represents the variation in the stress intensity factor K_R applied to an effective crack length a_{eff} as a function of the amount of stable crack growth $\Delta a_{eff} = (a_{eff} - a_0)$. What is meant by "effective" crack? It is a fictive crack loaded under purely elastic conditions which, under the effect of the same exterior forces, would cause the same strains, $2V_1$ and $2V_2$, in the specimen at the considered locations. If it is assumed for simplicity that the plastically deformed zone at the crack root has a circular contour of radius r_y, the "effective" crack length, a_{eff}, is, as a first approximation,

$$a_{eff} = a_0 + \Delta a \text{phys.} + r_y,$$ (17)

where $a_0=$ initial crack length (mechanical notch + fatigue crack), a phys. $=$ physical stable crack growth measured for example on one side of the specimen, and $r_y=$ radius of plastic zone at notch root.

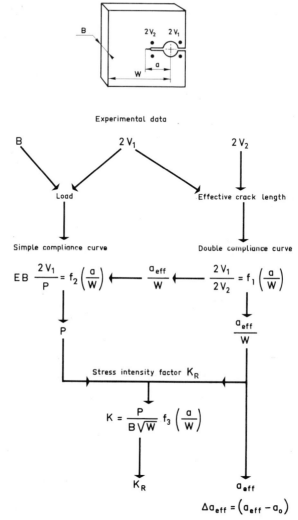

Fig. 7. Principle of calculating the *R*-curve by the elastic analysis method.

In the case of high-strength alloy steels, the dimension r_y of the plastic zone is often very small compared with the total length ($a_0 + \Delta a$ phys.) of the notch. Elastic analysis thus leads to an "effective" *R*-curve which is very similar to the "physical" *R*-curve (Fig. 8a).

In the case of medium-strength ferritic steels supplied in small thicknesses or austenitic stainless steels, the plastic zone at the notch root can develop over a greater extent ahead of the crack front. Elastic analysis then leads to

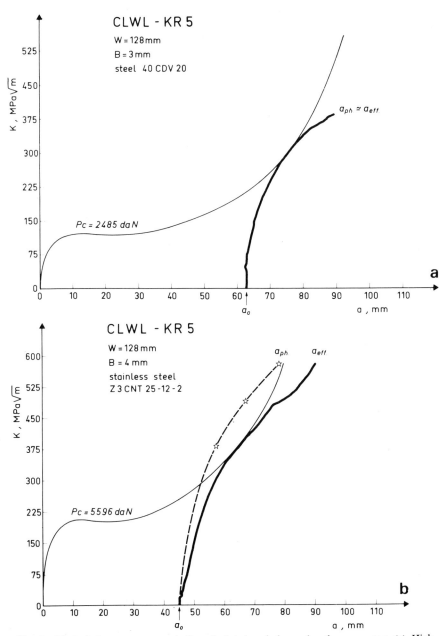

Fig. 8. Typical *R*-curves automatically calculated and drawn by the computer: (a) High strength steel 40 CDV 20; Sheet thickness 3 mm, $YS = 1150$ N/mm^2, $UTS = 1345$ N/mm^2, Elong.% = 13.7; (b) Stainless Z3 CNT 25-12-2, Sheet thickness 4 mm, $YS = 780$ N/mm^2, $UTS = 1070$ N/mm^2, Elong.% = 28; (c) Alloy INCONEL 718, Sheet thickness 4 mm, $YS = 1000$ N/mm^2, $UTS = 1250$ N/mm^2, Elong.% = 31.

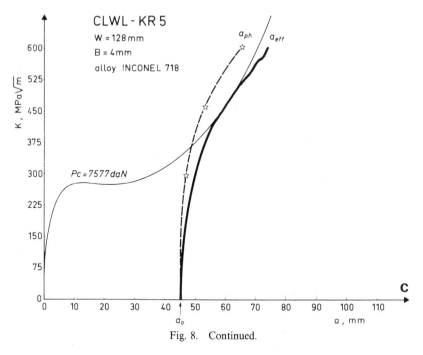

Fig. 8. Continued.

an "effective" R-curve which is very different from the "physical" R-curve (Figs. 8b and 8c).

3.3. Conditions regarding specimen dimensions

Elastic analysis is applicable to the extent that the plastic deformation remains confined to the crack tip (small-scale yielding). In general, it is thus necessary to use a specimen having sufficiently large dimensions so that the radius of the plastic zone can be regarded as negligible. According to Wang and McCabe (1976), the length $(W-a)$ of the ligament at the end of the test should meet the following condition:

$$(W-a) \geqslant \frac{4}{\pi} \left(\frac{K_{max}}{YS} \right)^2,$$
(18)

where K_{max} is the maximum stress intensity factor applied at the end of the test, and YS is the yield strength of the material.

To fulfill this condition, it is thus necessary to use specimens of varying size depending on the characteristics of the material, the thickness of the sheet, and so forth. The three models of CLWL specimens used at IRSID are presented in Fig. 9.

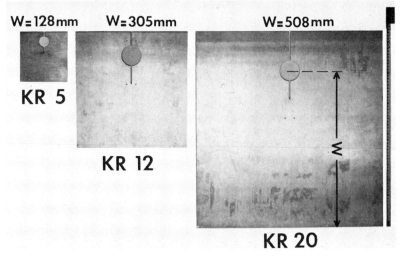

Fig. 9.   Different CLWL specimen models used at IRSID.

4.   Verification of basic hypothesis

The *R*-curve concept is based upon the hypothesis according to which, for a given material, thickness, temperature and loading rate, there is a unique relationship between the applied stress intensity factor, K_R, and the amount of stable crack growth, Δa. This means that the *R*-curve is a characteristic independent of the specimen type, initial crack length, and loading method; in other words, independent of the parameters related to the test. We are presenting below the results of experiments conducted in order to verify these basic hypotheses.

4.1.   Independence with respect to specimen type

This question is still the subject of a certain amount of controversy. It is quite evident that the nature and the distribution of elastic stresses, the shape and the dimensions of the plastic zone at the notch root are conditioned by specimen geometry. According to Walker (1970), the crack growth resistance of a material depends only on the fracture Mode (I, II or III). Experiments conducted more recently by Adams (1976) reveal, on the contrary, a fairly marked influence of geometry on the shape of the *R*-curve.

The center cracked panel (CCT) and the compact specimen (CLWL or CT) constitute two examples of radically different geometries. All the investigations published in the literature up to the present have dealt with the comparison of *R*-curves recorded by means of these two types of specimen.

The results obtained by Heyer and McCabe (1972a, b) on thin sheets (0.8 to 1.5 mm) in high-strength, low-toughness alloys (PH 14-8 Mo; 7075-T6; Ti-6Al-4V) show, for example, that in spite of certain deviations attributable to recording errors and to the heterogeneity of the material, there is satisfactory agreement between the *R*-curves recorded with specimens of the CCT and CLWL types. The same conclusion results from the experiments conducted recently by Wang and McCabe (1976) using CCT and CLWL specimens of various sizes in aluminum alloy (2024-T3; 7475-T761; 7075-T6, etc.). This is particularly well illustrated in Fig. 10 in which it is observed that the *R*-curves determined by means of CLWL specimens of the 4C (*W*=209 mm) or 7C (*W*=355 mm) types make it possible to predict with excellent precision the critical loads measured upon the fracture of center cracked panels having different widths (610 mm < *W* < 3048 mm) or different initial crack lengths (46 mm < a_0 < 381 mm). The same authors also note fairly good agreement between the experimentally determined values of K_c and those predicted from the *R*-curves.

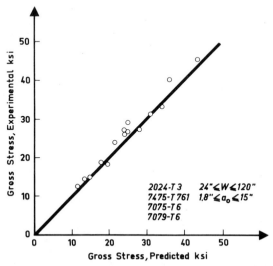

Fig. 10. Independence of the *R*-curve with respect to specimen type: The critical instability loads predicted from the *R*-curves recorded using CLWL specimens are in excellent agreement with the critical fracture loads of CCT specimens subjected to an increasing tensile load (Wang and McCabe, 1976).

4.2. Independence with respect to initial crack length

In 1966, Broek showed that modifying the initial crack length in a ratio of 1 to 2 had practically no effect on the *R*-curve or a CCT panel in aluminum alloy (2024-T3; 7075-T6). This is also the conclusion of Heyer and McCabe (1972a, b) who investigated the influence of the initial crack length on the *R*-curve of a thin sheet of high-strength stainless steel (PH 14-8 Mo). More recently, the tests performed by Marandet and Sanz (1977) using CLWL specimens in low-alloy, high-strength steel (40 CDV20) have made it possible to verify the independence of the *R*-curve with respect to initial crack length, in the interval $0.350 < a/W < 0.525$ (Fig. 11a). In agreement with the diagram proposed by Brown and Srawley (1965), it is thus sufficient to offset the *R*-curve on the scale of the abscissae to determine the critical

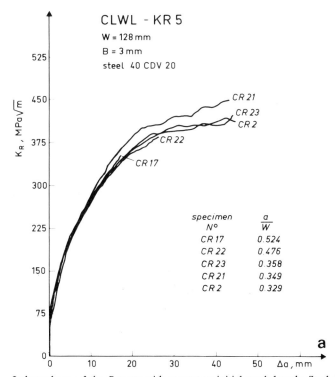

Fig. 11. Independence of the *R*-curve with respect to initial crack length: Steel 40 CDV 20 sheet thickness 3 mm, $YS = 1150$ N/mm^2, $UTS = 1345$ N/mm^2, Elong.% = 13.7; (a) *R*-curve is independent of initial crack length; (b) When the initial crack length varies, the *R*-curve is simply offset on the abscissae.

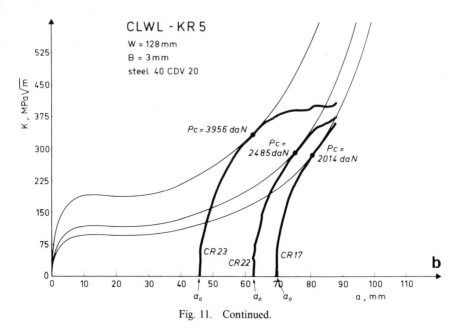

Fig. 11.　Continued.

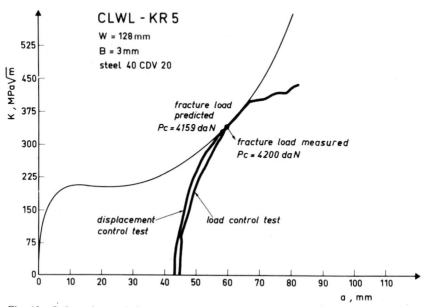

Fig. 12.　Independence of the *R*-curve with respect to loading method: Displacement- or load-controlled.

Table 1
Comparison between critical fracture loads calculated from the R-curve and measured on a tensile machine. CLWL specimens ($W=128$ mm; $B=3$ mm). Steel 40 CDV 20, $YS=1150$ N/mm², $UTS=1345$ N/mm², Elong.% = 13.7

$\left(\dfrac{a}{W}\right)$	P_{max} (measured) (daN)	P_{max} (predicted) (daN)
0.350	4159	4201
0.402	3678	3449
0.475	2810	2674
0.524	2335	2210
0.590	1683	1675

instability conditions, P_c, K_c, Δa_c, of a specimen of the same type having any initial crack length (Fig. 11b).

4.3. Independence with respect to loading method

If the test is conducted under the displacement-controlled condition, the crack propagates in a stable manner. On the other hand, if the test is carried out under the load-controlled condition, the crack first propagates in a stable manner and then in an unstable manner as soon as the conditions defined by the Irwin model are fulfilled.

According to Heyer and McCabe (1972a, b), the R-curve should not be dependent on the loading method. To check this, Marandet and Sanz (1977) have compared the maximum fracture load of a specimen loaded on a

Table 2
Comparison between critical fracture loads calculated from the R-curve (12T-CT) and measured (4T-CT and 1T-CT). Steel HY 130. Thickness 25 mm, $YS=965$ N/mm², $UTS=1040$ N/mm², Elong.% = 20, Reduction of area = 69% (D. E. McCabe and J. D. Landes, 1978)

W (in)	a_0 (in)	P_{max} (measured) (kips)	P_{max} (predicted) (kips)
2.0	1.252	13.850	13.000
2.0	1.234	15.200	13.200
2.0	0.777	35.350	32.000
2.0	0.770	35.150	32.000
8.0	4.772	59.000	53.000
8.0	4.819	53.800	50.000
8.0	2.855	153.500	137.000
8.0	2.827	153.625	137.000
24.0	9.300	378.500	378.500

tensile machine (test under the load-controlled condition) with the critical instability load predicted on the basis of the R-curve recorded elsewhere (test under the displacement-controlled condition), using identical specimens with the same initial crack length. The tests covered a series of CLWL specimens in high-strength, low-alloy steel (40 CDV20). The independence of the R-curve with respect to the loading method (Fig. 12) is proved by the excellent agreement obtained between the measured and calculated critical loads (Table 1).

In a very recent article, McCabe and Landes (1978) showed in the same manner that the R-curve recorded by means of a 12 T–CT specimen (W=610 mm) made it possible to predict to within less than 10 percent the critical fracture load of 4 T–CT (W=202 mm) and 1 T–CT (W=50.8 mm) specimens for very different crack lengths (Table 2).

5.   Application to the behavior of a flaw in a membrane under pressure

Does the Irwin model make it possible to predict the critical bursting conditions for a membrane under pressure, having a sharp flaw? It is to this question of considerable practical importance for the safe use of thin-walled vessels, that a team of researchers made up of mechanicians and metallurgists have recently attempted to reply[1].

The study has considered a thin disc (4 mm) in austenitic stainless steel, Z3 NCT25-12-2, traversed at its center by a fatigue crack having an initial length $a_0 = 22$ mm (Fig. 13a). The embedded disc was subjected to an increasing pressure on one side and on the other side was measured the stable propagation of the crack up to a critical pressure $P_c = 128$ bars corresponding to instability (Fig. 13b).

The computer programs developed at the CETIM were used to determine the curve representing the variation in the stress intensity factor K as a function of the length a of the flaw under a pressure P. It was found necessary to break down the analysis into several stages in order to take into account the general deformation of the disc in the form of a dome; these stages are:

(1) Loading of the uncracked disc and two-dimensional elastic-plastic calculation of stresses and strains by the axisymmetric finite element method.

(2) Unloading and introduction of the preceding residual stresses on the lips of a notch of length a at the pole of the disc having the form of a dome of radius R.

[1]DGRST contract No. 75-7-1284.

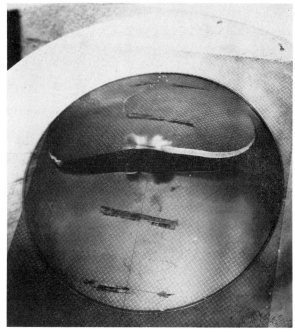

(a)

Fig. 13. Example of the application of the *R*-curve to the quantitative study of the critical fracture conditions of a membrane in the presence of a flaw: Stainless steel Z3 NCT25-12.2, Sheet thickness 4 mm, $YS = 780$ N/mm², $UTS = 1070$ N/mm², Elong.% = 28; (a) Embedded disc having a through-thickness flaw and subjected to an increasing pressure on one side; (b) Crack length as a function of pressure; (c) Prediction of instability conditions: critical burst pressure P_c and subcritical crack growth Δa_c, from the *R*-curve.

(3) Elastic loading of the cracked disc and three-dimensional elastic calculation by the integral equation method of the stress intensity factor K at the end of a notch of length a in a sphere of radius R subjected to the pressure P.

The calculations carried out for different values of the parameters R and a thus made it possible to determine the shape function, $f(a, R)$, which enters the formula,

$$K = Pf(a, R). \qquad (19)$$

The diagram presented in Fig. 13c shows, in agreement with experiment, that the through-crack of length $a_0 = 22$ mm can propagate in a stable manner over about 30 millimeters before becoming unstable at a critical pressure $P_c \sim 130$ bars, where the curve representative of the function (19) becomes tangent to the *R*-curve.

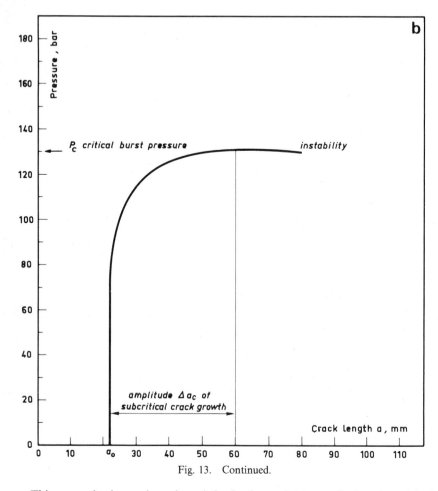

Fig. 13.   Continued.

This example shows the value of the Irwin model in predicting the critical instability conditions of a thin-walled structure. It shows that the quantitative analysis of in-service behavior calls for knowledge of two basic parameters:

—the *R*-curve characteristic of the toughness of the material for the thickness, temperature and loading rate considered; and

—the function $K=f$(geometry) under imposed stress for the considered structure.

6.   Conclusion

Significant progress has been achieved in the past decade in the quantitative analysis of stable or unstable ductile crack growth in cracked structures

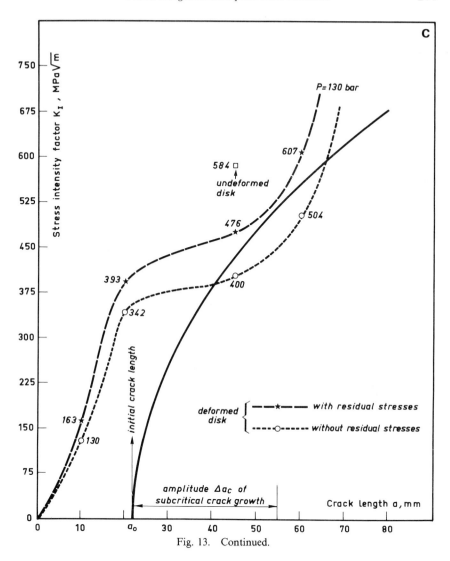

Fig. 13. Continued.

under predominantly plane stress loading conditions. Several investigators today consider with interest the hypothesis of an *R*-curve characteristic of the fracture toughness of a material under its utilization conditions: thickness, temperature, and the loading rate. This viewpoint appears to be supported by certain recent examples of practical applications, which have demonstrated the validity of the Irwin model for predicting the critical instability conditions of a thin-walled structure having a through-thickness crack.

Test techniques presently available are particularly suited to the study of high-strength steel or alloy sheets a few millimeters thick. The acquired experience must now be utilized to approach the much more difficult study of construction steel sheets (C–Mn) a few tenths of a millimeter thick. The investigations carried out at IRSID in this area are intended essentially to give fracture mechanics specialists the facilities necessary for calculating the optimum dimensions of welded elements to arrive at greater utilization safety.

References

Adams, N. J. (1976), "Influence of Configuration on R-curve Shape and G_c when Plane Stress Prevails," *American Society for Testing Materials, Special Technical Publication* 601, 330.

Boissenot, J. M. (1976), "Détermination de la Fonction Caractéristique d'une Éprouvette CLWL," Note technique CETIM, Senlis (France), communication privée.

Broek, D. (1966), "The Residual Strength of Aluminium Alloy Sheet Specimens Containing Fatigue Crack or Saw Cuts," NLR-TRM 2143, National Space Laboratory, Amsterdam, Netherlands.

Brown, W. F. Jr. and Srawley, J. E. (1965), "Fracture toughness testing methods," *American Society for Testing Materials, Special Technical Publication* 381, 133.

Contrat DGRST, no. 75-7-1284 (1975), "Etude Quantitative des Conditions de Rupture de Coques Minces Ductiles en Présence ou Non de Défauts."

Heyer, R. H. and McCabe, D. E. (1972a), "Crack growth resistance in plane stress fracture testing," *Eng. Fract. Mech.*, **4**, 413.

Heyer, R. H. and McCabe, D. E. (1972b), "Plane stress fracture toughness testing using a crack-line-loaded specimen," *Eng. Fract. Mech.*, **4**, 393.

Heyer, R. H. (1973), "Crack growth resistance curves (R-curves)," Literature Review, *American Society for Testing Materials, Special Technical Publication* 527.

Irwin, G. R. and Kies, J. A. (1954), *Welding Research Supplement*, **19**, 193.

Irwin, G. R. (1960), *ASTM Special Committee on Fracture Testing of High-Strength Materials*, ASTM Bulletin, Jan. 1960, 29.

Krafft, J. M., Sullivan, A. M. and Boyle, R. W. (1961), *Crack Propagation Symposium Proceedings*, College of Aeronautics, Cranfield, **1**, 8.

Marandet, B. and Sanz, G. (1977), "Caractérisation de la Ténacité des Matériaux en État de Contraintes Planes au Moyen des Courbes R," *Matériaux, Mécanique, Electricité*, numéro spécial, avril 1977, 19.

McCabe, D. E. and Heyer, R. H. (1973), "R-Curve Determination Using a Crack-Line-Wedge-Loaded (CLWL) Specimen," *American Society for Testing Materials, Special Technical Publication* 527, 17.

McCabe, D. E. (1976), "Determination of R-Curve for Structural Materials using Nonlinear Mechanics Methods," *American Society for Testing Materials, Special Technical Publication* 631, 245.

McCabe, D. E. and Landes, J. D. (1978), "Elastic-Plastic R-Curves," *J. Eng. Mat. Tech.* Transactions of ASME, **100**, 258.

Newman, J. C. (1976), "Crack Opening Displacements in Center Crack, Compact and Crack-Line-Wedge-Loaded Specimens," NASA Report, TN.D 8268.

Novak, S. R. (1976), "Resistance to Plane Stress Fracture (R-Curve Behavior) of A572 Structural Steel," *American Society for Testing Materials, Special Technical Publication* 590, 235.

Srawley, J. E. (1976), "Wide Range Stress Intensity Factor Expressions for ASTM E399 Standard Fracture Toughness Specimens," *Int. J. Fract.*, **12**, 475.

Tentative Recommended Practice for R-Curve Determination 1978, Annual Book of ASTM Standards, Part 10-E561-76T, 589.

Wang, D. Y. and McCabe, D. E. (1976), "Investigation of R-curve Using Comparative Tests with Center-Cracked-Tension and Crack-Line-Wedge-Loaded Specimens," *American Society for Testing Materials, Special Technical Publication*, 590, 169.

Walker, E. K. (1970), "A resistance slope concept for thin sheet fracture," Progress Report for Discussion at the October 1970 Meeting of Subcommittee 1, ASTM Committee E24.

S. Nemat-Nasser, Editor
THREE-DIMENSIONAL CONSTITUTIVE RELATIONS AND DUCTILE FRACTURE
North-Holland Publishing Company (1981) 275–288

MICROLOCALIZATIONS

Jes CHRISTOFFERSEN

The Technical University of Denmark, Lyngby, Denmark

Constitutive equations for a ductile material in a near-fracture state are proposed on the basis of a micromechanical model. In this, grains of sound material, assumed rigid, are separated by void-contaminated, deforming layers. Kinematical and statical relations pertaining to the mechanism are derived and may find applications in other fields, such as creep and deformation of granular media. Yield laws for the deforming layers are proposed, and their consequences, in terms of macroscopical behavior of the material, are investigated.

1. Introduction

Ductile fracture of metals is usually accompanied by an increase of volume, due to the nucleation and growth of voids. Indeed, fractured specimens have, at least locally, a characteristic spongy appearance.

The question naturally arises, how the sponginess is distributed through the material. The visible voidage is considerable and would, if extended uniformly through part of the specimen, account for volume increases well in excess of those actually encountered, cf. Parmar and Mellor (1980).

Thus we are led to the conclusion that voids are confined to local regions undergoing extreme deformations as a prelude to the final fracture. If, however, the voided regions contribute the major part of overall straining, the unvoided ones must necessarily consist of isolated blocks or granules interconnected only through the strongly deforming layers.

By these considerations we are led to propose a definitely idealized, micromechanical model of a material consisting of an agglomerate of rigid granules separated by plastically deforming material.

The kinematical and dynamical aspects of the model are applicable to other fields, obviously to those of creep and deformation of granular media.

Our model displays a number of features distinct from previous micromechanical models of metal plasticity, such as slip theory. Firstly, due to interlocking, closely packed rigid blocks cannot move relative to each other by pure slip. A separation, leading to an overall volume increase, is necessary. This leaves the model inapplicable to early stages of plastic deformation where no appreciable volume increase is evident.

275

Secondly, whereas most, if not all, earlier proposed micromechanical models predict the existence of a corner on the yield surface at the current loading point, the present one leads to a smooth surface. Corners may be ascribed to the simultaneous activity of several deformation systems, typically monocrystal glide. In the present instance there is but one system in action consisting of all the voided layers. The deformation is essentially strain controlled; microstresses build up according to the dictate of strains, and, while possible in some circumstances, it is hardly conceivable that microstresses corresponding to two different overall strain rates will in general add up to the same overall stress. Thus, our model cannot comply with a need for the destabilizing effect of a corner to explain bifurcations from a state of homogenous deformation. Again we emphasize that it is intended for application to the very last stages of deformation leading to fracture and not to structural failure, such as necking, in cases where the material remains sound. While a distinction between material and structural failure is thus attempted, the model deals with inhomogeneities, and the results are not in conflict with the often advocated view, that failure is always accompanied by structural changes on some, possibly microscopical, level.

Whereas our model has similarities with one investigated by Needleman and Rice (1978), it differs from it in predicting normality to apply. Nonnormality in the model of Needleman and Rice can be attributed to the nucleation of voids, their volume increasing effect to be distinguished from that due to growth of existing ones. Such a distinction is not attempted here.

Section 2 gives an account of the assumptions of a kinematical nature on which the model is based. A system of rigid polyhedral grains, closely fitting except for thin interconnecting layers of deformable material, is envisaged. Local deformation gradient rates, corresponding to a uniform overall rate, are determined on the basis of the assumption that the grain centroids experience velocities corresponding to those of a uniform field. An independent mode of deformation, corresponding to a uniform spin of the grains, is furthermore admitted. This is done for the purpose of an application of the principle of virtual work. In actual deformations the spin is assumed to correspond to the overall deformation gradient rate. Thus, micropolarity is excluded, although the model might be adjusted for the purpose of accommodating this effect.

Section 3 deals with such statical aspects of the microstresses that correspond to the modes of deformation admitted by the kinematical assumptions. The principle of virtual work is of course well suited as a tool for this purpose. Its application leads to expressions for overall stress in terms of microstresses, suitably averaged. The microstresses entering the expressions are, as might be expected, normal and shear stresses on planes separating the grains. A similar procedure has been applied successfully to problems of granular media by Christoffersen, Mehrabadi, and Nemat-Nasser (1980).

New results concerning the relations between microscopical and macroscopical stress rates are, furthermore, presented. We introduce, somewhat reluctantly, a new objective stress rate to join the considerable number currently in use. The objective nominal stress rate, which is the mean of convected (or Oldroyd) rate of contravariant Kirchhoff stress and rigid body (or Jaumann or Zaremba) rate of Kirchhoff stress, does, however, have a most direct interpretation in terms of rates of normal and shear stresses on the planes of the separating layers.

Section 4 contains suggestions concerning rate forms of constitutive laws pertaining to the deforming sheets. The combined effect of material and voids is accounted for by a Voigt estimate. The proposed laws are shown to be identical with a slightly simplified version of a model proposed by Gurson (1977). Refinements are possible, but the present model ought to be sufficient for the purpose of demonstrating characteristic features.

Results from the former sections are combined with the constitutive law in a standard fashion resulting in relations between macroscopical stresses, stress rates, and strain rates.

The investigation is completed in Section 5 where properties of the yield surface are demonstrated. Predictions of the model, in terms of material stability and eventual fracture, are presented, and a few suggestions concerning more particular forms are included.

2. Kinematics

Consider a system of rigid, convex polyhedral grains, closely fitting except for uniformly thin layers of deformable material. The following mode of deformation is assumed. Velocities of the grain centroids conform to a homogeneous field of deformation gradient rates ϕ_{ij}. Furthermore, the grains experience identical spins

$$\omega_{ij} = -\omega_{ji}. \tag{1}$$

It is assumed that the spin in actual deformations corresponds to the deformation gradient rate, i.e.

$$\omega_{ij} = \frac{1}{2}(\omega_{ij} - \omega_{ji}). \tag{2}$$

However, for the purpose of an application of the principle of virtual work, ω_{ij} will be retained as a free parameter.

The following velocity field v_i, compatible with the above specifications, is assumed, whereby x_i denotes fixed Cartesian coordinates. In a grain with centroid coordinates x_i^0,

$$v_i = \phi_{ij} x_j^0 + \omega_{ij}\left(x_j - x_j^0\right). \tag{3}$$

In the layer separating the grain from one of its neighbors,

$$v_i = \phi_{ij} x_j^0 + \omega_{ij}\left(x_j - x_j^0\right) + \gamma m_i n_j\left(x_j - x_j^*\right) + \varepsilon n_i n_j\left(x_j - x_j^*\right), \tag{4}$$

where x_j^* are coordinates of some point of the interface between grain and layer. Furthermore, n_j is the unit interface normal, outward with respect to the grain, while m_i is unit and parallel with the interface,

$$m_i m_i = 1; \qquad m_i n_i = 0; \qquad n_i n_i = 1. \tag{5a–c}$$

Finally, γ and ε denote constant shear and normal strain rates.

Continuity of velocities across the interface between layer and neighboring grain requires

$$\gamma m_i + \varepsilon n_i = \frac{1}{\delta}\left(\phi_{ij} - \omega_{ij}\right) d_j. \tag{6}$$

Here, d_j is the vector connecting the centroid of the grain with that of its neighbor, and δ is the thickness of the layer. The relative contribution of the layer to the total volume is the ratio between layer volume and the volume of a double pyramid with a layer plane as its basis and $n_i d_i$ as its height. Hence

$$c = 3\delta / n_i d_i. \tag{7}$$

Introducing c in (6) we find

$$\gamma m_i + \varepsilon n_i = \frac{3}{c}\left(\phi_{ij} - \omega_{ij}\right) p_j \tag{8}$$

where

$$p_j = d_j / n_i d_i. \tag{9}$$

Explicitly, cf. (5),

$$\gamma = \frac{3}{c}\left(\phi_{ij} - \omega_{ij}\right) m_i p_j; \qquad \varepsilon = \frac{3}{c}\left(\phi_{ij} - \omega_{ij}\right) n_i p_j. \tag{10a,b}$$

Naturally, m_i is indeterminate except for (5) if γ vanishes.

According to (3) and (4), the local deformation gradient rates are

$$\tilde{\phi}_{ij} = \omega_{ij} \tag{11}$$

in the grains and

$$\tilde{\phi}_{ij} = \omega_{ij} + \gamma m_i n_j + \varepsilon n_i n_j \tag{12}$$

in the separating layers. The corresponding strain rates,

$$\tilde{\varepsilon}_{ij} = \tfrac{1}{2}(\tilde{\phi}_{ij} + \tilde{\phi}_{ji}),$$ (13)

are

$$\tilde{\varepsilon}_{ij} = 0$$ (14)

in the grains and, in the layers,

$$\tilde{\varepsilon}_{ij} = \tfrac{1}{2}\gamma(m_i n_j + m_j n_i) + \varepsilon n_i n_j.$$ (15)

Note that the local spin in the layers is

$$\tilde{\omega}_{ij} = \omega_{ij} + \tfrac{1}{2}\gamma(m_i n_j - m_j n_i).$$ (16)

As mentioned earlier, effects of micropolarity are disregarded in that (2) is assumed to hold. Consequently, with

$$\varepsilon_{ij} = \tfrac{1}{2}(\phi_{ij} + \phi_{ji})$$ (17)

denoting the overall strain rate, Eqs. (8) and (10) simplify as follows,

$$\gamma m_i + \varepsilon n_i = \frac{3}{c}\varepsilon_{ij}p_j$$ (18)

and

$$\gamma = \frac{3}{c}\varepsilon_{ij}m_i p_j; \qquad \varepsilon = \frac{3}{c}\varepsilon_{ij}n_i p_j.$$ (19a–b)

3. Statics

An application of the principle of virtual work to the velocity field given by (3) and (4) leads to

$$\sigma_{ij}\varepsilon_{ij} = \overline{c(\tau\gamma + \sigma\varepsilon)},$$ (20)

where an overbar denotes volume averaging, and σ_{ij} is the overall stress. Furthermore, with

$$T_i = \sigma_{ij}n_j$$ (21)

denoting the tractions on planes of the layers,

$$\tau = T_i m_i; \qquad \sigma = T_i n_i$$ (22a,b)

are, respectively, the m_i-directed component of shear stress and the normal stress on the planes, averaged over the layer volume. Introducing (10) in (20) we find

$$\sigma_{ij}\varepsilon_{ij} = \overline{(3\tau m_i p_j + 3\sigma n_i p_j)}(\phi_{ij} - \omega_{ij}).$$ (23)

This is to hold for arbitrary ϕ_{ij} and ω_{ij} (note (17)). The ensuing equilibrium conditions are

$$\overline{3\tau m_i p_j} + \overline{3\sigma n_i p_j} = \overline{3\tau m_j p_i} + \overline{3\sigma n_j p_i} \tag{24}$$

and

$$\sigma_{ij} = \overline{3\tau m_i p_j} + \overline{3\sigma n_i p_j} \,. \tag{25}$$

These equations must hold for arbitrary sets of m_i compatible with some overall strain rate ε_{ij}.

The form $\dot{s}_{ij}\phi_{ij}$, where $\dot{s}_{ij}$ is the rate of nominal stress determined from ϕ_{ij} via appropriate constitutive equations, is central in investigations concerning stability and ellipticity. The objective rate of nominal stress

$$\dot{\tau}_{ij} = \dot{\tau}_{ji}, \tag{26}$$

which is related to $\dot{s}_{ij}$ by

$$\dot{s}_{ij} = \dot{\tau}_{ij} - \tfrac{1}{2}\sigma_{ik}\varepsilon_{kj} + \tfrac{1}{2}\sigma_{jk}\varepsilon_{ki} - \sigma_{jk}\omega_{ki}, \tag{27}$$

will prove convenient as a measure. It is related to the often encountered rigid body rate of Kirchhoff stress $\overset{\triangledown}{\tau}_{ij}$ through

$$\dot{\tau}_{ij} = \overset{\triangledown}{\tau}_{ij} - \tfrac{1}{2}\sigma_{ik}\varepsilon_{kj} - \tfrac{1}{2}\sigma_{jk}\varepsilon_{ki}. \tag{28}$$

In terms of $\dot{\tau}_{ij}$, the form $\dot{s}_{ij}\phi_{ij}$ becomes

$$\dot{s}_{ij}\phi_{ij} = \dot{\tau}_{ij}\varepsilon_{ij} + \sigma_{ij}\phi_{ki}\phi_{kj} - \sigma_{ij}\varepsilon_{ki}\varepsilon_{kj}. \tag{29}$$

Due to equilibrium conditions for $\dot{s}_{ij}$, the divergence theorem may be brought in use to obtain

$$\dot{s}_{ij}\phi_{ij} = \dot{s}_{ij}\omega_{ij} + \overline{c(\dot{T}_i m_i \gamma + \dot{T}_i n_i \varepsilon)}, \tag{30}$$

where

$$\dot{T}_i = \dot{s}_{ij} n_j \tag{31}$$

are rates of nominal tractions on the planes of the deforming layers, cf. (11) and (12). According to (22) we have

$$\dot{T}_i m_i = \dot{\tau} - T_i \frac{\mathrm{d} m_i}{\mathrm{d} t}; \qquad \dot{T}_i n_i = \dot{\sigma} - T_i \frac{\mathrm{d} n_i}{\mathrm{d} t}, \tag{32a,b}$$

where $\dot{\tau}$ and $\dot{\sigma}$ are rates of nominal shear and normal stress, both of them objective measures, while $\mathrm{d} m_i/\mathrm{d} t$ and $\mathrm{d} n_i/\mathrm{d} t$ denote rates of m_i and n_i. Decomposing the traction,

$$T_i = \tau m_i + \sigma n_i + \tau_i^*, \tag{33}$$

where τ_i^*, satisfying

$$\tau_i^* m_i = 0; \qquad \tau_i^* n_i = 0, \tag{34a,b}$$

is the shear stress component perpendicular to m_i, we find, noting (5) that

$$\dot{T}_i m_i = \dot{\tau} - \sigma n_i \frac{dm_i}{dt} - \tau_i^* \frac{dm_i}{dt}; \qquad \dot{T}_i n_i = \dot{\sigma} - \tau m_i \frac{dn_i}{dt} - \tau_i^* \frac{dn_i}{dt}.$$
$$\tag{35a,b}$$

Nanson's formula for the unit normal of a deforming surface is, in differential form,

$$\frac{dn_i}{dt} = -\tilde{\phi}_{ji} n_j + \varepsilon n_i. \tag{36}$$

In view of (12), this reduces to

$$\frac{dn_i}{dt} = \omega_{ij} n_j. \tag{37}$$

Hence

$$m_i \frac{dn_i}{dt} = \omega_{ij} m_i n_j; \qquad n_i \frac{dm_i}{dt} = -\omega_{ij} m_i n_j. \tag{38a,b}$$

Furthermore, in view of (34),

$$\tau_i^* \frac{dm_i}{dt} = -\dot{\tau}_i^* m_i; \qquad \tau_i^* \frac{dn_i}{dt} = -\dot{\tau}_i^* n_i, \tag{39a,b}$$

where $\dot{\tau}_i^*$ is the rate of τ_i^* when viewed as a nominal traction. We conclude that

$$\dot{T}_i m_i = \dot{\tau} + \sigma \omega_{ij} m_i n_j + \dot{\tau}_i^* m_i; \qquad \dot{T}_i n_i = \dot{\sigma} - \sigma \omega_{ij} m_i n_j + \dot{\tau}_i^* n_i,$$
$$\tag{40a,b}$$

which, when inserted in (30) with $\dot{s}_{ij}$ replaced by $\dot{\tau}_{ij}$ (Eq. (27)), leads to

$$\dot{\tau}_{ij} \varepsilon_{ij} = \overline{c(\dot{\tau}\gamma + \dot{\sigma}\varepsilon)} + \overline{c(\sigma\gamma - \tau\varepsilon)m_i n_j} \, \omega_{ij}$$
$$+ \overline{c\dot{\tau}_i^*(m_i\gamma + n_i\varepsilon)} + \sigma_{ik}\varepsilon_{kj}\omega_{ij}. \tag{41}$$

A further simplification may be obtained as follows. According to (25) and (8),

$$\sigma_{ik}\varepsilon_{kj} = \overline{c(\tau m_i + \sigma n_i)(\gamma m_j + \varepsilon n_j)}. \tag{42}$$

Twice the skew-symmetric part is

$$\sigma_{ik}\varepsilon_{kj} - \sigma_{jk}\varepsilon_{ki} = \overline{c(\tau\varepsilon - \sigma\gamma)(m_i n_j - m_j n_i)}. \tag{43}$$

Consequently,

$$\sigma_{ik}\varepsilon_{kj}\omega_{ij} = \overline{c(\tau\varepsilon - \sigma\gamma)m_i n_j} \, \omega_{ij}. \tag{44}$$

Utilizing (8) again,

$$\dot{\tau}_{ij}\varepsilon_{ij} = \overline{c(\dot{\tau}\gamma + \dot{\sigma}\varepsilon)} + \overline{3\dot{\tau}_i^* p_j \varepsilon_{ij}}. \tag{45}$$

The appearance of the last term is of course related to the kinematical restrictions. To see that it vanishes, use (8) to obtain

$$\overline{3\tau_i^* p_j (\phi_{ij} - \omega_{ij})} = \overline{c\tau_i^* m_i \gamma} + \overline{c\tau_i^* n_i \varepsilon}. \tag{46}$$

This expression vanishes identically, due to (34). Hence

$$\overline{3\tau_i^* p_j} = 0. \tag{47}$$

The local rate of area strain is taken to be

$$\alpha = \varepsilon_{ii} - \varepsilon_{ij} n_i n_j. \tag{48}$$

This is seemingly in conflict with the kinematical assumptions (12). Certainly, that expression leads to a locally vanishing α, but the expression (48) accounts consistently for area extensions concentrated to the layer intersections. We therefore have

$$\frac{d}{dt}(\tau_i^* p_j) = \dot{\tau}_i^* p_j - \alpha \tau_i^* p_j + \tau_i^* \frac{dp_j}{dt}. \tag{49}$$

With (9), (36), and

$$\frac{dd_j}{dt} = \phi_{jk} d_k \tag{50}$$

we realize that

$$\frac{dp_j}{dt} = \phi_{jk} p_k - \varepsilon_{kl} n_k n_l p_j. \tag{51}$$

Introducing this in (49) we find

$$\frac{d}{dt}(\tau_i^* p_j) = \dot{\tau}_i^* p_j + \tau_i^* p_k \phi_{jk} - \tau_i^* p_j \varepsilon_{kk}. \tag{52}$$

Volume averaging and noting that (47) is true at all times we finally conclude that

$$\overline{3\dot{\tau}_i^* p_j} = 0, \tag{53}$$

whence, according to (45),

$$\dot{\tau}_{ij}\varepsilon_{ij} = \overline{c(\dot{\tau}\gamma + \dot{\sigma}\varepsilon)}. \tag{54}$$

This, together with (19) and (29), determines the value of the form $\dot{s}_{ij}\phi_{ij}$, provided that constitutive equations, relating local stress rates $(\dot{\tau}, \dot{\sigma})$ to local strain rates (γ, ε), are available.

4. Constitutive equations

In a voided layer,

$$\tau=(1-f)\tau_0; \qquad \sigma=(1-f)\sigma_0, \tag{55a,b}$$

where f is the relative layer void volume, and (τ_0, σ_0) are average stresses in the material. An increase of f leads to an increase of volume resulting in a non-zero ε. As

$$\tilde{\varepsilon}_{kk} = \frac{1}{1-f}\frac{\mathrm{d}f}{\mathrm{d}t}, \tag{56}$$

and in view of (15),

$$\dot{\tau}=(1-f)\dot{\tau}_0-\tau\varepsilon; \qquad \dot{\sigma}=(1-f)\dot{\sigma}_0-\sigma\varepsilon. \tag{57a,b}$$

Hence

$$\dot{\tau}\gamma+\dot{\sigma}\varepsilon=(1-f)(\dot{\tau}_0\gamma+\dot{\sigma}_0\varepsilon)-(\tau\gamma+\sigma\varepsilon)\varepsilon. \tag{58}$$

The following simple yield law, relating (γ, ε) to $(\dot{\tau}_0, \dot{\sigma}_0)$, is proposed,

$$\tau_e \leqslant \tau_c, \tag{59}$$

where

$$\tau_e = \sqrt{\tau_0^2 + k\sigma_0^2}. \tag{60}$$

The parameters τ_c and k are subject to work hardening. Further,

$$\gamma=\gamma_e\frac{\tau_0}{\tau_e}; \qquad \varepsilon=k\gamma_e\frac{\sigma_0}{\tau_e}, \tag{61a,b}$$

where

$$\gamma_e = \frac{\tau_e}{G_{t0}} \quad \text{if} \quad \tau_e=\tau_c \quad \text{and} \quad \dot{\tau}_e>0; \quad \gamma_e=0 \quad \text{otherwise.} \tag{62}$$

Now

$$\dot{\tau}_e = \frac{\tau_0}{\tau_e}\dot{\tau}_0+k\frac{\sigma_0}{\tau_e}\dot{\sigma}_0, \tag{63}$$

and so it follows that

$$\gamma_e = \sqrt{\gamma^2 + \frac{1}{k}\varepsilon^2}. \tag{64}$$

Consequently,

$$\dot{\tau}_0\gamma+\dot{\sigma}_0\varepsilon=G_{t0}\left(\gamma^2 + \frac{1}{k}\varepsilon^2\right). \tag{65}$$

Moreover, from (61) and (55),

$$\tau\varepsilon = k\sigma\gamma, \tag{66}$$

and so, with (65), Eq. (58) becomes

$$\dot{\tau}\gamma + \dot{\sigma}\varepsilon = G_t\left(\gamma^2 + \frac{1}{k}\varepsilon^2\right) \tag{67}$$

where

$$G_t = (1-f)G_{t0} - k\sigma. \tag{68}$$

From (54), (67), and (19),

$$\dot{\tau}_{ij}\varepsilon_{ij} = \left(\frac{9}{c}G_t m_i\, p_j m_k\, p_l + \frac{9}{ck}G_t n_i\, p_j n_k\, p_l\right)\varepsilon_{kl}\varepsilon_{ij}. \tag{69}$$

The role of the voids should be obvious from (68) and (69). The parameter k increase from zero. This reduces the tangent modulus G_t when stresses are tensile. In fact, G_t may be negative even if G_{t0} is not. On the other hand, a hydrostatic pressure tends to increase the ductility.

We emphasize that f is the void ratio in the currently deforming layers. Continuity of matter requires that

$$fc + f_0(1-c) = \frac{J-1}{J}, \tag{70}$$

where f_0 is the void ratio in the grains while J is current (voided) volume relative to initial (unvoided) volume. Identifying

$$f = 1 \tag{71}$$

as the condition for fracture, in the fractured material

$$c^* + f_0^*(1-c^*) = \frac{J^*-1}{J^*}, \tag{72}$$

where starred symbols denote ultimate values. Assuming c to be a constant and f_0 to be proportional with f,

$$f = \frac{J^*}{J}\frac{J-1}{J^*-1} \tag{73}$$

during the deformation. That fracture may occur even at small increases of volume is evident from this expression.

Gurson (1977) has proposed a yield law for a voided Von Mises-material. With this, an estimate for the parameter k may be provided. According to Gurson's spherical void model, stresses $\tilde{\sigma}_{ij}$ must satisfy

$$\tfrac{1}{2}\tilde{\sigma}_{ij}\tilde{\sigma}_{ij} - \tfrac{1}{6}\tilde{\sigma}_{ii}\tilde{\sigma}_{jj} + 2f\tau_c^2\cosh\left\{\frac{\tilde{\sigma}_{ii}}{2\sqrt{3}\,\tau_c}\right\} - (1+f^2)\tau_c^2 \leqslant 0. \tag{74}$$

Deformation is in accordance with the normality rule. In the later stages of the deformation, the argument of the exponential function is relatively small. Replacing it by the first three terms in a Taylor expansion we arrive at

$$\tfrac{1}{2}\tilde{\sigma}_{ij}\tilde{\sigma}_{ij} - \frac{2-f}{12}\tilde{\sigma}_{ii}\tilde{\sigma}_{jj} - (1-f)^2\tau_c^2 \leq 0. \tag{75}$$

The corresponding strain rates are

$$\tilde{\varepsilon}_{ij} = \lambda\left(\tilde{\sigma}_{ij} - \frac{2-f}{6}\delta_{ij}\sigma_{kk}\right). \tag{76}$$

Taking for simplicity

$$m_i = (1,0,0); \quad n_i = (0,1,0), \tag{77a,b}$$

in view of (15) we have

$$\sigma_{11} = \frac{2-f}{2(1+f)}\sigma; \qquad \sigma_{22} = \sigma; \qquad \sigma_{33} = \frac{2-f}{2(1+f)}\sigma$$

$$\sigma_{23} = 0; \qquad \sigma_{31} = 0; \qquad \sigma_{12} = \tau. \tag{78a–f}$$

Introducing this in (75),

$$\tau^2 + \frac{3f}{4(1+f)}\sigma^2 - (1-f)^2\tau_c^2 \leq 0 \tag{79}$$

or, in view of (55),

$$\tau_0^2 + \frac{3f}{4(1+f)}\sigma_0^2 - \tau_c^2 \leq 0. \tag{80}$$

This is identical with the condition (59), (60) with

$$k = \frac{3f}{4(1+f)}. \tag{81}$$

According to this model, $0 \leq k \leq 3/8$. However, the case

$$k = 1 \tag{82}$$

will prove particularly interesting. This is obtained if the local state of stress in the layers is uniaxial. Perhaps an appropriate assumption for the very last stages of deformation where the material could be expected to develop a threadlike structure.

5. The yield surface

According to the present model, the voided material has a yield surface. Convexity and normality may be demonstrated by the following standard argument. The local yield surfaces are convex, and the normality rule applies, so

$$(\tau_0' - \tau_0)\gamma + (\sigma_0' - \sigma_0)\varepsilon \leq 0 \tag{83}$$

holds whenever (τ_0', σ_0') and (τ_0, σ_0) satisfy the yield condition (59), (60) and the strain rates (γ, ε) may be produced at (τ_0, σ_0). Let σ_{ij}' be an overall stress satisfying the macroscopic yield condition. This is the case if all local stresses (τ_0', σ_0') contributing to σ_{ij}' satisfy the local yield conditions. Also, let σ_{ij} be a stress at which a strain rate ε_{ij} is possible. With the aid of (18) and (25) we find the condition

$$(\sigma_{ij}' - \sigma_{ij})\varepsilon_{ij} \leq 0, \tag{84}$$

appropriate for convexity and normality.

Next we address the problem of finding the overall stress σ_{ij} required to produce a given overall strain rate ε_{ij}. According to (55) and (61)

$$\tau = \frac{\tau_c}{(1-f)\gamma_e}\gamma; \sigma = \frac{\tau_c}{(1-f)k\gamma_e}\varepsilon. \tag{85a,b}$$

We shall assume that c is constant apart from a random component. Hence (25) and (18) yield

$$\sigma_{ij} = \left\{ \overline{\frac{9\tau_c}{(1-f)g}m_i p_j m_k p_l} + \overline{\frac{9\tau_c}{(1-f)kg}n_i p_j n_k p_l} \right\} m_{kl}. \tag{86}$$

where m_{ij} is proportional with ε_{ij} but suitably normalized and

$$g^2 = \left\{ \overline{9m_i p_j m_k p_l} + \overline{(9/k)n_i p_j n_k p_l} \right\} m_{kl} m_{ij}. \tag{87}$$

It is not likely that different strain rate directions m_i will require identical stresses. The conclusion is that the yield surface is smooth. Perhaps an unexpected prediction from a micromechanical model. A corner effect may of course result from stipulating corners for the local surfaces. However, the general trend of the model is that it has a smoothening effect in striking contrast to slip theory.

The relation between strain rates and stress rates will be in the form

$$\varepsilon_{ij} = \frac{1}{h}m_{ij}m_{kl}\dot{\tau}_{kl}, \tag{88}$$

where m_{ij} is a yield surface normal. It is possible, at least in principle, to determine the yield surface from (86), (87). The work hardening parameter h is related to the parameter $\bar{c}$ through

$$m_{ij}\varepsilon_{ij} = \frac{m_{ij}m_{ij}}{h}m_{kl}\dot{\tau}_{kl} \tag{89}$$

and

$$\dot{\tau}_{ij}\varepsilon_{ij} = \frac{1}{h}(m_{ij}\dot{\tau}_{ij})^2 \tag{90}$$

leading to

$$h\left(\frac{m_{ij}\varepsilon_{ij}}{m_{kl}m_{kl}}\right)^2 = \dot{\tau}_{ij}\varepsilon_{ij}. \tag{91}$$

In view of (69), once again utilizing that c is distributed randomly around $\bar{c}$,

$$h\bar{c} = \left(\overline{9G_t m_i p_j m_k p_l} + \overline{\frac{9}{k}G_t n_i p_j n_k p_l}\right)m_{kl}m_{ij}. \tag{92}$$

A particularly simple result emerges in the case $k=1$ (Eq. (82)). According to (66) and (43)

$$\sigma_{ik}\varepsilon_{kj} = \sigma_{jk}\varepsilon_{ki} \tag{93}$$

in this case. This requires the expression for the yield surface to be in the form

$$F(I_1, I_2\,I_3) = 0, \tag{94}$$

where

$$I_1 = \sigma_{ii}; \qquad I_2 = \tfrac{1}{2}\sigma_{ij}\sigma_{ji}; \qquad I_3 = \tfrac{1}{3}\sigma_{ij}\sigma_{jk}\sigma_{ki} \tag{95a–c}$$

are stress invariants. We then have

$$m_{ij} = \frac{\partial F}{\partial I_1}\delta_{ij} + \frac{\partial F}{\partial I_2}\sigma_{ij} + \frac{\partial F}{\partial I_3}\sigma_{ik}\sigma_{kj}. \tag{96}$$

The further assumption, that

$$\frac{\partial F}{\partial I_3} = 0, \tag{97}$$

leads to a Levy-Von Mises type of yield condition with the yield stress depending on mean stress,

$$\tau_e \leqslant \tau_c(s), \tag{98}$$

where

$$\tau_e = \left(\tfrac{1}{2}\sigma_{ij}\sigma_{ij} - \tfrac{1}{6}\sigma_{ii}\sigma_{jj}\right)^{1/2}; \qquad s = \tfrac{1}{3}\sigma_{ii}. \tag{99a,b}$$

As mentioned, $k=1$ is perhaps a reasonable assumption in the immediate pre-fracture state

References

Christoffersen, J., Mehrabadi, M. M. and Nemat-Nasser, S. (1980), "A Micromechanical Description of Granular Material Behavior," to appear in *J. Appl. Mech.*

Gurson, A. L. (1977), "Continuum Theory of Ductile Rupture by Void Nucleation and Growth: Part 1—Yield Criteria and Flow Rules for Porous Ductile Media", *J. Eng. Mat. Tech.*, **99**, 2.

Needleman, A. and Rice, J. R. (1978), "Limits to Ductility Set by Plastic Flow Localization," *Mechanics of Sheet Metal Forming*, D. P. Koistinen and N. M. Wang (eds.), Plenum Press, 237.

Parmar, A. and Mellor, P. B. (1980), "Growth of Voids in Biaxial Stress Fields," *Int. J. Mech. Sci.*, **22**, 133.

S. Nemat-Nasser, Editor
THREE-DIMENSIONAL CONSTITUTIVE RELATIONS AND DUCTILE FRACTURE
North-Holland Publishing Company (1981) 289–306

ON CONSTITUTIVE RELATIONS IN THERMOPLASTICITY

Th. LEHMANN

Ruhr-University Bochum, Bochum, Fed. Rep. Germany

This paper deals with a phenomenological theory of large, nonisothermic inelastic deformations of solid bodies within the frame of classical continuum mechanics and thermodynamics. The general thermodynamical framework for such a theory is considered followed by a detailed discussion of an extended approach for the evolution law for strain. The capability of this approach is examined by comparison with some experimental results.

1. Introduction

This paper deals with a phenomenological theory of large, nonisothermic inelastic deformations of solid bodies within the frame of classical continuum mechanics and thermodynamics including internal processes like recrystallization, solid phase transformations, and others. The considerations are focussed on two different aspects:

(a) the general thermodynamical framework of such a theory, and

(b) a more detailed discussion of the stress–strain relations within this framework.

An extended discussion of the general thermodynamical framework is given by the author (1978, 1980). Therefore, we can restrict ourselves to the main aspects of these considerations. The reason for a new discussion of the inelastic stress–strain relations is due to the as yet unsolved discrepancy that in certain cases, e.g. (Hohenemser, 1931; Hohenemser and Prager, 1932; Pflüger, 1967), finite inelastic stress–strain relations lead to better results than an incremental stress–strain law, in spite of the fact that finite inelastic stress–strain relations are physically unsatisfactory. Sometimes the existence of corners in the yield surface is assumed for the explanation of the observed discrepancies, e.g. (Rice, 1976; Rudnicki and Rice, 1975; Sewell, 1974). It will be shown that another approach seems to be more satisfactory.

2. Some preliminary remarks

Considering constitutive relations it seems more natural to relate all quantities to the actual configuration of the body than to the original state.

In this way we do not only avoid some inconveniences concerning the elimination of rigid body rotations, but we also remove some difficulties in the description of elastic unloading processes which start at the actual state but do not end at the original state. Likewise for some other reason (see Lehmann, 1978, 1980) we are led to the following description of kinematics.

We introduce a body fixed coordinate system ξ^i with the metric $g_{ik}(\xi^r, t)$. The transition from the undeformed into the deformed configuration can be described by the metric transformation tensor

$$q_k^i = \mathring{g}^{ir} g_{rk} = q_k^i(\xi^r, t),\tag{1}$$

where the super-imposed open circle denotes the original (undeformed) state. Using q_k^i or its inverse $(q^{-1})_k^i$ we obtain

$$g_{ik} = q_i^r \mathring{g}_{rk} = \mathring{g}_{ir} q_k^r,$$

$$\mathring{g}_{ik} = (q^{-1})_i^r g_{rk} = g_{ir} (q^{-1})_k^r.\tag{2}$$

By means of isotropic tensor functions we may define any suitable strain tensor, e.g. Almansi strain tensor

$$\tilde{\varepsilon}_k^i = \tfrac{1}{2}\left\{ \delta_k^i - (q^{-1})_k^i \right\},\tag{3a}$$

or Hencky strain tensor

$$\varepsilon_k^i = \tfrac{1}{2}(\ln q)_k^i.\tag{3b}$$

The strain rate is given by

$$d_k^i = \tfrac{1}{2}(q^{-1})_r^i (\dot{q})_{.k}^r = -\tfrac{1}{2}(\dot{q}^{-1})_{.r}^i q_k^r.\tag{4}$$

With respect to Eq. (3a) we may also write

$$d_k^i = q_k^r (\dot{\tilde{\varepsilon}})_{.r}^i.\tag{5a}$$

For coaxial deformations furthermore we obtain

$$d_k^i = (\dot{\varepsilon})_k^i = \varepsilon_k^i|_0.\tag{5b}$$

Here $|_0$ denotes the covariant derivation with respect to time in the body-fixed coordinate system (with time $t = \xi^0$ introduced as 4th coordinate in a suitable manner (Lehmann, 1966). For an arbitrary second rank tensor it is defined by

$$a^i{}_{.k}|_0 = (\dot{a})_{.k}^i + d_r^i a_{.k}^r - d_k^r a_{.r}^i.\tag{6}$$

It coincides with the so-called Zaremab–Jaumann derivation. The total strain can be decomposed into an elastic and an inelastic part by a multiplicative splitting of the metric transformation according to

$$q_k^i = \overset{\circ}{g}^{im} \overset{*}{g}_{mr} g^{rn} \overset{*}{g}_{nk} = \underset{(i)}{q_{\cdot r}^i} \, \underset{(e)}{q_k^r}, \tag{7}$$

where the superimposed $*$ relates to the (imaginary) intermediate state. This leads to an additive decomposition of the strain rate

$$d_k^i = \tfrac{1}{2} \mathrm{sym}\left\{ \left(\underset{(e)}{q}^{-1} \right)^i_{\,r} \left(\underset{(e)}{\dot{q}} \right)^r_{\cdot k} \right\} + \tfrac{1}{2} \mathrm{sym}\left\{ (q^{-1})^i_{\,r} \left(\underset{(i)}{\dot{q}} \right)^r_{\cdot m} \underset{(e)}{q}^m_k \right\}$$

$$= -\tfrac{1}{2} \mathrm{sym}\left\{ \underset{(e)}{q}^r_k \left(\underset{(e)}{\dot{q}}^{-1} \right)^i_{\cdot r} \right\} - \tfrac{1}{2} \mathrm{sym}\left\{ \left(\underset{(e)}{q}^{-1} \right)^i_{\,r} \left(\underset{(i)}{\dot{q}}^{-1} \right)^r_{\cdot m} q^m_k \right\}$$

$$= \underset{(e)}{d^i_k} + \underset{(i)}{d^i_k}. \tag{8}$$

In these formulas sym$\{\ \}$ denotes the symmetric part of the respective tensor.

The rate of specific mechanical work is given by

$$\dot{w} = \frac{1}{\rho} \sigma_k^i \, d_i^k = \frac{1}{\overset{\circ}{\rho}} s_k^i \, d_i^k \tag{9}$$

with $\sigma_k^i = $ Cauchy stress tensor, $\rho = $ mass density, $s_k^i = (\rho/\overset{\circ}{\rho}) \sigma_k^i = $ weighted Cauchy stress tensor.

The rate of specific work can be decomposed into its elastic and its inelastic part according to

$$\dot{w} = \frac{1}{\overset{\circ}{\rho}} s_k^i \underset{(e)}{d^k_{\,i}} + \frac{1}{\overset{\circ}{\rho}} s_k^i \underset{(i)}{d^k_{\,i}}$$

$$= \underset{(e)}{\dot{w}} + \underset{(i)}{\dot{w}}. \tag{10}$$

If the elastic behavior of the body is isotropic we can also write the elastic part of the specific work rate in the form

$$\underset{(e)}{\dot{w}} = \frac{1}{\overset{\circ}{\rho}} \tilde{s}_k^i \tfrac{1}{2} \underset{(e)}{q}^k_i \big|_0 \qquad \text{with} \qquad \tilde{s}_k^i = s_r^i \left(\underset{(e)}{q}^{-1} \right)^r_k, \tag{11a}$$

$$= \frac{1}{\overset{\circ}{\rho}} s_k^i \underset{(e)}{\varepsilon}^k_i \big|_0 \qquad \text{with} \qquad \underset{(e)}{\varepsilon}^i_k = \tfrac{1}{2} \left(\ln \underset{(e)}{q} \right)^i_k. \tag{11b}$$

For simplicity this is supposed in the following.

One part $\dot{w}_{(d)}$ of the inelastic specific work rate is dissipated immediately. The remaining part $\dot{w}_{(h)}$ is correlated to changes of the internal structure of the material (hardening, solid phase transformations, etc.). This part is, in principle, reversible. Hence we may treat this part as pseudo-elastic. However, in most cases, at least a certain amount of this work is dissipated later on. Therefore we rank this part with the inelastic work writing

$$\dot{w} = \dot{w} + \dot{w} \,.$$
$$\,_{(i)} \quad _{(h)} \quad _{(d)}$$

(12)

3. Thermodynamic frame

The first law of thermodynamics states

$$\dot{u} = \dot{w} - \frac{1}{\rho}\left(q^i + h^i\right)\big|_i + r,$$

(13)

where u = specific internal energy, q^i = heat flux, h^i = other energy fluxes, r = specific energy supply by sources, $|_i$ = covariant derivation with respect to the body-fixed coordinates in the actual configuration.

Energy fluxes different from heat can be caused, for instance, by the migration of lattice defects which are not necessarily connected with macroscopic deformations. Analogously the energy supplied by sources may also comprehend energy different from heat as, for instance, generation of lattice defects by absorbed radiation. Following the concept of classical thermodynamics we assume that each material element can be considered as a local thermodynamical system whose state is uniquely determined by the actual values of a complete set of (external and internal) variables. This concept is certainly restricted to thermo-mechanical processes running not too far from thermodyamical equilibrium. This implies that the gradients of the state variables in space and time remain within certain bounds. Otherwise we have to use an extended concept. A capable one for such cases seems to be given by Müller (1973). We shall remain, however, within the frame work of classical thermodynamics.

Under this assumption the specific free energy must be expressible as a function of a set of thermodynamic state variables,

$$u = u\left(\varepsilon^{\,i}_{(e)\,k}, s, b, \beta^i_k \right).$$

(14)

In this relation, s = specific entropy, b = scalar valued internal variables like isotropic hardening parameters or mass fractions of solid phases, β^i_k = tensor valued internal variables (of arbitrary even rank) like anisotropic hardening parameters.

The quantities b and β^i_k may represent each a whole set of internal variables. For brevity, however, only one quantity of each group is introduced in (14).

The inelastic deformation represented by $q_{(i)}{}^i{}_{\cdot k}$ or the total strains, respectively, do not belong to the internal variables since they are not uniquely related to the internal state of the material in general. Considering, for instance, a dislocation passing through a single crystal, we finally obtain an inelastic deformation without a change of the internal state. Concerning polycrystalline materials we can also state that the same macroscopic inelastic deformation can be realized in different ways leading to different patterns of lattice defects, grain boundaries, etc., which determine the state of the material elements together with the elastic deformations of the crystal lattice and the entropy or the temperature, respectively.

With respect to many applications it is advantageous to replace the elastic deformations $\varepsilon_{(e)k}{}^i$ and the specific entropy s by the (weighted) stress s_k^i and the (absolute) temperature T. This can be achieved by a corresponding Legendre transformation leading to a particular form of the specific free enthalpy (Gibbs function)

$$\psi = u - \frac{1}{\overset{\circ}{\rho}} s_k^i \underset{(e)}{\varepsilon}{}_i^k - Ts = \psi\left(s_k^i, T, b, \beta_k^i\right). \tag{15}$$

Observing (13) we derive from (15)

$$\dot{\psi} = \dot{u} - \frac{1}{\overset{\circ}{\rho}} s_k^i \underset{(e)}{\varepsilon}{}_i^k\big|_0 - \frac{1}{\overset{\circ}{\rho}} \dot{s}_k^i\big|_0 \underset{(e)}{\varepsilon}{}_i^k - T\dot{s} - \dot{T}s$$

$$= \underset{(h)}{\dot{w}} + \underset{(d)}{\dot{w}} - \frac{1}{\rho}\left(q^i + h^i\right)\big|_i + r - \frac{1}{\overset{\circ}{\rho}} \dot{s}_k^i\big|_0 \underset{(e)}{\varepsilon}{}_i^k - T\dot{s} - \dot{T}s. \tag{16}$$

On the other hand $\dot{\psi}$ can be expressed by

$$\dot{\psi} = \frac{\partial \psi}{\partial s_k^i} \dot{s}_k^i\big|_0 + \frac{\partial \psi}{\partial T} \dot{T} + \frac{\partial \psi}{\partial b} \dot{b} + \frac{\partial \psi}{\partial \beta_k^i} \dot{\beta}_k^i\big|_0. \tag{17}$$

From the properties of the Legendre transformation we obtain immediately the thermic state equation:

$$\underset{(e)}{\varepsilon}{}_k^i = -\overset{\circ}{\rho}\frac{\partial \psi}{\partial s_i^k} = \underset{(e)}{\varepsilon}{}_k^i\left(s_k^i, T, b, \beta_k^i\right), \tag{18}$$

and caloric state equation:

$$s = -\frac{\partial \psi}{\partial T} = s\left(s_k^i, T, b, \beta_k^i\right). \tag{19}$$

Substituting the relations (18) and (19) into (16), and equating (16) and (17), we obtain an expression for the evolution of the specific entropy in the form:

$$T\dot{s} = \underset{(h)}{\dot{w}} + \underset{(d)}{\dot{w}} - \frac{1}{\rho}\left(q^i + h^i\right)\big|_i + r - \frac{\partial \psi}{\partial b} \dot{b} - \frac{\partial \psi}{\partial \beta_k^i} \dot{\beta}_k^i\big|_0. \tag{20}$$

This evolution law can by means of relation (19) also be written in the form

$$T\dot{s} = -T\left\{\frac{\partial^2 \psi}{\partial T^2} \dot{T} + \frac{\partial^2 \psi}{\partial s_k^i \partial T} \dot{s}_k^i\big|_0 + \frac{\partial^2 \psi}{\partial b \partial T} \dot{b} + \frac{\partial^2 \psi}{\partial \beta_k^i \partial T} \dot{\beta}_k^i\big|_0\right\}. \tag{21}$$

Comparing Eqs. (20) and (21) we derive the balance equation for the specific free enthalpy:

$$\dot{w}_{(h)} + \dot{w}_{(d)} - \frac{1}{\rho}(q^i + h^i)\big|_i + r = -T\left\{\frac{\partial^2\psi}{\partial T^2}\dot{T} + \frac{\partial^2\psi}{\partial s_k^i \partial T}s_k^i\big|_0\right\}$$

$$+ \frac{\partial}{\partial b}\left\{\psi - T\frac{\partial\psi}{\partial T}\right\}\dot{b} + \frac{\partial}{\partial \beta_k^i}\left\{\psi - T\frac{\partial\psi}{\partial T}\right\}\beta_k^i\big|_0,$$

$$(22)$$

The reversible part of entropy evolution is:

$$T\,\dot{s}_{(rev)} = \dot{w}_{(h)} - \frac{T}{\rho}\left(\frac{q^i}{T}\right)\bigg|_i - \frac{1}{\rho}h^i\big|_i + r - \frac{\partial\psi}{\partial b}\dot{b} - \frac{\partial\psi}{\partial \beta_k^i}\beta_k^i\big|_0 - T\dot{\eta}. \quad (23)$$

In this relation $\dot{\eta}$ represents that part of entropy production which is due to irreversible internal processes (connected with changes of the internal variables) and to the immediate dissipation of energy supplied by sources r and by the divergence of the fluxes h^i. $\dot{\eta}$ has to be specified in the constitutive law of the respective materials. Recrystallization, for instance, leading to changes of the internal parameter (b, β_k^i) is a transient (dissipative) process which contributes to the entropy production entering $\dot{\eta}$. The same is true for other annealing processes. Quasistatic solid phase transformations on the contrary are reversible processes. Dynamic solid phase transformations, however, are connected with dissipative processes, since they run through nonequilibrium states. Hence it depends on the particular processes under consideration how $\dot{\eta}$ has to be specified in the constitutive law.

From Eqs, (23) and (20) or (21), respectively, we obtain, for the total entropy production which is, according to the second law of thermodynamics, nonnegative,

$$0 \leqslant T\,\dot{s}_{(irr)} = T\dot{s} - T\,\dot{s}_{(rev)}$$

$$= \dot{w}_{(d)} + T\dot{\eta} - \frac{1}{\rho T}q^i T\big|_i$$

$$= -\dot{w}_{(h)} + \frac{T}{\rho}\left(\frac{q^i}{T}\right)\bigg|_i + \frac{1}{\rho}h^i\big|_i - r - T\left\{\frac{\partial^2\psi}{\partial T^2}\dot{T} + \frac{\partial^2\psi}{\partial s_k^i \partial T}s_k^i\big|_0\right\}$$

$$+ \frac{\partial}{\partial b}\left\{\psi - T\frac{\partial\psi}{\partial T}\right\}\dot{b} + \frac{\partial}{\partial \beta_k^i}\left\{\psi - T\frac{\partial\psi}{\partial T}\right\}\beta_k^i\big|_0 + T\dot{\eta}. \quad (24)$$

Equations (15), (22) and (24) represent the general thermodynamical frame for the formulation of the constitutive law for materials undergoing thermomechanical processes.

4. Some general remarks on the constitutive law

The constitutive law of the material comprehends:
— state function and balance equation for the specific free enthalpy;
— evolution laws for internal variables;
— flux laws of internal energy (heat and other);
— laws of entropy production.
Under the usual assumptions concerning regularity etc., the constitutive law forms a system of first order partial differential equations in space and time completed by some side conditions as, for instance, state function, yield condition, and recrystallization condition. Only if all energy fluxes are vanishing the constitutive law degenerates to a system of first order ordinary differential equations in time. This is true for so-called elementary processes which are homogeneous throughout the body.

In many cases, however, it can be assumed that for solid bodies also in inhomogeneous processes the energy fluxes apart from heat remain small. In such cases, correspondingly, the evolution laws for the total strain, for the internal variables, and for the entropy production terms $\dot{w}_{(d)}$ and $T\dot{\eta}$ reduce to first order ordinary differential equations in time. Only the flux law for heat entering both the balance equation for specific free enthalpy and the law of entropy production remains a field equation. Therefore the neglect of energy fluxes different from heat leads to important simplifications of the constitutive law. We shall proceed in this way. Then the constitutive law reduces to the following scheme:
specific free enthalpy:

$$\psi = \psi\left(s_k^i, T, b, \beta_k^i\right), \tag{25a}$$

evolution laws (with side conditions)
total strain:

$$d_k^i = d_k^i\left(s_k^i, T, b, \beta_k^i, s_k^i|_0, \dot{T}\right), \tag{25b}$$

internal variables:

$$\dot{b} = \dot{b}\left(s_k^i, T, b, \beta_k^i, s_k^i|_0, \dot{T}\right), \tag{25c}$$

$$\beta_k^i|_0 = \beta_k^i|_0\left(s_k^i, T, b, \beta_k^i, s_k^i|_0, \dot{T}\right), \tag{25d}$$

flux law for heat:

$$q^i = q^i\left(s_k^i, T, b, \beta_k^i, T|_i\right), \tag{25e}$$

laws for entropy production:

$$\underset{(d)}{\dot{w}} = \underset{(d)}{\dot{w}}\left(s_k^i, T, b, \beta_k^i, s_k^i|_0, \dot{T}\right), \tag{25f}$$

$$T\dot{\eta} = T\dot{\eta}\left(s_k^i, T, b, \beta_k^i, s_k^i|_0, \dot{T}\right). \tag{25g}$$

A more detailed discussion of the structure of the constitutive law and its position as a link between the proper process description and the description of the changes of the thermodynamical state is given by Lehmann (1978, 1980). Here we focus our considerations on the evolution law for strains. This shall be done in a general way in the next section followed by the discussion of a particular case with some applications.

5. Some general remarks on the evolution law for strain

In the first step we restrict our consideration to elastic-plastic bodies. For simplicity we disregard solid phase transformations. Annealing processes like recrystallization, recovery, or aging, however, may be included. The inelastic (or pseudo-elastic) deformations connected with solid phase transformations can be treated separately, though they influence, of course, the evolution laws for the remaining shares of strain. This influence, however, is similar to the effect, for instance due to other hardening and annealing phemonena. Therefore it is not necessary to include solid phase transformations into the following considerations (for more details see (Inoue and Raniecki, 1979; Lehmann, 1978)).

The total strain rate may be split according to Eq. (8). The evolution law for the elastic strain can be derived from the thermic state Eq. (18). This leads in the first step to:

$$\underset{(e)}{\varepsilon}{}^{i}_{k}\big|_0 = -\overset{\circ}{\rho}\left(\frac{\partial \psi}{\partial s_1^k}\right)\bigg|_0 = \underset{(e)}{\varepsilon}{}^{i}_{k}\big|_0\{s_k^i, T, b, \beta_k^i; s_k^i\big|_0, \dot{T}, \dot{b}, \beta_k^i\big|_0\}. \quad (26)$$

The derivations of the internal variables with respect to time can now be eliminated by means of the evolution laws for these quantities. Furthermore for small elastic deformations we can suppose

$$\underset{(e)}{\varepsilon}{}^{i}_{k}\big|_0 = \underset{(e)}{d}{}^{i}_{k} \quad (27)$$

even for noncoaxial elastic deformations and superimposed large inelastic deformations (Inoue and Raniecki, 1979; Lehmann, 1978). Thus we obtain

$$\underset{(e)}{d}{}^{i}_{k} = \underset{(e)}{d}{}^{i}_{k}\{s_k^i, T, b, \beta_k^i, s_k^i\big|_0, \dot{T}\}. \quad (28)$$

When we are using $q_{(e)k}^{i}$ instead of $\varepsilon_{(e)k}^{i}$ as state variables we are led to the same result.

In many cases we can approximate the elastic behavior by a simpler hypo-elastic law, e.g.,

$$\underset{(e)}{d}{}^{i}_{k} = \frac{1}{2G}t^i_k\big|_0 + \left\{\frac{1}{9K}s^r_r\big|_0 + \alpha\dot{T}\right\}\delta^i_k, \quad (29)$$

with $G=$ shear modulus, $K=$ bulk modulus, $\alpha=$ coefficient of thermal expansion, and $t_k^i = s_k^i - 1/3s_r^r \delta_k^i = $ (weighted) stress deviator.

The error is in most cases very small (Lehmann, 1973). It should be mentioned that neither relation (28) nor (29) is independent of the accompanying inealstic deformations since the total strain rate enters these formulas by the expression

$$s_k^i|_0 = (\dot{s})_{\cdot k}^i + d_r^i s_k^r - d_k^r s_r^i. \tag{30}$$

This coupling is by no means avoidable because the elastically unloaded configuration changes with inelastic deformations.

Concerning the inelastic (plastic) deformations we assume as usual that plastic deformations only occur if the state variables fulfill a yield condition

$$F(s_k^i, T, b, \beta_k^i)=0. \tag{31}$$

The quantities entering this yield condition, which is only a necessary condition, must be state variables in the sense of thermodynamics as indicated in (31). The continuation of plastic deformations requires

$$\dot{F} = \frac{\partial F}{\partial s_k^i}s_k^i|_0 + \frac{\partial F}{\partial T}\dot{T} + \frac{\partial F}{\partial b}\dot{b} + \frac{\partial F}{\partial \beta_k^i}\beta_k^i|_0 = 0. \tag{32}$$

Equations (31) and (32) together represent a sufficient condition for the occurrence of plastic deformations. In order to see this we have to express $\dot{b}$ and $\beta_k^i|_0$ by their respective evolution laws. If, for instance, the evolution laws for b and β_k^i take the form:

$$\dot{b}=\zeta_{(1)}(s_k^i, T, b, \beta_k^i)\frac{1}{\rho}s_k^i d_{(i)}^k - (b-\mathring{b})\vartheta_{(1)}(s_k^i, T, b, \beta_k^i),$$

$$\beta_k^i|_0 = \zeta_{(2)}(s_k^i, T, b, \beta_k^i) d_{(i)}^i - (\beta_k^i - \mathring{\beta}_k^i)\vartheta_{(2)}(s_k^i, T, b, \beta_k^i),$$

with

$$\zeta_{(i)}(s_k^i, T, b, \beta_k^i)>0;$$

$$\vartheta_{(i)}(s_k^i, T, b, \beta_k^i)\begin{cases} >0 & \text{for } T>T_R(s_k^i, b, \beta_k^i), \\ =0 & \text{for } T \leqslant T_R(s_k^i, b, \beta_k^i), \end{cases} \tag{33}$$

which represents the interaction between isotropic and anisotropic hardening on the one hand and annealing by recrystallization on the other hand

(Lehmann, 1980), we obtain:

$$0=\dot{F}=\frac{\partial F}{\partial s_k^i}s_k^i|_0 + \frac{\partial F}{\partial T}\dot{T} + \frac{\partial F}{\partial b}\left\{\frac{\mathring{\zeta}_{(1)}}{\mathring{\rho}}s_k^i\,d_{(i)}^k - (b-\mathring{b})\vartheta_{(1)}\right\}$$

$$+\frac{\partial F}{\partial \beta_k^i}\left\{\zeta_{(2)}\,d_{(i)}^i - \left(\beta_k^i - \mathring{\beta}_k^i\right)\vartheta_{(2)}\right\}. \tag{34}$$

Hence we conclude that inelastic deformations (i.e. $d_{(i)}^k \neq 0$) require

$$\frac{\partial F}{\partial s_k^i}s_k^i|_0 + \frac{\partial F}{\partial T}\dot{T} - \frac{\partial F}{\partial b}(b-\mathring{b})\vartheta_{(1)} - \frac{\partial F}{\partial \beta_k^i}\left(\beta_k^i - \mathring{\beta}_k^i\right)\vartheta_{(2)} > 0, \tag{35}$$

with

$$\vartheta_{(i)}\begin{cases} >0 & \text{for } T>T_R, \\ =0 & \text{for } T\leqslant T_R. \end{cases}$$

This is the so-called loading condition corresponding to the yield condition (31) and the evolution laws (33) for the internal variables. Analogously we can proceed in all other cases of yield conditions and evolution laws for the internal variables.

Mostly it is assumed that the plastic strain rate obeys the so-called normality rule,

$$d_{(i)}^i = \dot{\lambda}\frac{\partial F}{\partial s_i^k}, \tag{36}$$

taking the yield condition as the so-called plastic potential. This assumption is very often founded on the postulate of Ilyushin (1948) or the postulate of Drucker (1952). The postulate of Ilyushin is based on the consideration of a cycle in the total strain, which is not a thermodynamical state variable in general and the postulate of Drucker, which is based on the consideration of a cycle in stress, uses additional statements beyond the second law of thermodynamics. Therefore these postulates do not offer a general base for the formulation of the evolution law for the plastic strain. Particularly in general nonisothermic processes they loose their capability.

In fact, many experiments with nonproportional loading show slight deviations from the normality rule (see, for instance (Naghdi et al., 1955; Shiratori and Ikegami, 1969)). Furthermore, as already mentioned, incremental stress–strain relations based on the normality rule do not fit very well the experimental results in some particular problems as, for instance, in bifurcation problems (Pflüger, 1967; Rice, 1976; Rudnicki and Rice, 1975).

It seems that the agreement between experimental and theoretical results in such cases can be improved by an extended approach for the evolution law for the plastic strain based on the general approach (first introduced by

Hartung and Lehmann (1968), and Lehmann (1962),

$$d_k^i \underset{(i)}{=} \dot{\lambda} \frac{\partial F}{\partial s_i^k} + \kappa_{ks}^{ir} S_r^s |_0. \tag{37}$$

Such extended approaches are also used by other authors in connection with bifurcation problems (Pflüger, 1967; Rice, 1976; Rudnicki and Rice, 1975). These authors (see also (Sewell, 1974)) have based their approach on the assumption of the existence of vertices in the yield condition interpreted as a hypersurface in the stress space. At such vertices the normal to the yield surface becomes ambiguous, and this gives the possibility to introduce a second term in the evolution law for plastic strains which is, however, either undetermined or arbitrarily determined by the constitutive law.

This difficulty can be avoided by another physical interpretation of the second term in (37) which does not presuppose the existence of vertices in the yield condition. Plastic deformations can be considered thermodynamically as a sequence of equilibrium states. At constant temperature and constant values of the internal variables the stress increments can be interpreted as a disturbance of the actual equilibrium state. During loading processes the stress increments cause an increase of the internal energy. This increase may activate new slip processes (limited by hardening) which are described by the first term in (37). At the same time we can expect, however, that the response of the material, i.e. the plastic strain increment, contains a component which is oriented immediately in the direction of the disturbance, i.e. the stress increment. This component of the material response is described by the second term. κ_{ks}^{ir} becomes separately determined within the constitutive law. It may still depend on the whole set of state variables,

$$\kappa_{ks}^{ir} = \kappa_{ks}^{ir} (s_k^i, T, b, \beta_k^i). \tag{38}$$

In many cases it may be sufficient to simplify the approach (37) to

$$d_k^i \underset{(i)}{=} \dot{\lambda} \frac{\partial F}{\partial s_i^k} + \kappa t_k^i |_0, \tag{39}$$

with κ depending on the state variables or κ equal to constant. In this case, however, one difficulty arises. When we vary the direction of the stress increment in an actual stress state at the yield surface, we obtain a jump in the strain increment when the direction of the stress increment passes through the yield surface. This may cause some difficulties in the general investigation of the uniqueness of solutions. In most practical problems, however, this does not play an important role.

In contradiction to the definition of the quantity κ_{ks}^{ir} or κ, respectively, in the evolution law for plastic strain the quantity $\dot{\lambda}$ follows from the yield condition and the evolution laws for the internal variables. This may be demonstrated in a simple example in the next section. In this connection we shall also discuss how the second law can be fulfilled by our approach (see also (Lehmann 1979)).

The introduction of a second term into the evolution law for plastic strain opens further possibilities. In unloading we observe some slight deviations from the pure elastic behavior. These deviations start after a certain proportion of unloading as shown in Fig. 1. We suppose that the second term in the evolution law for plastic strain can also be used to describe this phenomena based on the following considerations. This second term, but not the first one, may be activated when in unloading the actual stress state differs sufficiently from the last state reached in loading. This leads us to the introduction of different yield conditions for the first and for the second term in the evolution law of elastic strain as sketched in Fig. 1. The separation of these two yield conditions may also be useful in the description of reloading processes. The concept of multiple yield conditions is not new (see (Eisenberg and Phillips, 1971; Mróz, 1969; Mróz and Lind, 1975; Phillips and Lee, 1979; Shiratori et al., 1979)). In the context, however, with an extended evolution law for plastic strain it becomes a clear physical meaning and seems to open new aspects.

The given approach for the evolution law for plastic strain can be easily extended to elastic-viscoplastic deformations in the following way (see also (Lehmann, 1977)). We suppose that the stress s_k^i can be decomposed into a plastic stress $\bar{s}_k^i$ and a viscous (over-) stress $\tilde{s}_k^i$ according to

$$s_k^i = \bar{s}_k^i + \tilde{s}_k^i = \bar{s}_k^i + \left(s_k^i - s_k^i\right). \tag{40}$$

Furthermore we assume that inelastic deformations only occur if the actual

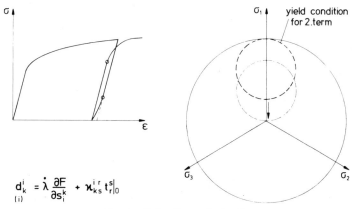

$$d_k^i \underset{(i)}{=} \dot{\lambda}\,\frac{\partial F}{\partial s_i^k} + \varkappa_{ks}^{i\,r}\,t_r^s\big|_0$$

Fig. 1.   Unloading and reloading.

stresses fulfill a yield condition,

$$F(s_k^i, T, b, \beta_k^i) = f^2(s_k^i, \beta_k^i) - k^2(b, T) > 0. \tag{41}$$

At the same time we demand during inelastic deformations,

$$\bar{F}(\bar{s}_k^i, T, b, \beta_k^i) = \bar{f}^2(\bar{s}_k^i, \beta_k^i) - k^2(b, T) = 0, \tag{42}$$

where $\bar{F}$ and $\bar{f}$ differ from F and f only by the different stresses entering these functions. Concerning the evolution law for the inelastic strain we assume,

$$\underset{(i)}{d_k^i} = \dot{\lambda} \frac{\partial \bar{F}}{\partial \bar{s}_i^k} + \kappa_{ks}^{ir} t_r^s|_0, \tag{43a}$$

$$\underbrace{\qquad}_{\substack{d_k^i \\ (i_1)}} \underbrace{\qquad}_{\substack{d_k^i \\ (i_2)}}$$

$$\underset{(i_1)}{d_k^i} = \dot{\varepsilon}_0 \Phi\left(\frac{f^2}{k^2} - 1\right) \frac{1}{\sqrt{f^2}} \frac{\partial F}{\partial s_i^k}, \tag{43b}$$

where Φ denotes a suitable function of the given argument and $\dot{\varepsilon}_0$ means a suitable reference deformation rate. Using (43b) the plastic stress $\bar{s}_k^i$ can be eliminated from the constitutive law. Thus finally the total strain rate results in the form,

$$d_k^i = \underset{(e)}{d_k^i} + \underset{(i_1)}{d_k^i} + \underset{(i_2)}{d_k^i} = d_k^i(s_k^i, T, b, \beta_k^i, s_k^i|_0, \dot{T}), \tag{44}$$

as in the case of elastic-plastic deformations. Therefore we can extend our general remarks stated before also to elastic-viscoplastic materials (Lehmann, 1977 and 1980).

6. Particular isothermic elastic-plastic processes

We consider isothermic deformations of an elastic-plastic body exhibiting a combination of isotropic and kinematic hardening. The specific free enthalpy may be given by

$$\psi = \psi(s_k^i, T, b, B) \qquad \text{with} \qquad B = \beta_k^i \beta_i^k. \tag{45}$$

The corresponding yield condition may read (for $T = \text{const.}$)

$$F(s_k^i, b, \beta_k^i) = (t_k^i - c\beta_k^i)(t_i^k - c\beta_i^k) - k^2(b) = 0$$

$$\text{with} \qquad c = \text{const.} \tag{46}$$

Concerning the evolution laws for the internal variables we assume (see Eq. (33)),

$$\dot{b} = \frac{\zeta^*}{\overset{\circ}{\rho}\frac{\partial}{\partial b}\left\{\psi - T\frac{\partial\psi}{\partial T}\right\}} s^i_k \, d^k_{(i)i} = \frac{\zeta}{\overset{\circ}{\rho}} t^i_k \, d^k_{(i)i}, \tag{47a}$$

$$\beta^i_k|_0 = \frac{(1-\zeta^*)c}{\overset{\circ}{\rho}2\frac{\partial}{\partial B}\left\{\psi - T\frac{\partial\psi}{\partial T}\right\}} d^i_{(i)k} = \xi \, d^i_{(i)k}. \tag{47b}$$

If ζ is constant or only a function of b then (47a) can be integrated independently of the particular process and we obtain:

$$b = b(w) \qquad \text{or} \qquad w = w(b). \tag{48a}$$

Furthermore when ξ is constant and when additionally the loading takes place proportionally $(t^i_k|_0 \sim t^i_k)$ the integration of (47b) yields:

$$\beta^i_k = \xi \, \varepsilon^i_{(i)k} \qquad \text{or} \qquad \varepsilon^i_{(i)k} = \frac{1}{\xi}\beta^i_k. \tag{48b}$$

This means that under the formulated conditions the plastic work $w_{(i)}$ and the plastic deformations $\varepsilon_{(i)k}^{\;i}$ can be used as state variables entering also into the yield condition. In general, however, the plastic work $w_{(i)}$ and the plastic deformations $\varepsilon_{(i)k}^{\;i}$ are not usable as hardening parameters not even in isothermic processes since ζ and ξ may depend on the whole set of state variables.

From (46) we derive by means of (47a, b),

$$\dot{F} = 0 = 2(t^i_k - c\beta^i_k)(t^k_i|_0 - c\beta^k_i|_0) - \frac{dk^2}{db}\dot{b}$$

$$= 2(t^i_k - c\beta^i_k)\left(t^k_i|_0 - c\xi \, d^k_{(i)i}\right) - \frac{dk^2}{db}\frac{\zeta}{\overset{\circ}{\rho}} t^i_k \, d^k_{(i)i}. \tag{49}$$

From hence the loading condition results in:

$$(t^i_k - c\beta^i_k)t^k_i|_0 > 0. \tag{50}$$

Concerning the evolution law for the inelastic strain we assume (according to Eq. (39))

$$d^i_{(i)k} = \dot{\lambda}\frac{\partial F}{\partial s^k_i} + \kappa t^i_k|_0. \tag{51}$$

Inserting this expression into Eq. (42) we can calculate $\dot\lambda$ and obtain finally

$$
\underset{(i)}{d^i_k} = \frac{\left\{ 2(t^r_s - c\beta^r_s) - \kappa\left[\frac{\zeta}{\overset{\circ}{\rho}} \frac{dk^2}{db} t^r_s - 2\xi c(t^r_s - c\beta^r_s) \right] \right\} t^s_r|_0}{\frac{\zeta}{\overset{\circ}{\rho}} \frac{dk^2}{db}(t^m_n - c\beta^m_n)t^n_m - 2\xi ck^2}(t^i_k - c\beta^i_k)
$$

$$
+ \kappa t^i_k|_0. \tag{52}
$$

From Eq. (52) we see that the terms containing κ cancel each other out in proportional loading. Therefore these terms only appear in nonproportional loading like, for instance, in shear processes where the principle axes of the stresses rotate against the material elements. This even holds in pure shear processes. Concerning the evolution law for the elastic strain we may adopt the hypo-elastic approach (Eq. (29)).

The constitutive law of the considered material has to be completed by the definition of the entropy production. One possible compatible approach is

$$
\underset{(d)}{w} = \frac{1-\zeta}{\overset{\circ}{\rho}}(t^i_k - c\beta^i_k)\underset{(i)}{d^k_i} \geq 0, \tag{53a}
$$

$$
T\dot\eta = 0. \tag{53b}
$$

From hence concerning $\dot w_{(h)}$ results

$$
\underset{(h)}{\dot w} = \underset{(i)}{\dot w} - \underset{(d)}{\dot w} = \frac{\zeta}{\overset{\circ}{\rho}} t^i_k \underset{(i)}{d^k_i} + \frac{1-\zeta}{\overset{\circ}{\rho}} c\beta^i_k \underset{(i)}{d^k_i} \leq 0. \tag{54}
$$

It should be mentioned, however, that the distribution of the inelastic work according to (53a) and (54) is not the only possible approach fulfilling the requirements of the second law of thermodynamics. An open question in particular is how far the inelastic work due to the second term in (51) can be considered as reversible.

In the following applications of the foregoing approaches the weighted stresses s^i_k are replaced by the stresses σ^i_k. The error of this approximation is very small (Lehmann, 1973).

The first example concerns the Poynting-effect in pure shear. The experimental results taken from Hecker (1967, p. 106) for brass are compared with an theoretical approach as indicated in Fig. 2. It can be seen that an approach with $\kappa=0$ leads to an unsatisfactory description of the Poynting-effect (see also (Thermann, 1969)).

The second example concerns a thin-walled steel tube in the first loading step twisted (twist angle ψ; length of the specimen 200 mm, outer diameter 26 mm) and in the second step extended with ψ held constant (Blix, 1980).

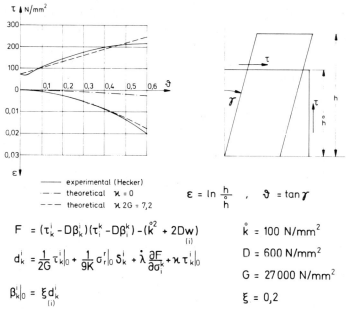

$$\varepsilon = \ln \frac{h}{\overset{\circ}{h}} \quad , \quad \vartheta = \tan \gamma$$

$$F = (\tau_k^i - D\beta_k^i)(\tau_i^k - D\beta_i^k) - (\overset{\circ}{k}{}^2 + 2Dw)_{(i)}$$

$$\overset{\circ}{k} = 100 \text{ N/mm}^2$$

$$d_k^i = \frac{1}{2G}\tau_k^i\big|_0 + \frac{1}{9K}\sigma_r^r\big|_0 \delta_k^i + \dot\lambda \frac{\partial F}{\partial \sigma_i^k} + \varkappa\tau_k^i\big|_0$$

$$D = 600 \text{ N/mm}^2$$

$$G = 27\,000 \text{ N/mm}^2$$

$$\beta_k^i\big|_0 = \xi d_k^i{}_{(i)}$$

$$\xi = 0,2$$

Fig. 2.   Poynting-effect in pure shear (brass).

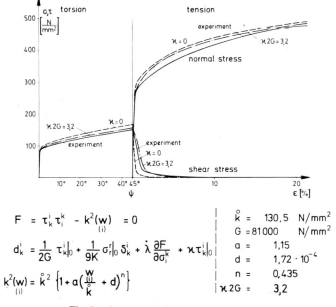

$$F = \tau_k^i \tau_i^k - k^2(w)_{(i)} = 0$$

$$d_k^i = \frac{1}{2G}\tau_k^i\big|_0 + \frac{1}{9K}\sigma_r^r\big|_0 \delta_k^i + \dot\lambda \frac{\partial F}{\partial \sigma_i^k} + \varkappa\tau_k^i\big|_0$$

$$k^2(w) = \overset{\circ}{k}{}^2 \left\{1 + a\left(\frac{w}{\overset{\circ}{k}} + d\right)^n\right\}_{(i)}$$

$$\overset{\circ}{k} = 130,5 \text{ N/mm}^2$$
$$G = 81000 \text{ N/mm}^2$$
$$a = 1,15$$
$$d = 1,72 \cdot 10^{-4}$$
$$n = 0,435$$
$$\varkappa 2G = 3,2$$

Fig. 3.   2-step strain history (steel Ck 15).

The experimental and theoretical results are shown in Fig. 3. We find again that an approach with $\kappa \neq 0$ leads to a better agreement between experimental and theoretical results.

7. Final remarks

The two examples treated in Section 6 represent only preliminary results which have to be improved in accuracy and have to be extended to more complex loading paths histories, to bifurcation problems, and to other problems with abrupt changes of the loading as in ductile fracture. One can expect, however, that an approach for the evolution law for plastic strain according to Eq. (37) or (39), respectively, will prove a good capability for problems with complex loading histories or problems with rotating principal axes of stress.

References

Blix, U. (1980), "Vergleich verschiedener Formänderungsgesetze der Plastizitätstheorie," Thesis, Ruhr-University, Bochum.

Drucker, D. C. (1952), "A More Fundamental Approach to Plastic Stress-Strain Relations," *Proc. 1st U.S. Nat. Congr. Appl. Mech.*, 487.

Eisenberg, M. A. and Phillips, A. (1971), "A Theory of Plasticity with Noncoincident Yield and Loading Surfaces," *Acta Mech.*, **11**, 247.

Hartung, Ch. and Lehmann, Th. (1968), "Vergleich einiger Formänderungsgesetze für Plastische Formänderungen," *ZAMM*, **48**, 138.

Hecker, F. W. (1967), "Die Wirkung des Bauschinger-Effektes bei großen Torsions-Formänderungen," Dissertation, TU Hannover.

Hohenemser, K. (1931), "Fließversuche an Rohren aus Stahl bei kombinierter Zug- und Torsionsbeanspruchung," *ZAMM*, **11**, 15.

Hohenemser, K. and Prager, W. (1932), "Beitrag zur Mechanik des bildsamen Verhaltens von Fließstahl," *ZAMM*, **12**, 1.

Ilyushin, A. A. (1948), "Foundations of the General Mathematical Theory of Plasticity," *Plasticity*, Moscow (orig. in Russian).

Inoue, R. and Raniecki, B. (1979), "Determination of Thermal-hardening Stress in Steels by use of Thermoplasticity," *J. Mech. Phys. Solids*, **26**, 187.

Lehmann, Th. (1962), "Zur Beschreibung großer plastischer Formänderungen unter Berücksichtigung der Werkstoffverfestigung," *Rheologica Acta*, **2**, 248.

Lehmann, Th. (1966), "Formänderungen eines Klassischen Kontinuums in vierdimensionaler Darstellung," *XI Int. Congr. Appl. Mech.*, München 1964, Springer-Verlag, Berlin.

Lehmann, Th. (1973), "On Large Elastic-plastic Deformations," Int. Symposium, Warsaw 1972, *Foundations of Plasticity*, A. Sawczuk (ed.), Noordhoff Int. Publ., Leyden, 571.

Lehmann, Th. (1977), "On the Theory of Large, Non-isothermic, Elastic-plastic and Elastic-viscoplastic Deformations," *Arch. Mech.*, **29**, 393.

Lehmann, Th. (1978), "Some Aspects of Non-isothermic Large Inelastic Deformations," *SM Arch.*, **3**, 261.

Lehmann, Th. (1980), "Some Aspects of Coupling Effects in Thermoplasticity," *Proc. Symp. on Large Deformations*, New Delhi, 1979, in press.

Lehmann, Th. (1980), "Coupling Phenomena in Thermoplasticity," *Nucl. Eng. Des.*, **57**, 323.

Mróz, Z. (1969), "An Attempt to Describe the Behavior of Metals Under Cyclic Loads Using a More General Work-hardening Model," *Acta Mech.*, **7**, 199.

Mróz, Z. and Lind, N. C. (1975), "Simplified Theories of Cyclic Plasticity," *Acta Mech.*, **22**, 131.

Müller, J. (1973), *Thermodynamik*, Bertelsmann Universitätsverlag, Düsseldorf.

Naghdi, P. M., Rowly, J. C. and Beadle, C. W. (1955), "Experiments Concerning the Yield Surface and the Assumption of Linearity in the Plastic Stress-strain Relations," *J. Appl. Mech.*, **22**, 416.

Pflüger, A. (1967), "Zur plastischen Beulung von Flächenträgern," *ZAMM*, **47**, 210.

Phillips, A. and Lee, Chong-Won (1979), "Yield Surfaces and Loading Surfaces. Experiments and Recommendations," *Int. J. Solids Struct.*, **15**, 715.

Rice, J. R. (1976), "The Localization of Plastic Deformation," *Proc. 14th IUTAM Congr.*, *Delft*, North-Holland Publ., Amsterdam, 207.

Rudnicki, J. W. and Rice, J. R. (1975), "Conditions for the Localization of Deformation in Pressure-sensitive Dilatant Materials," *J. Mech. Phys. Solids*, **23**, 371.

Sewell, M. J. (1974), "A Plastic Flow Rule at a Yield Vertex," *J. Mech. Phys. Solids*, **22**, 469.

Shiratori, E. and Ikegami, K. (1969), "Studies of the Anisotropic Yield Condition," *J. Mech. Phys. Solids*, **17**, 473.

Shiratori, E., Ikegami, K. and Yoshida, F. (1979), "Analysis of Stress-strain Relations by Use of an Anisotropic Hardening Potential," *J. Mech. Phys. Solids*, **27**, 213.

Thermann, K. (1969), "Zur elasto-plastischen Torsion kreiszylindrischer Körper bei endlichen Verzerrungen," Dissertation, TU Hannover.

S. Nemat-Nasser, Editor
THREE-DIMENSIONAL CONSTITUTIVE RELATIONS AND DUCTILE FRACTURE
North-Holland Publishing Company (1981) 307–310

DISCUSSION ON SESSION 5

Bertil STORÅKERS

Royal Institute of Technology, Stockholm, Sweden

Remarks on *B. Marandet and G. Sanz: A Quantitative Description of Fracture Toughness under Plane Stress Conditions by the R-Curve Method*

This contribution discusses with admirable clarity the *R*-curve approach to predict stable crack growth and instability of cracked members and in particular its relevance with respect to different geometric parameters. The conclusions drawn are based on earlier findings and on current experimental work, which seems to have been carried out with great care and skill.

The specimens utilized in the experimental study are of compact type, CLWL, for which the influence of initial crack length on the *R*-curve was studied in a considerable range. Subsequently instability predictions were made within very good (5%) accuracy. It was also found that in the preinstability range, the *R*-curve was independent of the mode of loading (controlled force or displacement).

The theoretical and experimental results achieved are consistent and strengthen the confidence in the *R*-curve as a useful tool in fracture mechanics. It seems valuable, however, to have a further discussion of its relevance when applied to other geometries and higher load levels. The method to evaluate the tests is based on linear fracture mechanics. It would be interesting to learn of the results of a similar evaluation based on the *J*-integral in particular as the investigated steel seems to possess a certain amount of ductility.

In the final part of the paper a successful prediction of crack growth in a pressurized membrane is described.

Remarks on *J. Christoffersen: Microlocalizations*

The growth of voids and its consequences as regards plastic flow has recently received much attention with the view of predicting significant features of the fracture process of ductile materials. Void growth has earlier been considered as a plausible second-order effect to accelerate localized necking at stretching of sheet metal. In this fashionable field a severe unbalance prevails between theory and experiment. The number of effects

307

proposed does not seem to deviate very much from the considerable number of theorists actively engaged. Still there seems to be no report so far of an actual experiment in which a metal sheet has been homogeneously and biaxially stretched under dynamically controlled conditions. Almost similar circumstances seem predominant in the discipline under discussion. Also Dr. Christoffersen's considerations of the mechanical properties of voided materials are restricted to pure theory and albeit for some cursory observations, the behavior of real materials passes unnoticed.

Dr. Christoffersen is concerned with near-fracture material behavior. The underlying physical model is based on rigid grains separated by plane deforming layers of finite thickness and containing voids. The assumed locally homogeneous deformation of the grain boundary layers is severely constrained being composed of transverse stretching (with no lateral flow admitted) combined with simple shear. The voids announce their presence by one scalar measure f (void volume ratio) and consequently their individual size and geometry are considered immaterial together with their distribution, interference and possible coalescence. No void nucleation mechanism is proposed and the (initially) present voids will grow purely due to the separation of neighboring grains. At predicted final fracture $f = 1$(Eq. [71]), which implies that voids of finite mean diameter will have evolved into (infinitely) long cracks *perpendicular* to the grain boundary. Such a finding would surprise any practical metallurgist and this feature does perhaps raise some doubts regarding the ability of the model to simulate the ductile fracture behavior of actual polycrystalline aggregates.

With admirable care kinematics and statics are dealt with on layer level. The connection with macro-variables is provided via a virtual work principle. A simple quadratic yield condition is proposed and it is shown that yield surface convexity and, if an apparently reasonable conjecture is accepted, smoothness on microlevel is reflected on macrolevel.

Once the macrobehavior of the material is established one would perhaps expect one or two applications of the theory dealing with common grain structures and generating results which could be tested against the outcome of, say, a few tensile tests with associated microscopic examinations. This desire, however, is not fulfilled. Instead it is shown that, when one constitutive variable (k) is suitably adjusted, the assumed layer yield condition is compatible with one proposed by Gurson for a continuum randomly contaminated by spherical voids. Whatever the shapes of the present model voids are, however, they will be spherical only temporarily due to the assumed kinematical constraints. It is further shown that for a second choice of k, the present model predicts isotropic macroscopic behavior.

In conclusion Dr. Christoffersen has presented a suggestive and seasonable contribution to a domain in which more knowledge is urgently needed.

Whether the proposed mechanisms are dominant for some particular material is hard to judge. So far a recipe, stimulating the appetite, has been provided and further appraisal will have to wait until the pudding is served.

Remarks on *Th. Lehmann: On Constitutive Relations in Thermoplasticity*

The topic chosen by Professor Lehmann is of great interest in particular regarding events such as bifurcation of inelastic deformation.

In the preliminaries Professor Lehmann discusses large strain kinematics and in particular the notorious question regarding a proper decomposition of the total deformation into its elastic and plastic parts, this matter being essential in case of anisotropy. There seems to be no convincing argument in favor of the present proposal, i.e. as regards the existence of variation principles for the solution of incremental boundary value problems, and the reviewer is left with the feeling that this option essentially remains a matter of taste. This feature is also reflected in the present decomposition of specific work rate into its elastic and inelastic parts.

After balance equations are laid down, within a thermo-mechanical framework, laws of evolution are proposed for internal state variables including yield and consistency conditions. The normality rule is abandoned and in Eq. (37) a nonassociated plastic strain-rate component is proposed. Professor Lehmann does not seem to appreciate the physical background of vertex theories (local Schmidt slip in polycrystalline aggregates) or that they may be unambiguous, cf., e.g. (Sewell, 1974), but proceeds to provide a "physical interpretation" of nonnormality in case of smooth yield functions. This is a main point in the paper and the motivation is given verbally as "we can *expect* (the italics are ours), however, that the response of the material, i.e. the plastic strain increment contains a component which is oriented immediately in the direction of the disturbance, i.e. the stress increment." Thus no appeal is made to the elaborate thermodynamic framework laid down and no further arguments are involved. The proposal remains a conjecture as regards the behavior of real materials.

A specific form of the constitutive equation is then proposed where a classical combination of isotropic and kinematic hardening is supplemented with the novel nonassociated flow rule. Two experimental investigations are singled out for comparison. The first illustration deals with the Poynting effect in shear and the second with extension followed by twist of a tube.

It is a long nourished wish by the reviewer that earnest writers, such as Professor Lehmann, laying down constitutive equations based on internal (nonmeasurable) state variables and having practical purposes in mind, would simultaneously propose fundamental experiments enabling the number and rank of the variables to be established in order to simulate characteristic features of particular materials. To the reviewer's knowledge there exist but a few efforts (Onat and Fardshisheh, 1972; Fardshisheh and Onat, 1974) to solve this nontrivial problem.

The present degenerate material model utilizes one second rank tensor and two scalars as state variables. In the first example, dealing with the Poynting effect, the experimental results exhibit a parabolic relation between extension and twist. This is a well-known feature from finite isotropic elasticity and thus one might expect any appropriate deformation theory of plasticity to explain the phenomenon as well. The second application mainly stays in a two-dimensional normal stress–shear stress space with a (small) excursion due to rotation of principal material directions. In such a subspace, however, state variables do not reveal their tensorial nature (Fardshisheh, 1973; Storåkers, 1979). Then the good agreement between theoretical and experimental results might be fortuitous and further applications, in which constitutive parameters are not evaluated from results which are to be predicted, are desirable. This might show how well Professor Lehmann's theory will stand up in comparison with competing theories, such as for instance Tvergaard's (1978) finite strain version of kinematic hardening, when confronted with the characteristics of particular materials.

References

Sewell, M. J. (1974), "A Plastic Flow Rule at a Yield Vertex," *J. Mech. Phys. Solids*, **22**, 469.
Onat, E. T. and Fardshisheh, F. (1972), "Representation of Creep of Metals," Oak Ridge National Laboratories, Report 4783.
Fardshisheh, F. and Onat, E. T. (1974), "Representation of Elastoplastic Behavior by Means of State Variables," *Foundations of Plasticity*, **2**, A. Sawczuk (ed.), Noordhoff Int. Publ., Leyden, 89.
Fardshisheh, F. (1973), Dissertation, Yale University.
Storåkers, B. (1979), "Yield Surface Characteristics Arising from Orthorhombic Symmetry," *J. Appl. Mech.*, **46**, 961.
Tvergaard, V. (1978), "Effect of Kinematic Hardening on Localized Necking in Biaxially Stretched Sheets," *Int. J. Mech. Sci.*, **20**, 651.

S. Nemat-Nasser, Editor
THREE-DIMENSIONAL CONSTITUTIVE RELATIONS AND DUCTILE FRACTURE
North-Holland Publishing Company (1981) 311

REPLY BY Th. LEHMANN

One basic requirement concerning the decomposition of the total deformation into its elastic and inelastic parts is that a vanishing elastic strain rate should result in a constant ratio ds/ds^* also in the case of non-vanishing inelastic strain rate; ds and ds^* denote the length of a line element in the deformed state and in the intermediate, elastically unloaded state, respectively. This requirement is, for instance, fulfilled by the presented proposal and by Mandel's (1971) approach, but not by an additive decomposition of the changes of the metric or of the deformation gradient.

The tensor κ_{ks}^{ir} in the general approach (37) can be specified in many different ways. The theoretical approaches by Sewell (1974) or by Christoffersen and Hutchinson (1979) can be considered as special cases of (37). The only difference is that the presented approach does not presuppose the existence of vertices in the yield condition. It should be emphasized once more that this approach does not contradict the thermodynamic requirements.

As mentioned, the presented comparison of theoretical and experimental results has only to be considered as a first preliminary step. Many further experimental studies improved in accuracy and extended to more complex loading histories have to be achieved in order to test the capability of the presented theoretical approach.

Additional references

Christoffersen, J. and Hutchinson, J. W. (1979), "A class of phenomenological corner theories of plasticity," *J. Mech. Phys. Solids*, **27**, 465.
Mandel, J. (1971), "Sur la décomposition d'une transformation élastoplastique", *Compt. Rendus Acad. Sci. Paris, série A*. **272**, 276.

SESSION 6

Chairman: P. Germain
                      Ecole Polytechnique, Palaiseau, France

Authors: Nguyen Q. S.
                      G. Rousselier

Discussion: H. D. Bui

S. Nemat-Nasser, Editor
THREE-DIMENSIONAL CONSTITUTIVE RELATIONS AND DUCTILE FRACTURE
North-Holland Publishing Company (1981) 315–330

A THERMODYNAMIC DESCRIPTION OF THE RUNNING CRACK PROBLEM

Quoc Son NGUYEN

Ecole Polytechnique, Palaiseau, France

The problem of plane crack extension in *dissipative continua* is studied within the framework of classical thermodynamics. By a global dissipation analysis the expression of the intrinsic dissipation is given and the crack tip force, G, associated with the crack extension velocity, is introduced. If $G \neq 2\gamma_0$ it is shown that the crack tip behaves like a moving heat source and the temperature is singular as $-\log r$. The obtained results are new and give a general framework for the study of coupled thermomechanical behavior near the crack tip. Usual models of elastic, viscoelastic and elastic-plastic materials are considered. Classical notions, such as the Rice–Eshelby integral, the energy flux of Freund, and the energy release rate, are extended for arbitrary continua.

1. Introduction

In fracture analysis, from a macroscopic point of view, the basic problem associated with the formulation of a criterion for crack extension is the characterization of the mechanical field near the crack tip, i.e. the local state of stress and strain.

This problem has been successfully studied in the context of the linear elasticity theory. Substantial results have been obtained, and such notions as the stress intensity factor, path independent integrals and conservation laws, and the energy release rate, have been introduced (Bui, 1977; Rice, 1968). The applicability of the linear fracture theory to brittle fracture of usual materials (metals, ceramics, etc.) has been extensively discussed; in particular, many design criteria are proposed, using this kind of analysis under static or dynamic response conditions.

However, the extension of the theory to the case of dissipative materials is not straightforward. Several authors, e.g. Cherepanov (1967), have suggested a systematic extension of the Griffith description to viscoelastic or plastic materials; this extension has been given in relation with thermodynamics. Although advances have been made, many open problems remain, especially in incremental theory of plasticity (Nguyen, 1980; Rice, 1968). For materials of this kind, the determination of the local state of stress and strain is a very difficult problem, because of the material nonlinearity. The stress and strain singularity near the crack tip has been only recently discussed,

(Slepyan, 1974; Rice *et al.*, 1979). The definition of a relevant fracture parameter in the description of the crack extension for ductile fracture is still an open problem.

The objective of this paper is to present a thermodynamic description of the running crack problem in *arbitrary dissipative continua*, (Nguyen, 1979, 1980; Bui *et al.*, 1979). This description illustrates the development of energy methods in the context of fracture. The proposed theory is different from Rice's (1978) or Gurtin's (1979) description devoted to the Griffith model of brittle fracture. It is shown here that in plasticity and in viscoelasticity, as in elasticity, the crack tip force can be derived in a consistent manner. The basic equations for the coupled thermomechanical behavior are established.

The principal idea is that crack extension is an irreversible thermodynamic process in a similar way as friction or viscosity or plasticity. The fracture process must be associated with dissipative mechanisms because crack extension is an energy consuming process. For example, we may consider the notion of surface energy, γ, in the classical theory of Griffith as a dissipation term and neglect the possible *reversible* surface energy.

For this, we present in the first section the thermodynamic analysis of a two dimensional problem of running line crack; the material is not necessarily elastic. The transformation is dynamic; assumed to be infinitesimal. The crack tip force, G, associated with the crack extension velocity, is introduced. The expression of the intrinsic dissipation is given. In particular, it is shown that if $G \neq 2\gamma_0$, the crack tip behaves like a moving heat source and the temperature is singular as $-\log r$.

The general theory is then illustrated by usual models of elastic, viscoelastic, and elastic-plastic materials. Significant results given in the literature are reviewed in the context of the present description. New results, such as the coupled thermomechanical singularity in thermoelasticity, are presented.

The estimate of the crack tip force is discussed in detail. Attention is focused on some paradoxes pointed out by Rice (1968) who gives the estimate $G=0$ in the elastic perfectly plastic case. Although our definition of G seems to be different, the estimate $G=0$ is obtained from the stress and strain singularity analysis (Slepyan, 1974), not only for the elastic perfectly-plastic model, but also for certain models in viscoelasticity, when the reversible surface energy is absent. The estimate of G in the presence of reversible surface energy is a rather complex problem; we give here a short discussion based upon a mechanical modeling of the surface energy.

In the last section, the crack tip force, G, is interpreted as an energy release rate in quasi-static transformation. It is shown that G is the partial derivative of a certain energy functional, derived from the free energy of the whole system.

2. Dissipation and force analysis

Let us consider the response of a simple body with a line crack as shown in Fig. 1.

The material, not necessarily elastic, is defined by a free energy density per unit mass, W, a function of the strain tensor, ε, the internal parameters, α, and the temperature, T. If S denotes the entropy density, we have,

$$S = -\frac{\partial W}{\partial T}(\varepsilon, \alpha, T).$$

The internal energy density is $e(\varepsilon, \alpha, S) = W(\varepsilon, \alpha, T) + TS$. For an arbitrary system of material points, occupying a volume V at time t, the two principles of thermodynamics give the following relations:

$$\dot{U}_V + \dot{C}_V = P_e + P_{cal}, \tag{1}$$

$$\frac{d}{dt}\int_V \rho S \, d\Omega + \int_{\partial V} \frac{qn}{T} \, ds \geqslant 0, \tag{2}$$

where P_e, P_{cal}, U, C, q denote, respectively, the external power, the caloric power, the internal energy, the kinematic energy, and the heat flux, where

$$P_e = \int_{\partial V} n\sigma \dot{u} \, ds, \qquad P_{cal} = -\int_{\partial V} qn \, ds. \tag{3}$$

It is well known that the second principle inequality leads to the definition of the intrinsic dissipation and the thermal dissipation. These dissipations are assumed to be separately positive. For regular responses, the intrinsic dissipation is,

$$\mathcal{D}_V = \int_V \rho T \dot{S} \, d\Omega - P_{cal} \geqslant 0, \tag{4}$$

with, (Germain, 1973; Halphen and Nguyen, 1975)

$$\mathcal{D}_V = \int_V D \, d\Omega, \qquad D = \sigma\dot{\varepsilon} - \rho(\dot{e} - T\dot{S}) = \sigma\dot{\varepsilon} - \rho(\dot{W} + S\dot{T}), \tag{5}$$

where D is the volumic dissipation.

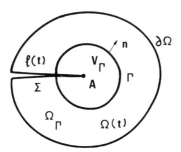

Fig. 1. Two-dimensional problem.

There is some difficulty when we take for V the whole body $\Omega(t)$. Indeed, the crack tip is a particular point where all thermomechanical fields may be singular. We introduce the following assumption:

Assumption H1. *The functions D, $T\dot{S}$, and $S\dot{T}$ are Lebesgue-integrable in $\Omega(t)$ for almost all t.*

In this manner we rule out the presence of shock waves, where both T and S may have finite discontinuities and $T\dot{S}$ and $S\dot{T}$ are not integrable functions. The integrability of D is a physical assumption since D corresponds to various volumic dissipations.

Assumption H1 enables us to write for the whole system, the second principle in the following form:

$$\mathcal{D}=\int_{\Omega}\rho T\dot{S}\,\mathrm{d}\Omega-P_{\mathrm{cal}}, \qquad \mathcal{D}\geqslant 0, \tag{6}$$

where

$$P_{\mathrm{cal}}=-\int_{\partial\Omega}qn\,\mathrm{d}s-\int_{\Sigma}qn\,\mathrm{d}s. \tag{7}$$

The internal energy, U, is

$$U=E+R=\int_{\Omega}\rho e\,\mathrm{d}\Omega+R, \tag{8}$$

where R denotes possible reversible surface energy. Although reversible surface energy is not necessarily a mechanical energy, we consider here the simple case of surface energy due to attractive surface forces with potential $\varphi([u])$, where $[u]$ is the opening displacement of the crack surface. In this case, the mechanism of exchange of energy is very clear and may serve as a model in order to discuss the general situation, when one wishes to study reversible surface energy. Then we have:

$$R=\int_{\Sigma}\varphi([u(s)])\,\mathrm{d}s. \tag{9}$$

For adhesion problem for example, it is interesting to consider the case of Fig. 2 where attractive forces $\varphi'([u])$ operate when $[u]\leqslant d$. In the limiting situation, $d\to 0$, and we simply obtain $R=2\gamma_0 l(t)$. Then

$$U=\int_{\Omega}\rho e\,\mathrm{d}\Omega+2\gamma_0 l. \tag{10}$$

Naturally, we can obtain Eq. (10) on the basis of other physical descriptions than the surface attractive forces. This proves that Eq. (10) is ambiguous, requiring in each situation a detailed analysis of the corresponding physical phenomenon. Here, we shall interpret it as the limiting case, $d\to 0$.

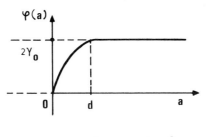

general case , d > 0

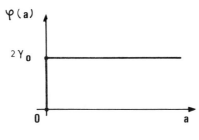

limit case , d → 0

Fig. 2. Reversible surface energy: $\varphi(a)=2\gamma_0$, if $a \geq d$, $\varphi(o)=0$ (top); $\varphi(a)=2\gamma_0, \forall a>0$, $\varphi(o)=0$ (bottom). ,

The first principle gives, for the whole system;

$$\dot{E}+\dot{C}+2\gamma_0\dot{l}=P_e+P_{cal}. \tag{11}$$

As $\rho\dot{e}$ and $\rho\ddot{u}\,\dot{u}$ are not necessarily integrable functions, the computation of $\dot{E}+\dot{C}$ is not straightforward. We introduce the following basic assumption:

Assumption H2 (Transport Condition of the Singularity). *The transport condition of the singularity for a physical quantity g is,*

$$\dot{g}=-\dot{l}g_{,1}+(more\ regular\ terms,\ integrable), \tag{12}$$

i.e. the dominant term of $\dot{g}$ is exactly $-\dot{l}g_{,1}$ for almost all t.

This assumption, identically verified for steady motions, means that in the general case, the singularity of the quantity g must have the same nature. Observe also that $\dot{g}+\dot{l}g_{,1}$ is the derivative of g with respect to time in the reference moving with the crack tip. For example, this assumption is verified for the energy density, W, if for $\dot{l}=0, \dot{W}$ is integrable; for $\dot{l}\neq0, \dot{W}+\dot{l}W_{,1}$ is integrable, although $\dot{W}$ is not.

Let us isolate the singularities by a closed curve Γ, delimiting a volume $V_\Gamma(t)$ in translation with the crack tip. We set,

$$E_\Gamma = \int_{\Omega(t)-V_\Gamma(t)} \rho e \, d\Omega \quad \text{and} \quad C_\Gamma = \int_{\Omega(t)-V_\Gamma(t)} \rho \dot{u}^2/2 \, d\Omega.$$

We obtain with Assumption H2, for $g = \rho(e + \dot{u}^2/2)$,

$$\lim_{\Gamma \to 0} (\dot{E}_\Gamma + \dot{C}_\Gamma) = \dot{E} + \dot{C}. \tag{13}$$

Indeed, we have,

$$\frac{d}{dt}[(E+C)-(E_\Gamma+C_\Gamma)] = \frac{d}{dt}\int_{V_\Gamma(t_0)} \rho(e + \dot{u}^2/2)$$

$$\times [x + l(t) - l(t_0), y, t] \, d\Omega,$$

$$\lim_{\Gamma \to 0}[(\dot{E}+\dot{C})-(\dot{E}_\Gamma+\dot{C}_\Gamma)] = \lim_{\Gamma \to 0}\int_{V_\Gamma(t_0)} \{\rho(e + \dot{u}^2/2)_{,t}$$

$$+ \dot{l}\rho(e + \dot{u}^2/2)_{,1}\} \, d\Omega = 0,$$

because the integrant is an integrable function.

Relation (13) may be written as,

$$\dot{E} + \dot{C} = \lim_{\Gamma \to 0}\left(\int_{\Omega-V_\Gamma} \rho(\dot{e} + \dot{u}\ddot{u}) \, d\Gamma + \int_\Gamma \rho(e + \dot{u}^2/2)\dot{l}n_1 \, ds \right). \tag{14}$$

But the two principles give, for the system of material points occupying the volume $\Omega(t) - V_\Gamma(t)$ at time $t' = t$,

$$\int_{\Omega-V_\Gamma} \rho(\dot{e} + \dot{u}\ddot{u}) \, d\Omega = \int_{\partial\Omega} n\sigma\dot{u} \, ds - \int_\Gamma n\sigma\dot{u} \, ds + \int_\Gamma qn \, ds \tag{15}$$

$$- \int_{\partial\Omega} qn \, ds - \int_{\Sigma_\Gamma} qn \, ds,$$

$$\int_{\Omega-V_\Gamma} D \, d\Omega = \int_{\Omega-V_\Gamma} \rho T\dot{S} \, d\Omega$$

$$- \int_\Gamma qn \, ds + \int_{\partial\Omega} qn \, ds + \int_{\Sigma_\Gamma} qn \, ds. \tag{16}$$

If G is defined by,

$$G = \lim_{\Gamma \to 0}\int_\Gamma \{\rho(e + \dot{u}^2/2)n_1 + n\sigma\dot{u}/\dot{l}\} \, ds, \tag{17}$$

then it follows from Eqs. (11), (14) and (15) that,

$$(G-2\gamma_0)\dot{l}= \lim_{\Gamma \to 0} \int_\Gamma qn \, ds, \tag{18}$$

and from Eqs. (17), (6), and (16) that,

$$\mathcal{D}= \int_\Omega D \, d\Omega +(G-2\gamma_0)\dot{l}. \tag{19}$$

The crack tip force, G, is thus introduced in a natural manner. Relation (18) proves that the crack tip behaves like a heat source, $(G-2\gamma_0)\dot{l}$. For this reason G is a finite quantity. Equation (19) shows that the global dissipation consists of a volumic dissipation, D, and a dissipation concentrated at the crack tip.

From (4) and (18), the thermal equation for the whole body can be written as,

$$\text{Div } q+\rho T\dot{S}=D+(G-2\gamma_0)\dot{l}\,\delta(A), \tag{20}$$

where $\delta(A)$ denotes the Dirac measure at the crack tip. For example, if the Fourier law of conduction is assumed, then we obtain in $\Omega(t)$ the following equation:

$$-k\Delta T+\rho T\dot{S}=D+(G-2\gamma_0)\dot{l}\,\delta(A). \tag{21}$$

There is thermomechanical coupling at two levels: one by the classical terms D and $\rho T\dot{S}$, and the other by the heat source $(G-2\gamma_0)\dot{l}\delta(A)$. The thermal Eq., (21), has been given by Bui *et al.* (1979).

Relation (17) gives G in terms of the singularity of the thermomechanical fields. We remark, from Assumption H1, that $(TS)^{\cdot}=T\dot{S}+S\dot{T}$ is an integrable function. If TS verifies Assumption H2, then $-\dot{l}(TS)_{,1}$ is also integrable. The following relation then holds:

$$\int_\Gamma TSn_1 \, ds= \int_{V_\Gamma} (TS)_{,1} \, ds,$$

and hence

$$\lim_{\Gamma \to 0} \int_\Gamma \dot{l}TSn_1 \, ds=- \lim_{\Gamma \to 0} \int_{V_\Gamma} (T\dot{S}+S\dot{T}) \, d\Omega=0.$$

If the displacement also satisfies Assumption H2, we have, near the tip, $\dot{u}\simeq -\dot{l}u_{,1}$. Under these conditions, G can also be written as,

$$G= \lim_{\Gamma \to 0} \int_\Gamma \{\rho(W+\dot{u}^2/2)n_1 -n\sigma u_{,1}\} \, ds. \tag{22}$$

The second principle gives, for an arbitrary volume V_Γ,

$$\mathcal{D}_V = \int_{V_\Gamma} D \, d\Omega + (G - 2\gamma_0)\dot{l} \geq 0.$$

The following thermodynamic restrictions are then derived:

$$D \geq 0, \qquad G - 2\gamma_0 \geq 0. \tag{23}$$

Note that the asymptotic behavior of the temperature is completely determined by the two assumptions, H1 and H2. We prove that the *temperature is singular as* $-\log r$ *if and only if* $G \neq 2\gamma_0$ *otherwise T is finite* at the crack tip.

Indeed, if the temperature can be expanded as $T = H(t)r^\alpha (\log r)^\beta \cdot \hat{T}(\theta, \dot{l})$ we obtain necessarily, $\alpha = 0, \beta = 1, \hat{T} = \hat{T}(\dot{l})$, and since *the heat source must be finite*, $q = -k \, \text{grad} \, T \sim r^{-1}$. The temperature is then $T = H(t) \log r \cdot \hat{T}(\dot{l}) +$ more regular terms $r^m (\log r)^n \, f(\theta, \dot{l}), m \geq 0$. But from Eq. (21), the integrability of $\rho T \dot{S}$ and D shows that,

$$T = -\frac{G - 2\gamma_0}{2k\pi} \dot{l} \log r + (\text{more regular terms}). \tag{24}$$

This result is new. In elasticity, it has been given recently by Bui *et al.* (1979). We stress the fact that the temperature obtained here applies to a large class of materials.

3. Elasticity, viscoelasticity and plasticity

The theory is here illustrated by usual models of materials in elasticity, viscoelasticity, and plasticity, and compared with different results given in the literature, almost all of which have been established for the mechanical problem of crack extension with the stress free condition over the cracked surface Σ. These published results are thus related to a purely mechanical analysis without reversible surface energy.

We assume first that reversible surface energy is neglected. The case of reversible surface energy, (9) and (10), is considered in the last paragraph. We wish to verify by singularity analysis that for usual materials, the two assumptions, H1 and H2, are fulfilled and the proposed theory is consistent. It furnishes a general framework for the discussion of the thermomechanical behavior near the crack tip.

3.1. Elasticity

For elastic materials, the free energy density is a function of the state variables (ε, T), and $\sigma = \rho(\partial W/\partial \varepsilon)$. There is no volume dissipation $(D = 0)$.

If the transformation is isothermal and quasi-static, Rice (1968) and Cherepanov (1967) have discussed the stress and strain singularities when $W(\varepsilon, T_0)$ is a homogeneous polynomial of ε. It has been shown that $W = r^{-1} H(t) \hat{W}(\theta)$ and thus Assumption H2 is verified. If we assume that $W(\varepsilon, T_0)$ is also a homogeneous polynomial of degree $(n + 1/n)$ in ε and T_0, then we verify easily that, $S \simeq r^{-(1/n+1)}$ and $T_0 \dot{S} \simeq r^{-(n+2/n+1)}$ are integrable functions for every $n > 0$. Since $D = 0$ and $S\dot{T} = 0$, Assumption H1 is verified.

Equation (23) reduces to the classical Rice–Eshelby integral,

$$G = \lim_{\Gamma \to 0} \int_\Gamma \{ W n_1 - n\sigma u_{,1} \}\, \mathrm{d}s = \int_\Gamma \{ W n_1 - n\sigma u_{,1} \}\, \mathrm{d}s.$$

If the transformation is isentropic and quasi-static, we can obtain the same conclusions by replacing W by e.

We can remark that $\sigma\dot{\varepsilon} \simeq r^{-2}$, so the function $\sigma\dot{\varepsilon}$ is not Lebesgue integrable, although its Cauchy principal value exists.

If the transformation is isothermal and dynamic, Yoffe (1951) or Achenbach and Bazant (1975) have studied *smooth* solutions in linear elasticity. We verify without difficulty Assumptions, H1 and H2. The quantity

$$G\dot{l} = \lim_{\Gamma \to 0} \int_\Gamma \{ \rho(W + \dot{u}^2/2)\dot{l}n_1 - n\sigma\dot{u} \}\, \mathrm{d}s$$

has been introduced in this case by Freund (1972) as the energy flux into the crack tip.

The general case of dynamic thermoelasticity has been recently discussed in the context of our theory, Bui *et al.* (1979). From the thermal equation, (21), it is shown in linear elasticity that the temperature is singular when the crack propagates $(T \simeq -(G\dot{l}/2k\pi)\log r)$. However, this field does not change the dominant singularity of the mechanical fields which are the same as that obtained in isothermal elasticity. This new result illustrates the coupled behavior.

For elastic materials, Eq. (19) gives, $\mathcal{D} = G\dot{l}$; the dissipation is a finite measure at the crack tip. This expression enables us to construct systematically "standard" laws of crack propagation, if we assume that the dissipation is normal (Germain, 1973; Halphen and Nguyen, 1975; Nguyen, 1979) in a similar way as standard laws of plasticity and viscoplasticity. These laws correspond to the existence of a dissipation potential, $\Omega^\star(\dot{l})$, such that,

$$G = \frac{\partial \Omega^\star}{\partial \dot{l}}(\dot{l}). \tag{25}$$

For example, the Griffith law corresponds to $\Omega^\star = G_c\langle \dot{l} \rangle$ and (25) may be written as,

$$\text{if } G < G_c \quad \text{then} \quad \dot{l} = 0,$$
$$\text{if } G = G_c \quad \text{then} \quad \dot{l} \geqslant 0, \tag{26}$$

where G_c is a critical dissipative surface energy.

3.2. Viscoelasticity

Let us consider firstly the Kelvin model of viscoelasticity. In perfect viscoelasticity (i.e. without internal parameters), the free energy is $W = W(\varepsilon, T)$. The quantity $\sigma_R = \rho \partial W / \partial \varepsilon$ is the reversible stress, the irreversible stress is $\sigma_{IR} = \sigma - \sigma_R$ and we obtain simply, $D = \sigma_{IR} \dot{\varepsilon}$.

In isothermal transformation and for linear models, the constitutive equations are $\sigma_R = L\varepsilon$ and $\sigma_{IR} = M\dot{\varepsilon}$. These equations show that if tensors L and M are positive definite, σ_{IR} is more singular than σ_R.

A singularity analysis shows that $\sigma_{IR} \simeq r^{-1/2}$ and $\varepsilon \simeq r^{1/2} \simeq \sigma_R$, and hence (22) gives $G = 0$.

The perfect Maxwell model is discussed in a similar way. The free energy density is a function of the elastic strain, $\varepsilon - \varepsilon^{Vp}$, and the temperature, $W = W(\varepsilon - \varepsilon^{Vp}, T)$, with $\sigma = \rho \partial W / \partial \varepsilon = -\rho \partial W / \partial \varepsilon^{Vp}$. The volume dissipation, D, is $D = \sigma \dot{\varepsilon}^{Vp}$.

In isothermal transformation and for linear models, the constitutive equations are $\sigma = L(\varepsilon - \varepsilon^{Vp})$ and $\sigma = N\dot{\varepsilon}^{Vp}$. These equations show that $\sigma \simeq L\varepsilon$ and thus a singularity analysis gives $\sigma \simeq \varepsilon \simeq r^{-1/2}$ and $G \neq 0$. We obtain, for example in plane strain, $G = (1 - \nu^2 / E)(K_I^2 + K_{II}^2)$ if we assume isotropy.

Nonlinear models have been discussed recently (Hui and Riedel, 1980; Kachanov, 1978). Let us assume that

$$\dot{\varepsilon}_{ij}^{Vp} = C \left(\frac{|\sigma|}{\sigma_0} \right)^{m-1} \frac{\sigma_{ij}}{\sigma_0} \quad \text{where} \quad |\sigma|^2 = \sigma_{ij}\sigma_{ij}.$$

The integrability of $D = \sigma \dot{\varepsilon}^{Vp}$ implies that β necessarily verifies $(m+1)\beta > -2$ when $\sigma \simeq K(t) r^\beta \hat{\sigma}(\theta, \dot{l})$ near the crack tip.

If ε^{Vp} is more regular than ε^e, i.e. if $m\beta > \beta - 1$, a singularity analysis like that in elasticity gives $\beta = -\frac{1}{2}$. Thus, if $m < 3$, the stress singularity is that of linear elasticity ($\sigma \simeq r^{-1/2}$).

If $m > 3$, we have $-(1/m - 1) \geqslant \beta > -(2/m + 1)$. A numerical result obtained by Hui and Riedel (1980) seems to show that in fact one obtains $\beta = -(1/m - 1)$. This can be easily established from the dynamic equation $\text{Div } \sigma - \rho \ddot{u} = 0$.

We remark that if $\beta = -(1/m - 1)$, $m > 3$, then (22) gives $G = 0$.

Assumption H1 is verified by the proposed solution ($\beta = -\frac{1}{2}$ if $m<3$, $\beta = -(1/m-1)$ if $m>3$), except for $m=3$.

3.3. Plasticity

In perfect plasticity, the free energy is $W = W(\varepsilon - \varepsilon^p, T)$ with $\sigma = \rho \partial W / \partial \varepsilon = -\rho \partial W / \partial \varepsilon^p$, $D = \sigma \dot{\varepsilon}^p$. Let us consider the case of linear elastic strain–stress relation and quasi-static, isothermal transformation. The constitutive equations are $\sigma = L(\varepsilon - \varepsilon^p)$ and $\dot{\varepsilon}^p = \lambda \partial f / \partial \sigma$, where $\lambda \geq 0$ in the plastic range.

Although crack extension in plastic material has been extensively discussed, few significant results have been obtained [Chitaley and McClintock, 1971; Slepyan, 1974; Rice *et al.*, 1979). In Mode I, plane strain analysis shows that the stress distribution can be compared to the Prandtl solution, but an unloading region near the stress free surface is observed (Slepyan, 1974; Rice *et al*, 1979; Nguyen and Rahimian, 1980). The stress is bounded and the velocity $\dot{u} \sim \log r$ gives an opening displacement $[u] \sim r \log r$. Assumption H1 is fulfilled since $D = \sigma \dot{\varepsilon}^p \sim r^{-1}$. The basic assumption, H2, is naturally verified. Formula (22) gives $G=0$.

The last estimate of G has been suggested by Rice (1968) and discussed in many published papers. But the definition of G used by Rice *et al.* (1979) is not the same; we denote it here by $G^\star$,

$$G^\star = \lim_{\Gamma \to 0} \int_\Gamma \{W^\star n_1 - n\sigma u_{,1}\} \, ds, \tag{27}$$

where $W^\star = \int_0^\varepsilon \sigma \, d\varepsilon$ is the deformation energy which is *not* a state function in incremental plasticity. Except in elasticity, $G^\star$ has no energetic interpretation. From the proposed solution (σ bounded, $\dot{u} \sim \log r$), one verifies easily that $G^\star = 0$.

The estimate, $G=0$, means that the dissipation of the whole body is simply the volume dissipation, $\int_\Omega D \, d\Omega$; there is no other dissipative term. Note also that this estimate is obtained when reversible surface energy is neglected.

From the remark of the preceding paragraph, it is expected that the temperature distribution is regular near the crack tip, although the thermo-mechanical behavior is not completely established.

For hardening materials, there exists a discussion from Amazigo and Hutchinson (1977) in the case of linear isotropic hardening. It has been shown that the stress is singular as r^s with $-\frac{1}{2} \leq s \leq 0$, $s \to -\frac{1}{2}$ if $E_t/E \to 1$ and $s \to 0$ if $E_t/E \to 0$. Formula (22) gives $G=0$, from this singularity analysis.

The estimate, $G=0$, does not result from bounded stress distribution near the crack tip. It is expected that in incremental plasticity (Nguyen, 1980) the function $\sigma \dot{\varepsilon}$ behaves like $\sigma \dot{\varepsilon}^p$; thus it is integrable and gives $G=0$.

3.4. Reversible surface energy

We give here a short discussion on the estimate of G in presence of reversible surface energy. For clarity, we consider only quasi-static, isothermal transformations. It is necessary to distinguish the two limits:

$$G = \lim_{\Gamma \to 0} \lim_{d \to 0} G_\Gamma(d), \tag{28}$$

$$G_0 = \lim_{d \to 0} \lim_{\Gamma \to 0} G_\Gamma(d). \tag{29}$$

If body force is neglected, we obtain exactly as in elasticity,

$$G_{\Gamma_1}(d) = \lim_{\Gamma_0 \to 0} G_{\Gamma_0}(d) + 2\gamma_0,$$

where Γ_0 and Γ_1 are closed curves, V_{Γ_1} contains the part of Σ with attractive force.

When $d \to 0$ and $\Gamma_1 \to 0$ we obtain,

$$G = G_0 + 2\gamma_0. \tag{30}$$

For example, if the material is isotropic and linearly elastic, we obtain in plane strain,

$$\lim_{\Gamma_0 \to 0} G_{\Gamma_0}(d) = \frac{1-\nu^2}{E} \left(K_I^2(d) + K_{II}^2(d) \right)$$

and thus,

$$G - 2\gamma_0 = \lim_{d \to 0} \frac{1-\nu^2}{E} \left(K_I^2(d) + K_{II}^2(d) \right) \geqslant 0.$$

In the general case, we have,

$$G_{\Gamma_1}(d) = G_{\Gamma_0}(d) - \int_{\Sigma_{\Gamma_1} - \Sigma_{\Gamma_0}} \varphi' \, ds - \int_{V_{\Gamma_1} - V_{\Gamma_0}} (\sigma \varepsilon_{,1} - \rho W_{,1}) \, d\Omega,$$

and we obtain when $\Gamma_0 \to 0$,

$$G_{\Gamma_1}(d) = \lim_{\Gamma_0 \to 0} G_{\Gamma_0}(d) + 2\gamma_0 - \int_{V_{\Gamma_1}} (\sigma \varepsilon_{,1} - \rho W_{,1}) \, d\Omega.$$

Let us assume that the solution exists when $d \to 0$ (note that the perfect plastic model is not compatible with this procedure since the attractive force is greater than the limit load). We can expect that the last term behaves like

$$\int_{V_{\Gamma_1}} (\sigma \dot{\varepsilon} - \rho \dot{W}) \, d\Omega = \int_{V_{\Gamma_1}} D \, d\Omega$$

by virtue of the basic assumption, H2, and thus it vanishes with Γ_1. We obtain again (30).

Although more rigorous proofs must be given, this short discussion shows that the estimate of G in the presence of reversible surface energy, (10), is a rather complex problem. In particular, it should be emphasized that the physical background of this energy must be understood in order to complete the thermomechanical description.

4. Energy release rate

In this section, we present the extension of the well-known relation,

$$G = -\frac{\partial P}{\partial l}$$

given by Rice (1968) in quasi-static elasticity. This extension has been obtained in quasi-static transformation (Nguyen, 1979 and 1980). Let us introduce the following notation:

$$\sigma_R = \rho \frac{\partial W}{\partial \varepsilon}, \qquad \sigma_{IR} = \sigma - \sigma_R, \qquad A = -\rho \frac{\partial W}{\partial \alpha}. \tag{31}$$

The volume dissipation is $D = \sigma \dot{\varepsilon} - \rho(\dot{W} + S\dot{T}) = \sigma_{IR}\dot{\varepsilon} + A\dot{\alpha} \geq 0$.

In quasi-static transformation, we have the possibility of defining a global parametric description of the system of cracked body. The system is completely defined by the state variables $(\xi, \alpha, l, T, \sigma_{IR})$ which are respectively the displacement over $\partial\Omega$, the internal parameter field, the crack length, the temperature field, and the irreversible stress field.

Indeed, if we assume that the fields α, T, σ_{IR} are given, then the relation $\varepsilon \to \sigma = \sigma_{IR} + \rho \partial W / \partial \varepsilon(\varepsilon, \alpha, T)$ defines everywhere an elastic law with initial stress. We can obtain the strain ε as a function of the state variables by the solution of a boundary value problem of elastic response with unilateral constraints and attractive forces over Σ:

Find the fields u and σ such that:

(i) $u = \xi$ on $\partial\Omega$ (kinematic condition);

         $[u] \geq 0$ on Σ

(ii) $\text{Div } \sigma = 0$ in Ω

         $\sigma_{12} = 0, \sigma_{22} = \varphi'[u]$ if $[u] > 0$ on Σ (static condition);   (32)

         $\sigma_{22} \leq \varphi'[u]$ if $[u] = 0$

(iii) $\sigma = \sigma_{IR} + \rho \dfrac{\partial W}{\partial \varepsilon}(\varepsilon, \alpha, T)$ (constitutive equation).

Solution of this purely elastic boundary value problem gives the displacement field and the strain ε. We recall that if the functional,

$$\int_{\Omega(l)} \{\rho W(\varepsilon^\star, \alpha, T) + \sigma_{IR}\varepsilon^\star\} \, d\Omega + \int_\Sigma \varphi[u^\star] \, ds$$

is locally convex with respect to $u^\star$ for fixed $\alpha, T, l, \sigma_{IR}, \xi$, then the solution u realizes a local minimum of the potential energy,

$$\phi_d\{u^\star\} = \int_{\Omega(l)} \{\rho W(\varepsilon^\star, \alpha, T) + \sigma_{IR}\varepsilon^\star\} \, d\Omega + \int_\Sigma \varphi[u^\star] \, ds, \quad (33)$$

among all kinematically admissible displacement fields.

Since the solution u is a function of the state variables $(\xi, \alpha, l, T, \sigma_{IR})$, the functional $\phi\{u\}$ is a function of these variables,

$$\phi_d\{u\} = \phi_d(\xi, \alpha, l, T, \sigma_{IR}).$$

Again, we must define the limit $\phi = \lim_{d \to 0} \phi_d$ to obtain the case of reversible surface energy, (10).

To simplify the presentation, we neglect all possible reversible energy and consider only the functional,

$$\Phi(\xi, \alpha, l, T, \sigma_{IR}) = \int_{\Omega(l)} \{\rho W(\varepsilon, \alpha, T) + \sigma_{IR}\varepsilon\} \, d\Omega. \quad (34)$$

It is interesting to study the dependence of Φ with respect to each state variable. The partial derivative $\partial\Phi/\partial\xi$ represents the surface force, F, acting on $\partial\Omega$; this follows simply from the classical Clapeyron theorem. The partial derivative $\partial\Phi/\partial\alpha$ represents the generalized force field $-A$; the partial derivative $-\partial\Phi/\partial T$ represents the field ρS; the derivative $\partial\Phi/\partial\sigma_{IR}$ is the field ε; these results follow simply from (34) and from the equilibrium equation.

Let us denote by $\mathfrak{g}$ the derivative $-\partial\Phi/\partial l$,

$$\mathfrak{g} = -\frac{\partial\Phi}{\partial l}(\xi, \alpha, l, T, \sigma_{IR}). \quad (35)$$

Note that if the energy density $W(\varepsilon, \alpha, T)$ is a convex function of ε for fixed α, T, then the functional $\Phi\{u\}$ is convex. From the minimum of potential energy functional, (33), it is shown (Nguyen, 1980) that,

$$\mathfrak{g} \geqslant 0. \quad (36)$$

The functional $\phi(\xi, \alpha, l, T, \sigma_{IR})$ is Gateaux-differentiable if for arbitrary virtual variations $\xi[\lambda], \alpha[\lambda], l[\lambda], T[\lambda], \sigma_{IR}[\lambda]$ defined by a kinematic parameter λ such that $\xi[0] = \xi, \ldots, \sigma_{IR}[0] = \sigma_{IR}$, the virtual rate $\delta\Phi = \Phi'[0]$ can be obtained from the virtual rates $\delta\xi = \xi'[0], \ldots, \delta\sigma_{IR} = \sigma'_{IR}[0]$ in the following form:

$$\delta\Phi = \int_{\partial\Omega} F\delta\xi \, ds - \int_\Omega (A\delta\alpha + \rho S \, \delta T - \varepsilon\delta\sigma_{IR}) \, d\Omega - \mathfrak{g} \, \delta l. \quad (37)$$

The differentiability of ϕ is a quite restrictive property. It depends on the regularity of the introduced functions $\xi[\lambda], \ldots, \sigma_{IR}[\lambda]$. In particular, it can be established for the linear Maxwell model. Let us assume this property in order to obtain, for the actual evolution,

$$\dot{\Phi} = \int_{\partial\Omega} F\dot{\xi}\, ds - \int_{\Omega} \left(A\dot{\alpha} + \rho S\dot{T} - \varepsilon\dot{\sigma}_{IR} \right) d\Omega - \mathfrak{g}\dot{l}. \tag{38}$$

But we can also compute directly $\dot{\Phi}$ as indicated in Section 2. If, in accordance with Assumption H1, we assume that

$$\varepsilon\dot{\sigma}_{IR}, \sigma_{IR}\dot{\varepsilon} \text{ are Lebesgue-integrable,} \tag{39}$$

then we obtain,

$$\dot{\Phi} = \int_{\partial\Omega} F\dot{\xi}\, ds - \int_{\Omega} A\dot{\alpha}\, d\Omega - \int_{\Omega} \rho S\dot{T}\, d\Omega - G\dot{l} + \int_{\Omega} \varepsilon\dot{\sigma}_{IR}\, d\Omega. \tag{40}$$

It follows from (38), (40), that $\mathfrak{g} \equiv G$.

As an illustration, we give here a formal expression of the functional Φ in isothermal, perfect viscoelasticity when the energy density W is quadratic. The relation $\sigma = \sigma_{IR} + \rho \partial W / \partial \varepsilon$ reduces to $\sigma = L(\varepsilon - \varepsilon^p)$ and the functional Φ is,

$$\Phi = \int_{\Omega(l)} \tfrac{1}{2}(\varepsilon - \varepsilon^p) L(\varepsilon - \varepsilon^p)\, d\Omega = \tfrac{1}{2} \| \varepsilon - \varepsilon^p \|^2_{\Omega(l)},$$

where $\|.\|$ denotes the energy norm. To compute Φ as a function of the state variables, (ξ, ε^p, l), let us introduce the kinematically admissible set of displacements,

$$I(\xi, l) = \{ u \mid [u] \geqslant 0 \text{ on } \Sigma(l), u = \xi \text{ on } \partial\Omega \}.$$

The solution u minimizes $\tfrac{1}{2} \| \varepsilon(u) - \varepsilon^p \|^2_{\Omega(l)}$ in the set $I(\xi, l)$, thus the functional Φ represents,

$$\Phi(\xi, \varepsilon^p, l) = \tfrac{1}{2} \left[\text{Dist}(\varepsilon^p, I(\xi, l)) \right]^2,$$

where $\text{Dist}(\varepsilon^p, I(\xi, l))$ denotes the distance of the plastic strain field to the set $I(\xi, l)$.

5. Conclusion

Our discussion shows that for several models of materials, the energy method does not provide a simple fracture parameter, except in elasticity. In small scale plasticity, the global plastic dissipation may be a good parameter since it reduces to the dissipation $G\dot{l}$ concentrated at the crack tip for very small plastic zone size. In ductile fracture however, this parameter does not characterize the local state and perhaps more physical descriptions will be necessary.

Acknowledgment

The author wishes to thank Professors P. Germain and J. Mandel for stimulating discussions. The private communication "Sur certaines définitions liées à l'énergie en Mécanique des Milieux Continus" from Professor P. Germain contributed greatly to the presentation of Section 4.

References

Achenbach, J. D. and Bazant, Z. P. (1975), "Elastodynamic Near Tip Stress and Displacement Fields for Rapidly Propagation Cracks in Orthotropic Materials," *J. Appl. Mech.*, **97**, 183.

Amazigo, J. C. and Hutchinson, J. W. (1977), "Crack Tip Fields in Steady Crack Growth with Linear Strain Hardening," *J. Mech. Phys. Solids*, **25**, 81.

Bui, H. D., Ehrlacher, A. and Nguyen, Q. S. (1979), "Propagation de Fissure en Thermoélasticité Dynamique," *C. R. Acad. Sci. Paris*, **289**, 211.

Bui, H. D. (1977), *La Mécanique de la Rupture Fragile*, Masson, Paris.

Cherepanov, G. P. (1967), "Crack Propagation in Continuous Media," *P. M. M.*, **31**, 476.

Chitaley, A. D. and McClintock, F. A. (1971), "Elastic Plastic Mechanics of Steady Crack Growth under Antiplane Shear," *J. Mech. Phys. Solids*, **19**, 147.

Freund, L. B. (1972), "Energy Flux into the Tip of an Extending Crack in an Elastic Solid," *J. Elasticity*, **2**, 341.

Germain, P. (1973), *Cours de Mécanique des Milieux Continus*, Masson, Paris.

Gurtin, M. E. (1979), "Thermodynamics and the Griffith Criterion for Brittle Fracture," *Int. J. Solids Struct.*, **15**, 553.

Halphen, B. and Nguyen, Q. S. (1975), "Sur les Matériaux Standards Généralisés," *J. Mécanique*, **14**, 39.

Hui, C. Y. and Riedel, H. (1980), "The Asymptotic Stress and Strain Field Near the Tip of a Growing Crack under Creep Conditions," to be published.

Kachanov, L. M. (1978), "Crack Growth under Creep Condition," *M. Tver. Tela.*, **13**, 97.

Nguyen, Q. S. (1979), "Une Description Thermodynamique du Problème de Fissure Mobile," *C. R. Acad. Sci.*, **288**, 201.

Nguyen, Q. S. (1980), "Méthodes Energétiques en Mécanique de la Rupture," *J. Mécanique*, **19**, 363.

Nguyen, Q. S. and Rahimian, M. (1980), "Mouvement Permanent d'une Fissure en Milieu Elastoplastique," *J. Mécanique Appliquée*, to be published.

Rice, J. R. (1968), "Mathematical Analysis in the Mechanics of Fracture," *Fracture*, **2**, 191.

Rice, J. R. (1978), "Thermodynamics of the Quasi Static Growth of Griffith Cracks," *J. Mech. Phys. Solids*, **16**, 61.

Rice, J. R., Drugan, W. J. and Sham, T. L. (1979), "Elastic-Plastic Analysis of Growing Cracks," Report no. 65, Brown University, to be published.

Slepyan, L. I. (1974), "Growing Crack During Plane Deformation of an Elastic-Plastic Body," *Mek. Tve. Tela.*, **9**, 57.

Yoffe, E. H. (1951), "The Moving Griffith Crack," *Phil. Mag.*, **42**, 739.

S. Nemat-Nasser, Editor
THREE-DIMENSIONAL CONSTITUTIVE RELATIONS AND DUCTILE FRACTURE
North-Holland Publishing Company (1981) 331–355

FINITE DEFORMATION CONSTITUTIVE RELATIONS INCLUDING DUCTILE FRACTURE DAMAGE

G. ROUSSELIER

Electricité de France, Moret-sur-Loing, France

Constitutive relations are developed for finite deformation of plastically dilatant materials. These relations, which model the ductile fracture of metals, are derived from the macroscopic viewpoint, in the framework of generalized standard materials. An exponential dependence of ductile fracture damage on stress triaxiality is demonstrated and the occurrence of material instability, $\dot{\sigma}\dot{\varepsilon}^{P} < 0$, is shown by various examples. In finite element applications to cracked specimens, stable crack growth takes place naturally by localization of deformation without it being necessary to postulate a local fracture criterion nor to release the nodes.

1. Introduction

The examination of a crack tip or the minimum section of a tensile specimen points out that ductile fracture in metals involves considerable damage, via the nucleation and growth of voids, which should be taken into account in the mathematical modelling in order to predict the conditions at fracture and also to characterize the stress–strain relations before the ultimate stage of damage. A damage function may be used with usual stress-strain relations to define the conditions at fracture, but this is only an approximation, a reasonable one for fatigue and creep damage, but not for ductile fracture. Usual theories of plasticity imply plastic incompressibility, which is inconsistent with the dilatancy evident in ductile fracture mechanisms.

For the construction of a new plasticity theory including ductile fracture damage, which reduces to usual theories of plasticity if damage is negligible, two approaches may be taken: microscopic and macroscopic.

In the first approach, taken by Gurson (1977), the macroscopic constitutive relations are constructed from microscopic components of material: matrix, particles, voids, and their individual and interactive behavior. The transition from nonhomogeneous microscopic to homogeneous macroscopic material is the main difficulty here, requiring a number of simplifying assumptions. Moreover the microscopic components are not well known. McClintock (1968), Rice and Tracey (1969) have developed models of the growth of voids, that give an exponential dependence on stress triaxiality, σ_m / σ_0, where $\sigma_m = \sigma_{kk} / 3$ is the hydrostatic tension and σ_0 the yield stress. The growth rate of a single spherical void in a rigid-plastic material is approximated by Rice and Tracey as,

$$\frac{\dot{R}_0}{R_0} = 0.283 \dot{\varepsilon}_{eq}^P \exp\left(\frac{3\sigma_m}{2\sigma_0}\right), \tag{1}$$

where R_0 is the void radius and $\dot{\varepsilon}_{eq}^P$ the remote equivalent strain rate. But as far as we know, there are no acknowledged models of void interaction and coalescence, that apply to the last stages of the ductile fracture of metal.

The second approach, used here, is macroscopic. It requires no description of micromechanisms and will be considered applicable to the whole process of damage, on condition that the predictions of the model are consistent with the current understanding of ductile fracture. The theory is motivated by the microscopic models but is not deduced from them.

We suppose that the thermodynamical state of the material, the hardening and also damage, are characterized by internal variables in the framework of generalized standard materials (GSM), i.e. existence of a quasi-potential of dissipation and normality rule for the plastic strain rate *and* the internal variables rates, as proposed by Nguyen (1973). The characterization of damage by internal variables is a natural consequence of the choice of GSM. The same hypothesis was formulated quite independently by Kachanov (1958) and Rabotnov (1968) for creep damage and developed recently by Lemaitre and Chaboche (1978), who establish creep, fatigue, and creep-fatigue cumulation models.

Ductile rupture is generally preceded by large plastic deformations Lautridou and Pineau (1978), by recrystallization technique, have measured a strain of 100% at the tip of a blunted crack in A508 Cl 3 steel. Hence constitutive relations are established for *finite deformations*.

First an account of GSM is given for finite deformation. Then application to ductile fracture damage is performed. Finally, the initiation and stable growth of a crack in a bend specimen are analyzed by a finite element model which takes into account some finite deformation effects.

2.   Generalized standard materials (GSM) in finite deformation

Consider the configurations (o) of a macroscopic element of material at time 0, in an unstressed state, and (a) at time t, under stress σ. Introduce (Mandel, 1973) an actual intermediate relaxed configuration (κ), by supposing the element of material unloaded at time t^1. Let G be the gradient of the

---

[1] *Virtual* unloading according to the elastic properties, in order to avoid plastic deformations in the opposite direction of the deformation experienced during loading.

total transformation (o)→(a), P the gradient of the plastic transformation (o)→(κ) and E the gradient of the elastic transformation (κ)→(a). As $G=EP$, the velocity gradient tensor grad $\vec{v}=(DG/Dt)G^{-1}$ is,

$$\text{grad } \vec{v}=\mathcal{D}+\omega=\frac{DE}{Dt}E^{-1}+E\frac{DP}{Dt}P^{-1}E^{-1}, \tag{2}$$

where the symmetric and antisymmetric parts are the deformation rate tensor, $\mathcal{D}$, and spin tensor, ω. The objective derivative D/Dt may be a convected derivative, the Jaumann derivative, or the derivative introduced by Mandel (1973) relative to the rotation of the director frame which specifies the physical orientation of the macroelement.

Consider only isothermal quasistatic transformations. The thermodynamic state of the macroelement is defined by $\Delta^e=(E^tE-1)/2$, the Green deformation tensor between (κ) and (a), and α_i, scalar or generally tensorial variables that characterize damage and strain-hardening of the material; including the orientation of the director frame (Mandel, 1973).

The local form of the second principle of thermodynamics is,

$$\Phi=\text{tr}\left(\frac{\sigma \mathcal{D}}{\rho}\right)-\frac{D\varphi}{Dt}\geq 0, \tag{3}$$

where $\varphi(\Delta^e, \alpha_i)$ is the specific free energy. Equations (2) and (3) give,

$$\Phi=\text{tr}\left(E^{-1}\frac{\sigma}{\rho}E\frac{DP}{Dt}P^{-1}\right)-\text{tr}\left(\frac{\partial\varphi}{\partial\Delta^e}-\frac{\pi}{\rho_\kappa}\right)\frac{D\Delta^e}{Dt}-\frac{\partial\varphi}{\partial\alpha_i}\frac{D\alpha_i}{Dt}\geq 0, \tag{4}$$

where $\rho=1/\det G$ and $\rho_\kappa=1/\det P$ are the densities in the configurations (a) and (κ) ($\rho=1$ in the configuration (o)) and π is the Kirchhoff stress tensor relative to the configuration (κ),

$$\frac{\sigma}{\rho}=E\frac{\pi}{\rho_\kappa}E^t. \tag{5}$$

In the (virtual) reversible elastic transformation from (κ) to (a), the dissipated power Φ is zero, and $DP/Dt=D\alpha_i/Dt=0$, so,

$$\frac{\pi}{\rho_\kappa}=\frac{\partial\varphi}{\partial\Delta^e}. \tag{6}$$

The generalized forces,

$$\Sigma=\left\{E^t\frac{\sigma}{\rho}E^{t-1}\right\}, \qquad A_i=-\frac{\partial\varphi}{\partial\alpha_i}, \tag{7}$$

are the work conjugates of the plastic deformation rate, $\mathcal{D}^P=\{(DP/Dt)P^{-1}\}$, and the internal variables rates, $D\alpha_i/Dt$ (Nguyen, 1973).

The resulting inequality is,

$$\Phi = \mathrm{tr}(\Sigma \mathcal{D}^P) + A_i \frac{\mathrm{D}\alpha_i}{\mathrm{D}t} \geq 0. \tag{8}$$

The constitutive relations of the material express the rates in terms of the generalized forces. Normal dissipativity is assumed, i.e. (Moreau, 1970): there exists a convex quasi-potential of dissipation $\Psi(\Sigma, A_i)$ and the conjugated rates are given by the normality rule.

Considering nonviscous plastic materials only, we further assume that the yield surface is defined by the single differentiable plastic potential, $F(\Sigma, A_i)$. It implies[2] that

$$\mathcal{D}^P = \lambda \frac{\partial F}{\partial \Sigma}, \qquad \frac{\mathrm{D}\alpha_i}{\mathrm{D}t} = \lambda \frac{\partial F}{\partial A_i}, \tag{9}$$

where $\lambda \geq 0$ if $F = \dot{F} = 0$, otherwise $\lambda = 0$; i.e., the rates are oriented in the direction of the external normal to the yield surface.

These postulates characterize GSM. In the usual theories of plasticity, only the first equation (9) applies[3].

The configuration (κ) may be chosen so that the elastic transformation is a pure deformation, $E = E^t$. Since elastic strains in metals are small, they may be neglected compared to unity. Then $E = 1 + \Delta^e = 1 + \varepsilon^e$ and $\rho = \rho_\kappa$; $\sigma = \pi$, $\Sigma = \sigma/\rho$. With these hypotheses we have,

$$\mathcal{D} = \frac{\mathrm{D}\varepsilon^e}{\mathrm{D}t} + \mathcal{D}^P, \tag{10}$$

and the constitutive relations become,

$$\frac{\sigma}{\rho} = \frac{\partial \varphi}{\partial \varepsilon^e}, \qquad A_i = -\frac{\partial \varphi}{\partial \alpha_i}, \tag{11}$$

$$\mathcal{D}^P = \lambda \frac{\partial F}{\partial(\sigma/\rho)}, \qquad \frac{\mathrm{D}\alpha_i}{\mathrm{D}t} = \lambda \frac{\partial F}{\partial A_i}, \tag{12}$$

where $\lambda \geq 0$ if $F(\sigma/\rho, A_i) = \dot{F}(\sigma/\rho, A_i) = 0$, otherwise $\lambda = 0$.

3. Application to ductile fracture damage

The hardening of the metal is supposed to be isotropic, characterized by a single scalar internal variable $\alpha_1 = \alpha$. This is a reasonable assumption as chiefly monotonously increasing loadings are considered in ductile fracture,

---

[2] As a general rule, multiple potentials may be considered (Nguyen, 1973). $\Psi(\Sigma, A_i)$ is the indicatory function of the convex bounded by the yield surface.

[3] In the general anisotropic case, a constitutive equation for the plastic spin tensor ω^P has to be added to Eqs. (9) (Mandel, 1973). It defines the orientation of the director frame.

with no noticeable reverse plastic deformation. However, the generalization to anisotropic hardening may be made, in the same way as in the theories of plasticity without damage.

The ductile fracture damage is also supposed to be isotropic, characterized by the scalar $\alpha_2 = \beta$. This is a simplifying hypothesis, because an initially spherical void usually grows into an ellipsoid, the characterization of which requires a symmetric tensorial internal variable $\beta_{ij} = \beta_{ji}$. On the other hand, the isotropy is derived from Rice and Tracey's (1969) results of symmetric void growth under large stress triaxiality, σ_m / σ_0.

The following form of the specific free energy is considered:

$$\varphi(\varepsilon^e, \alpha, \beta) = \frac{1}{2}\varepsilon^e L \varepsilon^e + \varphi_1(\alpha) + \varphi_2(\beta). \tag{13}$$

The first term is the elastic recoverable energy. $\varphi_1 + \varphi_2$ is the "locked" free energy, related to dislocations, residual stresses, voids, etc. The split of the free energy into three terms means that the elastic moduli tensor L does not depend on hardening and damage. Only for very large plastic strains, due to high hardening, which creates a texture in the metal, or high damage, the elastic moduli may be altered. The isotropic part of this alteration is certainly taken into account, according to the elastic constitutive relation (11): $\sigma/\rho = \partial\varphi/\partial\varepsilon^e$, that gives $\sigma = \rho L \varepsilon^e$. The density ρ is related to the damage, so are the apparent elastic moduli, $L_a = \rho L$. In fact, measurements of the variation of density and Young's modulus are performed as an indirect assessment of damage (Lemaitre and Chaboche, 1978); an isotropic damage D has been first introduced in the constitutive relations by Kachanov (1958) and Rabotnov (1968) with the notion of effective stress $\sigma_a = \sigma/(1-D)$, that gives $L_a = (1-D)L$. In the case of ductile fracture damage, and with the present formulation[4], it leads to the definition of damage $\beta = D = 1 - \rho$.

Finally, the split of the terms $\varphi_1(\alpha)$ and $\varphi_2(\beta)$ in the specific free energy means that the texture created by hardening and the porosity resulting from damage do not affect the characteristics of each other.

The Von Mises form of the plastic potential is,

$$F\left(\frac{\sigma}{\rho}, A\right) = \left[J_2\left(\frac{\sigma}{\rho}\right)\right]^{1/2} + \frac{A}{\sqrt{3}}, \tag{14}$$

where $J_2(T)$ is the second invariant of the deviatoric part of a second-order tensor T. The generalized force, $A = -d\varphi_1(\alpha)/d\alpha$, characterizes the hardening curve of the metal.

---

[4]Lemaître and Chaboche (1978) consider the *volumic* free energy $\rho\varphi$ instead of the specific free energy φ. The damage D only appears in the elastic term in the form,

$$\rho\varphi = \tfrac{1}{2}(1-D)\varepsilon^e L \varepsilon^e + \rho\varphi_1(\alpha_i),$$

where D has to be independent of ρ.

Damage is introduced with a third term, depending only on the first invariant, σ_m, of the stress tensor, according to theoretical and experimental results on ductile fracture up to date (Rice and Tracey, 1969; Hancock and Mackenzie, 1976; Auger and François, 1977),

$$F\left(\frac{\sigma}{\rho}, A, B\right) = \left[J_2\left(\frac{\sigma}{\rho}\right)\right]^{1/2} + \frac{A}{\sqrt{3}} + Bg\left(\frac{\sigma_m}{\rho}\right). \tag{15}$$

The generalized force B is the work conjugate of β. Let d^P and s be the deviatoric parts, $\mathcal{D}_m^P$ and σ_m the spherically symmetric parts of $\mathcal{D}^P$ and σ, respectively. The constitutive relations (11) and (12) give,

$$\sigma = \rho L \varepsilon^e, \tag{16}$$

$$A = -\varphi_1'(\alpha), \qquad B = -\varphi_2'(\beta), \tag{17}$$

$$d^P = \lambda \frac{s}{2[J_2(\sigma)]^{1/2}}, \qquad \mathcal{D}_m^P = \lambda \frac{B}{3} g'\left(\frac{\sigma_m}{\rho}\right), \tag{18}$$

$$\dot{\alpha} = \frac{\lambda}{\sqrt{3}}, \qquad \dot{\beta} = \lambda g\left(\frac{\sigma_m}{\rho}\right). \tag{19}$$

The equivalent deformation rate is usually defined as,

$$\mathcal{D}_{eq}^P = \left[\tfrac{4}{3}J_2(\mathcal{D}^P)\right]^{1/2} = \left[\tfrac{2}{3}d_{ij}^P d_{ij}^P\right]^{1/2}. \tag{20}$$

Taking the second invariant of both sides of the first equation (18), it follows that

$$\mathcal{D}_{eq}^P = \frac{\lambda}{\sqrt{3}} = \dot{\alpha}, \tag{21}$$

$$d^P = \frac{3\mathcal{D}_{eq}^P}{2\sigma_{eq}} s, \tag{22}$$

where $\sigma_{eq} = [3J_2(\sigma)]^{1/2}$ is the equivalent stress. In the absence of damage, the hardening curve is $\sigma_{eq} = \varphi_1'(\alpha)$, where $\alpha = \int \mathcal{D}_{eq}^P \, dt$.

Equation (22) is the usual constitutive relation with the Von Mises yield criterion. In addition, the second equations in (18) and (19) characterize the volumetric plastic deformation rate and the development of damage.

It may be argued that damage coincides with incipient plastic deformation, as the same threshold $F = 0$ (15) is considered for plasticity and damage. It is a reasonable approximation when the stress triaxiality, σ_m/σ_0, is large (Beremin, 1979). In any case damage indeed starts with the formation and piling up of dislocation loops at the matrix–particle interface, at incipient plastic strains, and not only with the subsequent nucleation of voids (Henry and Hortsmann, 1979). Besides, if the stress triaxiality is low, the plastic potential (15) will reduce to the usual form (14), as will be demonstrated below.

With the hypotheses of isotropic hardening and damage, from an isotropic unstressed initial configuration[5], the orientation of the director frame, which is then corotational with the macroelement, does not intervene any longer in the constitutive relations (16)–(19) (Mandel, 1973), and the Jaumann and Mandel derivatives are identical. As α and β are scalars, it seems that these derivatives, and the spin tensor $\omega = \omega^P$, do not enter the constitutive relations. Actually, the time derivative of the elastic relation $\sigma = \rho L \varepsilon^e$ is required, and according to Eq. (10),

$$\frac{D}{Dt}\left(\frac{\sigma}{\rho}\right) = L(\mathcal{D} - \mathcal{D}^P); \tag{23}$$

The spin tensor appears in the derivative,

$$\frac{D}{Dt}\left(\frac{\sigma}{\rho}\right) = \frac{d}{dt}\left(\frac{\sigma}{\rho}\right) - \frac{\omega\sigma}{\rho} + \frac{\sigma\omega}{\rho}. \tag{24}$$

The function $g(\sigma_m/\rho)$ has not been explicitly defined as yet.

In ductile fracture, the damage parameter β is directly related to the change of density of the metal; thus $\beta = \beta(\rho)$. Expressing $\mathcal{D}_m^P$ and $\dot{\beta} = \beta'(\rho)\dot{\rho}$ in the mass conservation law $\dot{\rho} + 3\rho\mathcal{D}_m^P = 0$[6], according to (18) and (19), we find,

$$\frac{g'(\sigma_m/\rho)}{g(\sigma_m/\rho)} = -\frac{1}{\rho\beta'(\rho)B}, \tag{25}$$

where $B = -\varphi_2'(\beta(\rho))$ is a function of ρ only. Therefore, the two sides of (25) are constant of dimension $1/\sigma$, say C/σ_0, where σ_0 is the yield stress. The integration of the left-hand side gives,

$$g\left(\frac{\sigma_m}{\rho}\right) \equiv D \exp\left(\frac{C\sigma_m}{\rho\sigma_0}\right), \tag{26}$$

where D is the constant of integration; C and D are supposed to be positive, in order that damage increases with stress triaxiality and that $\dot{\beta} > 0$; damage is irreversible, since λ cannot be negative.

The general hypotheses—generalized standard materials—and plastic potential depending only on the two first invariants of the stress tensor— yield an exponential dependence of damage on stress triaxiality. This result is similar to that obtained by McClintock (1968), Rice and Tracey (1969).

---

[5] The elastic moduli tensor is also isotropic, $L_{ijhk} = \lambda^* \delta_{ij}\delta_{hk} + \mu(\delta_{ih}\delta_{jk} + \delta_{ik}\delta_{jh})$.

[6] The elastic deformations are small, so $\rho \simeq \rho_\kappa$ and div $\vec{v} = 3\mathcal{D}_m \simeq 3\mathcal{D}_m^P$. As for an exact measure of damage, ρ_κ should be considered instead of ρ, i.e. the elastic volumetric deformation is excluded.

Still there is a difference, the consequence of which will be shown; the density ρ appears in the exponential. This is due to the fact that, in the change of configuration, σ is a material 2—contravariant tensorial density. In usual plasticity theories, ρ is assumed to be constant (unity), but for ductile fracture damage this is not a good assumption.

The functions $\beta(\rho)$ and $\varphi_2(\beta)$ are related by

$$C\rho\beta'(\rho)\,\varphi_2'(\beta(\rho))=\sigma_0. \tag{27}$$

One of these two functions, just as the constants C and D, must be chosen to match the theoretical and experimental results for given materials and micromechanisms. Simple choices are:

$$\beta=1-\rho \;\Rightarrow\; \varphi_2(\beta)=\frac{\sigma_0}{C}\,\ln(1-\beta), \tag{28}$$

$$\beta=\frac{1}{\rho}-1 \;\Rightarrow\; \varphi_2(\beta)=-\frac{\sigma_0}{C}\,\ln(1+\beta), \tag{29}$$

$$\beta=f-f_0 \;\Rightarrow\; \varphi_2(\beta)=\frac{\sigma_0}{C}\,\ln(1-f_0-\beta). \tag{30}$$

In the latter example (not really distinct from $\beta=1-\rho$), f is the void volume fraction, including the particles, the volume fraction of which is f_0. As the metal of the matrix is supposed to be incompressible in the plastic deformation, according to (18) without damage ($g\equiv0$), the relation between f and ρ is,

$$\rho=\frac{1-f}{1-f_0}. \tag{31}$$

An alternative choice for $\beta(\rho)$ is suggested by the Rice and Tracey formula (1) for the growth of a spherical void in case of high stress triaxiality,

$$\frac{\dot{R}_0}{R_0}=0.283\mathcal{D}_{eq}^P\exp\!\left(\frac{3\sigma_m}{2\sigma_0}\right), \tag{32}$$

(with the present notations). This formula is only valid for incipient void growth ($R=R_0$ and $\rho=1$) but it may be extended to the whole damage process. As $\lambda=\mathcal{D}_{eq}^P\sqrt{3}$ and

$$\frac{\dot{f}}{f(1-f)}=3\frac{\dot{R}}{R}, \tag{33}$$

(19) and (32) give, if $C=3/2$,

$$\frac{\dot{f}}{f(1-f)\dot{\beta}}=\frac{0.283\sqrt{3}}{D}. \tag{34}$$

It thus follows that,

$$C = 3/2, \qquad D = 0.49, \tag{35}$$

$$\beta = \ln \frac{f(1-f_0)}{f_0(1-f)} = \ln\left(1 + \frac{1-\rho}{\rho f_0}\right), \tag{36}$$

$$\varphi_2(\beta) = \frac{\sigma_0}{C} \ln \frac{1-f}{1-f_0} = -\frac{\sigma_0}{C} \ln(1 - f_0 + f_0 \exp \beta), \tag{37}$$

$$B(\beta) = -\varphi_2'(\beta) = \frac{\sigma_0 f}{C} = \frac{\sigma_0}{C} \frac{f_0 \exp \beta}{1 - f_0 + f_0 \exp \beta}. \tag{38}$$

The yield criterion, $F = 0$, Eq. (15), becomes,

$$\frac{\sigma_{eq}}{\rho} + 0.57\sigma_0 f \exp\left(\frac{3\sigma_m}{2\rho\sigma_0}\right) - \varphi_1'(\alpha) = 0. \tag{39}$$

This can be compared with the Gurson approximation (1977) resulting from his microscopic analysis, with the matrix material idealized as rigid—perfectly-plastic,

$$\sigma_{eq}^2 + 2\sigma_0^2 f \cosh\left(\frac{3\sigma_m}{2\sigma_0}\right) - (1 + f^2)\sigma_0^2 = 0. \tag{40}$$

If the stress triaxiality σ_m/σ_0 is large, $2 \cosh = \exp$. As the Gurson analysis is similar to that of Rice and Tracey, the correspondence between (39) and (40) is not unexpected[7].

4. Infinitesimal strain approximation

Assume both elastic and plastic deformations are small. The constitutive relations become,

$$\sigma = L\varepsilon^e, \tag{41}$$

$$\dot{e}^P = \lambda \frac{\sqrt{3} \, s}{2\sigma_{eq}}, \qquad \dot{\varepsilon}_m^P = \lambda \frac{DC}{3\sigma_0} B(\beta) \exp\left(\frac{C\sigma_m}{\sigma_0}\right), \tag{42}$$

$$\dot{\alpha} = \frac{\lambda}{\sqrt{3}}, \qquad \dot{\beta} = \lambda D \exp\left(\frac{C\sigma_m}{\sigma_0}\right). \tag{43}$$

---

[7] The consideration of σ_{eq}^2 instead of σ_{eq} in the plastic potential would make it difficult to give a physical meaning to the hardening variable α. With the present formulation, $\alpha = \int \mathcal{D}_{eq}^P \, dt$. For large σ_m/σ_0 and incipient damage (small f and f exp), an approximate form of Eq. (40) is, $\sigma_{eq} + 0.5\sigma_0 f \exp(3\sigma_m/2\sigma_0) - \sigma_0 = 0$, which is similar (39).

Since $\mathcal{D}^P$ is now the time derivative, $\dot{\varepsilon}^P = d\varepsilon^P/dt$, of plastic strain tensor, ε^P, with deviatoric and spherical symmetric parts denoted by e^P and $\varepsilon_m^P = \varepsilon_{kk}^P/3$, respectively.

The plastic potential is,

$$F(\sigma, \alpha, \beta) = \sigma_{eq} - \varphi_1'(\alpha) + DB(\beta) \exp\left(\frac{C\sigma_m}{\sigma_0}\right). \tag{44}$$

The plastic multiplier λ is expressed in terms of the stress rate $\dot{\sigma}$,

$$\lambda = \frac{1}{H} \frac{\partial F}{\partial \sigma} \dot{\sigma}, \tag{45}$$

where

$$H = \frac{\partial F}{\partial A_i} \frac{\partial^2 \varphi}{\partial \alpha_i \partial \alpha_j} \frac{\partial F}{\partial A_j}. \tag{46}$$

The elastic energy on the one hand, hardening ($\alpha_1 = \alpha$) and damage ($\alpha_2 = \beta$) on the other hand, are supposed to be uncoupled in the free energy (13), $\partial^2 \varphi/\partial \alpha_i \partial \varepsilon^e = 0$. Thus,

$$\dot{A}_i = \frac{d}{dt}\left(-\frac{\partial \varphi}{\partial \alpha_i}\right) = -\frac{\partial^2 \varphi}{\partial \alpha_i \partial \alpha_j} \dot{\alpha}_j = -\frac{\partial^2 \varphi}{\partial \alpha_i \partial \alpha_j} \lambda \frac{\partial F}{\partial A_j}. \tag{47}$$

In plastic loading, $F = \dot{F} = 0$, and hence,

$$\frac{\partial F}{\partial \sigma} \dot{\sigma} + \frac{\partial F}{\partial A_i} \dot{A}_i = 0. \tag{48}$$

Equations (47) and (48) can then be combined to give (45). Equation (45) yields,

$$\dot{\sigma}\dot{\varepsilon}^P = \lambda \frac{\partial F}{\partial \sigma}\dot{\sigma} = H\lambda^2. \tag{49}$$

If the matrix $Z_{ij} = \partial^2 \varphi/\partial \alpha_i \partial \alpha_j$ is positive, then $H \geq 0$ and the Drucker inequality holds (Nguyen, 1973; Nguyen and Bui, 1974),

$$\dot{\sigma}\dot{\varepsilon}^P \geq 0, \tag{50}$$

the equality implying $\dot{\varepsilon}^P = 0$. In the present theory,

$$Z = \begin{bmatrix} \varphi_1''(\alpha) & 0 \\ 0 & \varphi_2''(\beta) \end{bmatrix}. \tag{51}$$

Assume the hardening of the matrix material is positive, then $\varphi_1''(\alpha) > 0$.

If the damage is defined by $\beta = 1/\rho - 1$, (29), then $\varphi_2''(\beta) = \sigma_0/C(1+\beta)^2$ is also positive. In that case, as stated by Rousselier (1979), instability (in Drucker's sense), and therefore rupture, is impossible. It is still necessary to postulate a fracture criterion. In the next section it will be demonstrated that instability takes place if the infinitesimal strain hypothesis is relaxed.

The positiveness of Z is not a general rule. The choice of $\beta = 1 - \rho$, (28), or $\beta = f - f_0$, (30), yields $\varphi_2''(\beta) = -\sigma_0/C(1-\beta)^2$ or $\varphi_2''(\beta) = -\sigma_0/C(1-f_0 -\beta)^2$, both negative. Equation (36), suggested by the Rice and Tracey formula (1), gives,

$$\varphi_2''(\beta) = -\frac{\sigma_0}{C} \frac{(1-f_0)f \exp \beta}{(1-f_0 +f_0 \exp \beta)^2},$$ (52)

which is also negative. In infinitesimal strain numerical applications, the damage β should be defined so as to avoid a positive $\varphi_2''(\beta)$ and allow instability of the material.

5. Instability and fracture: example

Consider the homogeneous transformation without rotation of an element of material subjected to a triaxial stress state,

$$\sigma_{22} = \sigma_{33} = k\sigma_{11}, \qquad k \text{ constant} < 1,$$
$$\sigma_{12} = \sigma_{23} = \sigma_{31} = 0.$$ (53)

The plastic potential and the constitutive relations for $\dot{\alpha}$ and $\mathcal{D}_m^P = -\dot{\rho}/3\rho$ give,

$$\frac{1-k}{\sqrt{3}} \frac{\sigma_{11}}{\rho} - \frac{\varphi_1'(\alpha)}{\sqrt{3}} + DB(\beta) \exp\left[\frac{C(1+2k)}{3\sigma_0} \frac{\sigma_{11}}{\rho}\right] = 0,$$ (54)

$$\dot{\rho} = -\dot{\alpha} \frac{DC\sqrt{3}}{\sigma_0} \rho B(\beta) \exp\left[\frac{C(1+2k)}{3\sigma_0} \frac{\sigma_{11}}{\rho}\right] = 0.$$ (55)

According to Eq. (27), $B(\beta) = -\sigma_0/C\rho\beta'(\rho)$, and the combination of the two latter equations gives,

$$\frac{\dot{\rho}}{\rho\dot{\alpha}} - \frac{3(1-k)}{1+2k} \ln\left[\frac{\beta'(\rho)}{D\sqrt{3}} \frac{\dot{\rho}}{\dot{\alpha}}\right] + C\frac{\varphi_1'(\alpha)}{\sigma_0} = 0,$$ (56)

$$(1-k)\frac{\sigma_{11}}{\rho\sigma_0} - \frac{D\sqrt{3}}{C\rho\beta'(\rho)} \exp\left[\frac{(1+2k)C}{3} \frac{\sigma_{11}}{\rho\sigma_0}\right] - \frac{\varphi_1'(\alpha)}{\sigma_0} = 0.$$ (57)

Equation (56) is an algebraic equation for $\dot{\rho}/\dot{\alpha}$. From an actual state $(\sigma_{11}, \rho, \alpha)$, its numerical solution, considering an increment $\Delta\alpha$, yields the new values $\alpha + \Delta\alpha$, $\rho + (\dot{\rho}/\dot{\alpha})\Delta\alpha$. The new stress σ_{11} is given by the numerical solution of the algebraic Eq. (57).

The strain rates are,

$$d_{11}^P = -2d_{22}^P = -2d_{33}^P = \mathcal{D}_{eq}^P = \dot{\alpha},$$
$$d_{12}^P = d_{23}^P = d_{31}^P = 0, \qquad \mathcal{D}_m^P = -\frac{\dot{\rho}}{3\rho}.$$ (58)

We shall consider the "logarithmic plastic deformation tensor", $\hat{\varepsilon}^P = \int \mathfrak{D}^P \, dt$, the first component of which is,

$$\hat{\varepsilon}_{11}^P = \alpha - \tfrac{1}{3} \ln \rho. \tag{59}$$

The results are shown in Figs. 1–8. In Fig. 1 a power-hardening law $\varphi_1'(\alpha) = \sigma_0(1 + \sqrt{\alpha})$ is considered. The damage is defined by $\beta = 1/\rho - 1$, (29), or by $\beta = f - f_0$, (30). In both cases, after a certain amount of plastic deformation depending on stress triaxiality, *instability takes place*: $\dot{\sigma} \dot{\varepsilon}^P < 0$. The damage-related softening overcomes the strain-hardening of the matrix material.

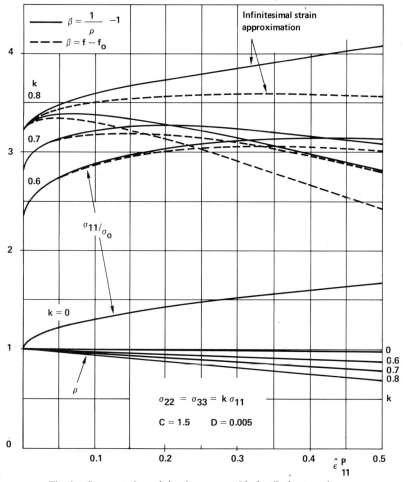

Fig. 1.   Stress–strain and density curves with ductile fracture damage.

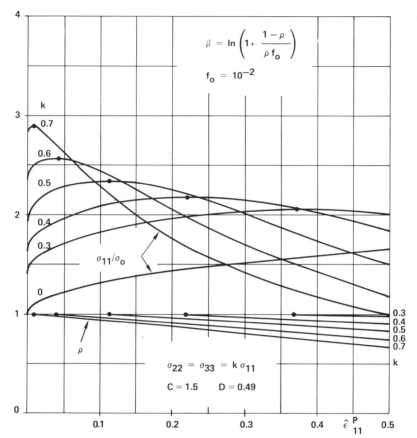

Fig. 2. Stress–strain and density curves with ductile fracture damage.

If the infinitesimal strain constitutive relations, (41)–(44), are used instead of the finite strain ones, instability is not revealed if $\beta=1/\rho-1$, even with a very high stress triaxiality ($k=0.8$). It was demonstrated in Section 4 that instability was impossible indeed in that case. If $\beta=f-f_0$, instability is still possible but is dramatically postponed (Fig. 1).

In Figs. 2 to 5 damage defined by $B=\sigma_0 f/C$ ((35) to (38)) is considered, as suggested by the Rice and Tracey formula, with the power-hardening law of Fig. 1. The effects of various stress triaxialities and various initial void volume fractions f_0 are investigated. As expected, the smaller the volume fraction f_0 the higher the stress triaxiality and plastic deformation required to yield instability. This is emphasized in Fig. 6, where the plastic strain at instability is plotted versus $\sigma_m/\sigma_{eq}\equiv(1+2k)/3(1-k)$ (that has to be distinguished from σ_m/σ_0). The dependence on f_0 is slightly stronger than

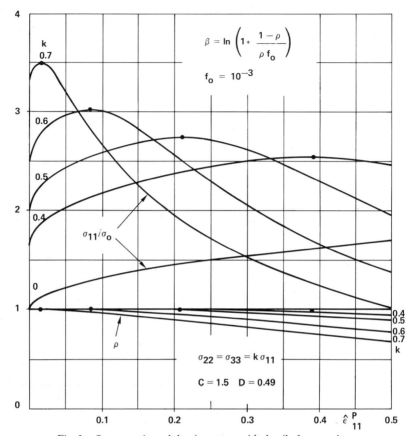

Fig. 3.   Stress–strain and density curves with ductile fracture damage.

the squaring of the initial void fraction required to double the critical strain obtained by McClintock (1968).

With $B=\sigma_0 f/C$, the infinitesimal strain curves, shown in Fig. 5, do not differ from the finite strain ones until instability. But the final decohesion resulting from these curves is entirely different as shown in Fig. 7 with a reduced strain scale (in Fig. 7, the power-hardening law is changed for an exponential one: $\varphi_1'(\alpha)=\sigma_0[2-\exp(-20\alpha)]$).

The finite strain curves of Fig. 7 clearly show that the present theory, with damage depending exponentially on σ_m/ρ, is perfectly fitted to the modelling of instability and decohesion in ductile fracture. *This is not the case with an exponential dependence on σ_m alone.*

Finally, it should be noticed that the use of fracture criteria, here $\beta=\beta_c$ or $\rho=\rho_c$ or $f=f_c$, is not consistent with the results of the present analysis. The

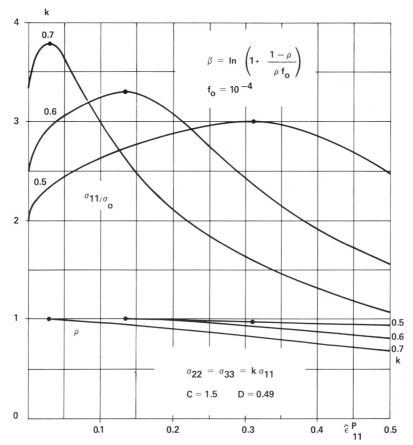

Fig. 4.   Stress–strain and density curves with ductile fracture damage.

critical values of β, ρ or f, corresponding to material instability, depend on the mean stress σ_m; see Fig. 8. This is in agreement with the experimental results of Beremin (1979) where a decrease of the critical void growth at instability with increasing stress triaxiality is obtained.

6.   Finite element analysis of a cracked specimen

The analysis of ductile fracture and stable crack growth in a three-point bend specimen has been performed with a 2D infinitesimal strain finite element model (plane strain). As pointed out in Sections 4 and 5, the infinitesimal strain approximation is a serious limitation of the present

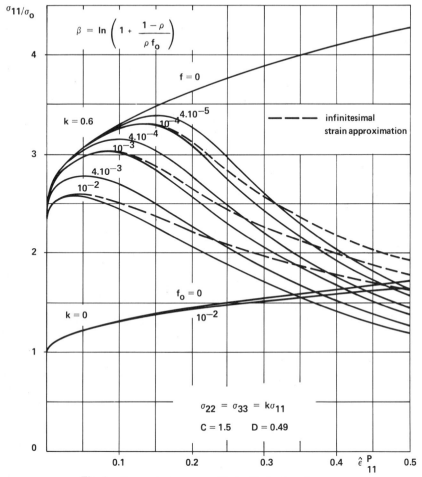

Fig. 5.   Stress–strain curves with ductile fracture damage.

theory. That is why Eqs. (41)–(44) are modified in order to *include some finite deformation effect*: namely, the ratio σ_m/ρ is substituted for σ_m into all the exponentials, according to the finite deformation relation (26), and σ_{eq}/ρ for σ_{eq} into the plastic potential (44).

A constant stiffness method is used. It requires the iterative computation of the "initial stress" $S = L\Delta\varepsilon^P$ that appears in the linear elastic relation between the increments of stress and strain: $\Delta\sigma = L\Delta\varepsilon - S$, which is the approximate form of Eq. (23). The initial stress, S, is computed with an implicit algorithm (Nguyen, 1973); see Appendix. This algorithm eliminates the systematic numerical errors usually found in the explicit method and is

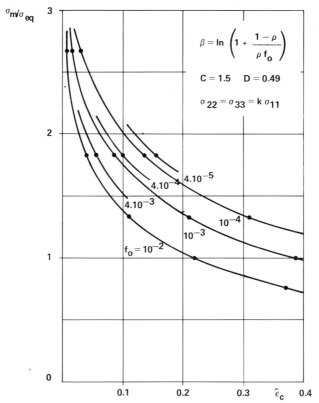

Fig. 6. Plastic strain at instability versus σ_m/σ_{eq}.

particularly suited to the present analysis where material instability $\dot{\sigma}\dot{\varepsilon}^P < 0$ takes place.

The geometry of the specimen is specified in Fig. 9. The increments of the displacement $u_1 = d$ of node A are prescribed; the corresponding load is $F_1 = P/2B$. The specimen is discretized into 377 constant strain triangular elements, 420 nodes and 840 degrees of liberty. The size of the finite elements at the crack tip is shown in Fig. 10.

Note that at the tip of a crack, where steep gradients of stresses and especially strains take place, the critical conditions for instability shall be achieved over some characteristic length l_c, related to interparticle spacing. Otherwise void coalescence and material decohesion will not occur. In the numerical modelling, with constant strain elements, l_c is the length of the finite elements at the crack tip. If the specimen size or geometry are changed, the elements at the crack tip will keep the same absolute size.

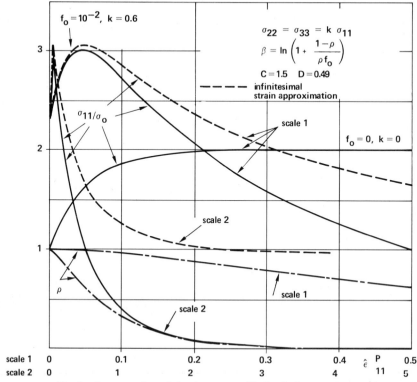

Fig. 7.   Stress–strain and density curves with ductile fracture damage.

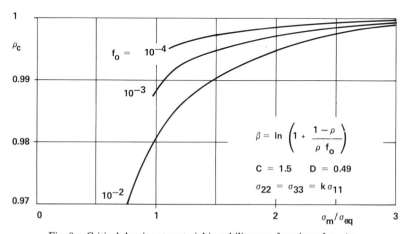

Fig. 8.   Critical density at material instability as a function of σ_m/σ_{eq}.

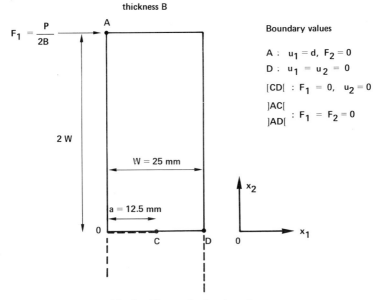

thickness B

$$F_1 = \frac{P}{2B}$$

A

2 W

W = 25 mm

a = 12.5 mm

0

C

D

0

x_2

x_1

Boundary values

A : $u_1 = d$, $F_2 = 0$

D : $u_1 = u_2 = 0$

[CD[: $F_1 = 0$, $u_2 = 0$

]AC[
]AD[: $F_1 = F_2 = 0$

Fig. 9.   Three-point bend specimen.

	d (mm)	P/B (N/mm)
– – –	0	0
——	0.548	686
——	0.748	652

$$\beta = \ln \left(1 + \frac{1-\rho}{\rho\, f_o}\right)$$

Displacements x 10

$l_c = 0.5$ mm

C = 1.5   D = 0.49

$f_o = 10^{-2}$

1

0.5

0.05
0

ω

ω

ω

11.4   12   13   13.5   14

C   12.5

Fig. 10.   Localization of deformation in the crack tip elements, by ductile fracture damage, resulting in stable crack growth.

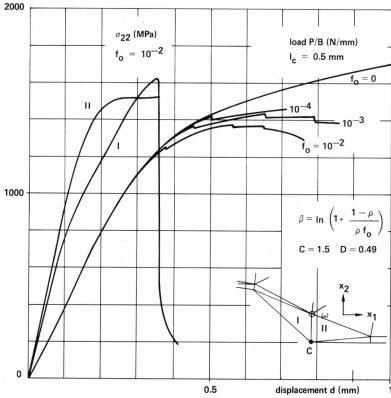

Fig. 11.   Load-displacement curves of a three point bend specimen, with ductile fracture damage. Stress curves in two crack tip elements showing the local instability (crack tip geometry different from that of load curves).

Moreover, in order to model stable crack growth, the length of the elements on the crack prolongation are uniform.

The same exponential strain-hardening law as in Fig. 7 is considered; $\varphi'_1(\alpha) = \sigma_0[2 - \exp(-20\alpha)]$ where $\sigma_0 = 500$ MPa. The elastic constants are $E = 200$ GPa and $\nu = 0.3$. According to the discussion in Sections 4 and 5, the definition of damage (35)–(38) is used. So, *the ductile fracture properties of the metal are defined by two parameters*: (1) l_c, related to the *interparticle spacing*; and (2) f_0, related to the *particle volume fraction*. These are required for the characterization of the particle size distribution[8].

The deformation of the crack tip zone is shown in Fig. 10. In the two most deformed elements, undergoing stress triaxialities σ_m / σ_0 of nearly 3, the damage increases rapidly. At some point, corresponding to the maximum of

---

[8]Note that in a tensile specimen, with no strain gradients, the ductility is dependent on the volume fraction f_0 only, as observed by Edelson and Baldwin (1962).

the local stress–strain curves, as in Fig. 7, the deformation of these elements increases abruptly, and the stresses decrease according to curves I and II of Fig. 11. This quasi-rupture of the two elements may be identified with the coalescence stage of the ductile fracture process. The node ω (Fig. 10) is no longer bound to node C (initial crack tip) nor to the symmetric node ω'. So *stable crack growth occurs naturally, by localization of deformation, resulting from the constitutive relations only, without it being necessary to define a critical state nor to release the nodes* as in usual models. A further stage of stable crack growth (four nodes, $d = 0.748$ mm) is shown in Fig. 10; the strains in the most deformed elements exceed unity.

As constant strain elements are used, the localization of deformation takes place over the whole element. That is why very thin elements are needed along the crack path. This drawback should be avoided with nonlinear elements that would give a better localization of deformation and allow crack propagation in any direction. Note that the blunting of the initial crack tip, that may be of importance for the initiation of crack growth, is not modelled by the simple constant strain elements of Fig. 10.

The load-displacement curves are given in Fig. 11 for various initial void volume fractions $f_0 = 10^{-2}$, 10^{-3}, 10^{-4} and 0 (no damage). In spite of strain hardening, up to $2\sigma_0$ at the most, the load rapidly stops increasing when ductile fracture and stable crack growth take place. After the maximum load the ductile tearing of the specimen goes on under decreasing load, in agreement with the acknowledged experimental evidence.

7. Conclusions

Constitutive relations have been developed for finite transformation of plastic dilatant materials. The dilatancy is related to the growth of voids nucleated at particles present in the matrix material. These constitutive relations model the ductile fracture of metals. They have been derived, in a macroscopic approach, from the internal variables hypothesis within the framework of generalized standard materials.

An exponential dependence of ductile fracture damage on stress triaxiality is demonstrated, and the occurrence of material instability $\dot{\sigma}\dot{\varepsilon}^P < 0$ shown for various examples. The finite transformation effects are also discussed.

Finite element analyses of a cracked specimen have been performed. Stable crack growth occurs naturally by localization of deformation, which is in better agreement with the physical behavior of metals, and the method is simpler, and probably cheaper than usual node relaxation models. An advantage of the present approach is that nonsymmetrical complex loading conditions (the angled crack extension) and 3D-problems could be handled, without special finite element modelling in the crack tip zone.

In further applications nonlinear elements must be considered, as constant strain triangular elements are not suitable for localization of deformation. The necessity of modelling the initial crack tip blunting and of a complete finite transformation formulation, according to the equations of Section 3, should also be investigated.

As for the parameters involved in the theory, C and D are derived from microscopic models, and could be checked by experiments. The testing and computing of circumferentially notched tensile specimens give f_0, which should be consistent with the observed inclusions and/or precipitates volume fractions, according to the micromechanisms involved. The characteristic length l_c that intervenes in crack problems, though relevant to microstructural aspects (a few interparticle spacings), must be regarded as essentially an empirically obtained quantity. Its determination requires experiments on appropriate specimen geometries, which can be modelled with 2D-finite element analyses, like side grooved CT specimens or circumferentially cracked tensile specimens. Such an experimental program, in collaboration with the French research group Beremin, is in progress on A508 Cl 3 steel.

Appendix. Implicit algorithm for the constitutive relations with ductile rupture damage

In the numerical solution of the elastic-plastic problem, for the increment $t-t+\Delta t$, the initial stress $S=L\Delta\varepsilon^P$ has to be expressed as a function of $\sigma(t)$, $\alpha_i(t)$, and $\Delta\varepsilon$. The following form of the constitutive relations is used:

$$\Delta s = 2\mu(\Delta e - \Delta e^P), \tag{A.1}$$

$$\Delta\sigma_m = (3\lambda^* + 2\mu)(\Delta\varepsilon_m - \Delta\varepsilon_m^P), \tag{A.2}$$

$$\Delta e^P = \lambda \frac{s + \Delta s}{2\sqrt{J_2(s + \Delta s)}}, \tag{A.3}$$

$$\Delta\varepsilon_m^P = \lambda \frac{DC}{3\sigma_0} B(\beta + \Delta\beta) \exp\left(\frac{C}{\rho\sigma_0}(\sigma_m + \Delta\sigma_m)\right), \tag{A.4}$$

$$\Delta\alpha = \frac{\lambda}{\sqrt{3}}, \tag{A.5}$$

$$\Delta\beta = \lambda D \exp\left(\frac{C}{\rho\sigma_0}(\sigma_m + \Delta\sigma_m)\right), \tag{A.6}$$

$$\sqrt{3J_2\left(\frac{s + \Delta s}{\rho}\right)} + \sqrt{3}\, DB(\beta + \Delta\beta) \exp\left(\frac{C}{\rho\sigma_0}(\sigma_m + \Delta\sigma_m)\right)$$
$$- \varphi_1'(\alpha + \Delta\alpha) = 0. \tag{A.7}$$

These relations are written at time $t+\Delta t$ instead of t in the explicit algorithm, and the plastic potential $F=0$, (A.7), is used instead of Eq. (45). If Δs and $\Delta \sigma_m$ are eliminated between (A.1) and (A.3), (A.2) and (A.4) respectively,

$$\Delta \varepsilon_m^P = \Delta \alpha \frac{DC}{\sigma_0 \sqrt{3}} B(\beta + \Delta \beta)$$

$$\times \exp\left\{ \frac{C}{\rho \sigma_0} \left[\sigma_m + (3\lambda^* + 2\mu)(\Delta \varepsilon_m - \Delta \varepsilon_m^P) \right] \right\} \quad (A.8)$$

$$s + \Delta s = s + 2\mu(\Delta e - \Delta e^P) = s + 2\mu \Delta e - \mu \Delta \alpha \sqrt{3} \frac{s + \Delta s}{\sqrt{J_2(s + \Delta s)}},$$

then,

$$\left(1 + \frac{\mu \Delta \alpha \sqrt{3}}{\sqrt{J_2(s + \Delta s)}} \right)(s + \Delta s) = s + 2\mu \Delta e. \quad (A.9)$$

Taking the second invariant of this equation, we obtain,

$$\sqrt{J_2(s + \Delta s)} + \mu \Delta \alpha \sqrt{3} = \frac{\gamma}{\sqrt{3}}, \quad (A.10)$$

where

$$\gamma^2 = \frac{3}{2}(s + 2\mu \Delta e)(s + 2\mu \Delta e). \quad (A.11)$$

Equation (A.9) gives,

$$\frac{\gamma(s + \Delta s)}{\sqrt{3 J_2(s + \Delta s)}} = s + 2\mu \Delta e.$$

Finally,

$$\Delta e^P = \frac{3\Delta \alpha}{2\gamma}(s + 2\mu \Delta e). \quad (A.12)$$

Equations (A.7) and (A.10) give,

$$\frac{\gamma - 3\mu \Delta \alpha}{\rho} - \varphi_1'(\alpha + \Delta \alpha) + \sqrt{3} \, DB(\beta + \Delta \beta)$$

$$\times \exp\left\{ \frac{C}{\rho \sigma_0} \left[\sigma_m + (3\lambda^* + 2\mu)(\Delta \varepsilon_m - \Delta \varepsilon_m^P) \right] \right\} = 0. \quad (A.13)$$

The exponential is eliminated between (A.8) and (A.13),

$$\Delta \varepsilon_m^P = \frac{C\Delta \alpha}{3\sigma_0}\left(\varphi_1'(\alpha + \Delta \alpha) + \frac{3\mu \Delta \alpha - \gamma}{\rho} \right), \quad (A.14)$$

354   G. Rousselier

and (A.6) is written as

$$\Delta\beta = D\Delta\alpha\sqrt{3} \ \exp\left\{\frac{C}{\rho\sigma_0}\left[\sigma_m + (3\lambda^* + 2\mu)\left(\Delta\varepsilon_m - \Delta\varepsilon_m^P\right)\right]\right\}. \quad (A.15)$$

The substitution of (A.14) for $\Delta\varepsilon_m^P$ into (A.13) and (A.15) and of (A.15) for $\Delta\beta$ into (A.13) yields a *simple algebraic equation for $\Delta\alpha$ alone*. The numerical solution of this equation gives $\Delta\alpha$ as a function of σ, α, β and $\Delta\varepsilon$. The initial stress $S = L\Delta\varepsilon^P = (3\lambda^* + 2\mu)\Delta\varepsilon_m^P + 2\mu\Delta e^P$ is then given by (A.12) and (A.14).

Note: With the infinitesimal strain hypothesis, $\rho \equiv 1$. In the above equations the density ρ is given by (A.15) and $\beta + \Delta\beta = \beta(\rho)$ (see Section 3).

References

Auger, J. P. and François, D. (1977), "Variation of Fracture Toughness of a 7075 Aluminum Alloy with Hydrostatic Pressure and Relationship with Tensile Ductility," *Int. J. Fract.*, **13**, 321.

Beremin (1979), *Rapport d'Activité 1979*, Framatome, Paris, TM/C DC/79.095.

Edelson, B. I. and Baldwin, W. M., Jr. (1962), "The Effect of Second Phases on the Mechanical Properties of Alloys," *Trans. Am. Soc. Metals*, **55**, 230.

Gurson, A. L. (1977), "Continuum Theory of Ductile Rupture by Void Nucleation and Growth: Part I—Yield Criteria and Flow Rules for Porous Ductile Media," *J. Engg. Mat. Tech.*, **99**, 2.

Hancock, J. W. and Mackenzie, A. C. (1976), "On the Mechanisms of Ductile Failure in High-strength Steels Subjected to Multi-axial Stress-states," *J. Mech. Phys. Solids*, **24**, 147.

Henry, G. and Hortsmann, D. (1979), *De Ferri Metallographia*, V, Verlag Stahleisen, Düsseldorf.

Kachanov, L. M. (1958), "Time of the Rupture Process under Creep Conditions," *Izv. Akad. Nauk. S.S.R. Otd. Tekh. Nauk.*, **8**, 26.

Lautridou, J. C. and Pineau, A. (1978), private communication, published in Beremin (1979); see above.

Lemaitre, J. and Chaboche, J. L. (1978), "Aspect Phénomènologique de la Rupture par Endommagement," *J. Mécanique Appl.*, **2**, 317.

McClintock, F. A. (1968), "A Criterion for Ductile Fracture by the Growth of Holes," *J. Appl. Mech.*, **35**, 363.

Mandel, J. (1973), "Equations Constitutives et Directeurs dans les Matériaux Plastiques et Viscoplastiques," *Int. J. Solids Struct.*, **9**, 725.

Moreau, J. J. (1970), "Sur les Lois de Frottement de Plasticité et de Viscosité," *C.R. Acad. Sci.*, série A, **271**, 608.

Nguyen, Q. S. (1973), "Contribution à la Théorie Macroscopique de l'Élastoplasticité avec Écrouissage," Thèse Doctorat-ès-Sciences, Paris, CNRS A.O. 9317.

Nguyen, Q. S. and Bui, H. D. (1974), "Sur les Matériaux Élastoplastiques à Écrouissage Positif et Négatif," *J. Mécanique*, **13**, 321.

Rabotnov, Y. N. (1968), "Creep Rupture," *Proc. XII Int. Cong. Appl. Mech.*, Springer Verlag, Stanford, 342.

Rice, J. R. and Tracey, D. M. (1969), "On the Ductile Enlargement of Voids in Triaxial Stress Fields," *J. Mech. Phys. Solids*, **17**, 201.

Rousselier, G. (1979a), "Contribution à l'Étude de la Rupture des Métaux dans le Domaine de l'Élastoplasticité," Thèse Doctorat-ès-Sciences, Paris, CNRS.

Rousselier, G. (1979b), "Numerical Treatment of Crack Growth Problems," *Advances in Elasto-Plastic Fracture Mechanics*, L. H. Larsson (ed.), Applied Science Publishers, London, 165.

S. Nemat-Nasser, Editor
THREE-DIMENSIONAL CONSTITUTIVE RELATIONS AND DUCTILE FRACTURE
North-Holland Publishing Company (1981) 357–362

DISCUSSION ON SESSION 6

H. D. BUI

Ecole Polytechnique, Palaiseau, France

The contributions to this session deal with some aspects of crack propagation. The lecture presented by Nguyen is focused on ductile fracture. The subject relates closely to Cherepanov's work. Since I am familiar with the main lines of his paper, I think that it is interesting to discuss this afternoon's papers in connection with Cherepanov's published work. Let me remark that the contribution by Rousselier makes a link between the two subjects of the symposium: Constitutive relationships and ductile fracture.

1. In relation to Nguyen's lecture, I would like to make some comments. The paper is firstly focused on the interpretation of the energy release rate, G. There are many well-known interpretations of G: decrease of potential energy, generalized force, fracture energy rate $\dot{a}G$, entropy production (Rice, 1978). Neglecting true surface energy, acoustic emission or radiation, the new interpretation is the point heat source at the moving crack tip, $\dot{a}G\delta(A)$. The consequence of this new result is the logarithmic singularity of the positive temperature field in a heat conducting material which satisfies the Fourier law (Bui, Ehrlacher and Nguyen, 1979, 1980),

$$T \simeq -(G\dot{a}/2k\pi)\log r + \cdots. \tag{1}$$

Unbounded temperature, or high temperature at the tip of a crack can be obtained from the literature. While the earlier works focused on the local plastic flow (Rice and Levy, 1969; Weichert and Schönert, 1978), the work by Nguyen and others (1979) shows that unbounded temperature occurs even in linear elasticity, without any reference to plasticity at the tip, because of the dissipative nature of fracture. The analysis presented by Nguyen is noticeably different from Gurtin's analysis (1979) which did not consider unbounded temperature.

One can object that an unbounded temperature is meaningless, since no fusion has been observed, at least on a macroscopic scale. In fact, experimental works, (Weichert, Schönert, 1978), have shown evidence of a very high temperature in fast fracture of glass ($T \simeq 3000$ K), and steel, in a very small zone ($\simeq 20$ Å) of the same order as the characteristic length, which can be derived from Eq. (1). Light emission was observed in the fracture of glass. Fuller and others (1975) measured a temperature of 500 K in polymers. The high temperature observed in the fast fracture of steel can be explained by the local plastic deformation. However, the point heat source does not theoretically exist in perfect plasticity, because $G=0$. The temperature is finite, although its value is high at the tip. This is only true in a heat conducting material satisfying Fourier's law, because the singularity of the term $\mathrm{div}(q)=-k\Delta T$ is stronger than that of $\dot{T}$. In the case of nonconducting material, the singularities of $\dot{T}$ and the plastic power are the same. I just mention Fourier's law. I am not sure that the analysis still applies to real materials, the coefficient of conduction of which probably depends on T. A realistic analysis must also take into account the latent heat of fusion of the solid.

2. The second point of the first lecture concerns the paradox $G=0$, pointed out by Rice some time ago. In order to clarify this point, and in connection with Cherepanov's works and Rousselier's lecture, I would like to present a model for the discussions.

Let us introduce the path-dependent integral of Eshelby (1968),

$$J_\Gamma = \int_\Gamma \left\{ \overline{W}(\varepsilon - \varepsilon^\mathrm{P})n_1 - T_i u_{i,1} \right\} \mathrm{d}s, \tag{2}$$

where the elastic energy density $\overline{W}$ is used, instead of the stress working density $W = \int \sigma \, \mathrm{d}\varepsilon$. Eshelby denoted the integral by F_1 and gave the interpretation of the generalized force acting on the "inhomogeneities" or "sources of internal stress" lying within the path Γ. Plastic deformation is indeed a "source of internal stress". Recently, the J-integral (2) was used by Ehrlacher and myself (1980) in connection with a model of fracture which took account of the process zone *or* the damaged zone. Briefly speaking, for any contour Γ, the J-integral (times the velocity $\dot{a}$) in the steady state condition, is equal to the energy rate dissipated through the contour Δ of the process zone, i.e. $\dot{a}J_\Delta$, plus the plastic power of the domain between Γ and Δ,

$$\dot{a}J_\Gamma = \dot{a}J_\Delta + \int_\Omega \sigma\dot{\varepsilon}^\mathrm{P} \, \mathrm{d}v. \tag{3}$$

The last equation is similar to the energy equation of Cherepanov. Replacing the left-hand side of Eq. (3) by the effective surface energy $2\gamma_*$ (the Griffith–Orowan concept) and the terms of the right-hand side respectively by the separation energy rate of order AK^2, and the plastic power of order $BK^3 \mathrm{d}K/\mathrm{d}a$, where K is the "stress intensity factor" of the "fine structure," one obtains Cherepanov's model,

$$2\gamma_* = AK^2 + BK^3 \frac{\mathrm{d}K}{\mathrm{d}a}. \tag{4}$$

Integrating Eq. (4) to get his R-curve (K^2 versus a), one has,

$$a - a_0 = -\frac{BK^2}{2A} - \frac{B\gamma_*}{A^2} \log\left(1 - \frac{AK^2}{2\gamma_*}\right). \tag{5}$$

Equation (5) has various applications, such as monotonic crack growth, fatigue crack growth. One observes that, without the separation energy rate ($A = 0$ or $J_\Delta = 0$), the R-curve increases without limit.

3. Instead of a large contour Γ, let us consider a small one and *neglect* the process zone. Let Γ and Δ tend to zero. Then the J_Γ or J_Δ integral vanishes. Equivalently, the crack extension force is zero.

This kind of paradox is frequently observed in other fields of continuum mechanics, e.g. the well-known paradox of d'Alembert. The null drag coefficient $F_1 = 0$ of a body in the irrotational flow of a perfect fluid, results from the infinite Reynolds number, $\mathfrak{R} = \infty$, and the absence of wake or cavitation zone.

Similarly, in perfect plasticity, the null generalized force, $G = 0$, results from the fact that the ductility of metal has been assumed to be infinite, $\varepsilon_R = \infty$, in the crack model.

Consequently, we do not have the paradox $G = 0$ in models in which both stress and strain are finite. Otherwise, there is a change of state of the material which has so many cavities, microcracks, etc., that it cannot support any stress, just like the cavitation in fluid dynamics where the pressure is zero. The zero-pressure criterion, or the constant velocity $|V|$, determine the cavitation zone.

As a result of the assumption on finite values of stress and strain in the constitutive relations, we obtain a finite characteristic dimension Δ, and completely determine the process zone.

In the literature, the characteristic dimension was generally introduced in the models of cracks, via the fracture strain, ε_R (or ε_f), criterion, the mean distance between voids etc.

Also, the concept G^Δ of Kfouri and Rice (1977) has illustrated the necessity of the process zone of finite length.

4. Let us show two analytical examples of the process zones obtained for Mode III loading (Bui, 1980). The first example corresponds to the elastic-brittle model. We obtain a strip of finite thickness relating to the external load and the fracture stress, and the front in the shape of a cusped cycloid.

The second example takes account of plasticity. The plastic-damage front is a cusped cycloid. The elastic-plastic boundary is a curled cycloid (Fig. 1).

In the particular case where the ductility of metal is unbounded, we recover the model of a crack without a process zone and the well-known circular plastic zone of Hult and McClintock.

FLOW PASSED A BODY   PROCESS-ZONES IN MODE III LOADING

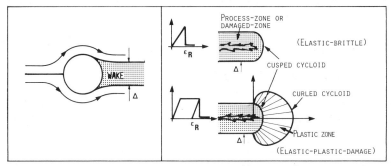

Fig. 1.   An analogy between fluid dynamics and ductile fracture mechanics. Wake zone in the potential flow and process zone in ductile fracture. Examples of process-zone or damaged-zone in Mode III (Bui, 1980, and Bui and Ehrlacher, 1980).

5. We have just discussed some aspects of ductile fracture in relation to damage. Rousselier's contribution is central to our discussion. Rousselier develops a model of plasticity with damage effects, by introducing the plastic parameter α and the damage parameter β. The theory is confined to isotropic and uncoupled effects. There is some similarity, in the choice of internal parameters, with the model independently developed by Dragon and Mroz (1979) who also introduce two kinds of parameters, hardening parameters and softening parameters. In many direct approaches (Nguyen and Bui, 1974; Sidoroff, 1975; Dafalias, 1977) no distinction has been made between these internal parameters, although the softening behavior has been analyzed in the mentioned works.

Rousselier confines himself to time-independent effects and is successful in his analysis. His first result is related to the exponential coefficient of the damage component of the potential F. The result agrees with the Gurson's conclusion (1977) obtained from a different point of view. Also it agrees with the growing cavity model of Rice and Tracey. The second result is the instability criterion of the material, in the Drucker's sense $\dot{\sigma}\dot{\epsilon}^{p} < 0$, which is obtained by various and simple choices of the function $\beta(\rho)$. The thermodynamic approach of Rousselier has many implications, however, perhaps one may object to the uncoupling between plasticity and damage. In fact, the model can be generalized to more complex situations. The open question is how to effectively determine the physical constants introduced in the theory.

In particular, the experimental characterization of the damage state by a measure of the density does not seem to be straightforward. Recent work done by Lemaitre et al. (1979), shows a possible characterization of damage by the elastic property changes. As to the question of the coupling effects, I think that any attempt at their analysis should be of great interest.

6. In connection with the question about plasticity and damage interaction, I would like to recall the constitutive equations in plasticity given by Mandel (1966),

$$\dot{\varepsilon}^{R}=\tfrac{1}{2}\sum_{r}\left(n^{r}\times m^{r}+m^{r}\times n^{r}\right)\dot{\gamma}^{r}, \tag{6a}$$

$$\tfrac{1}{2}\dot{\sigma}:\left(n^{r}\times m^{r}+m^{r}\times n^{r}\right)=\sum_{s}h^{rs}\dot{\gamma}^{s}, \tag{6b}$$

where n^{r}, m^{r} are respectively the normal of the glide plane and the direction of glide associated with the slip mechanism (r), and $\dot{\gamma}^{r}$ is the shear strain rate. If one assumes that damage is due to microcrack formation, one can write down similar equations. Consider the family (r) of randomly distributed microcracks, the plane of which is characterized by the normal N^{r}. The mean opening rate is $\dot{\delta}^{r}$, in the direction M^{r}. Then the damage rate can be defined by the tensor of Mroz (1977),

$$\dot{D}=\tfrac{1}{2}\sum_{r}\left(N^{r}\times M^{r}+M^{r}\times N^{r}\right)\dot{\delta}^{r}. \tag{7}$$

The spherical part of $\dot{D}$ contributes to the volumetric part of the plastic strain rate, as in Rousselier's model. The static counterpart of Eq. (7) is:

$$\tfrac{1}{2}\dot{\sigma}:\left(N^{r}\times M^{r}+M^{r}\times N^{r}\right)=\sum_{s}p^{rs}\dot{\delta}^{s}+\sum_{s}q^{rs}\dot{\gamma}^{s} \tag{8}$$

The coupling effect can be also introduced in Eq. (6) by adding the term $g^{rs}\dot{\delta}^{s}$ to Eq. (6b).

Referring to the work of Zarka, (1972), I mention that the interaction matrix h of Eq. (6), can be determined by simple models of dislocations. I wonder whether similar models of interactions between dislocations and microcracks, or cavities, can be worked out in order to theoretically derive the matrices h, p and q.

Perhaps the microscopic approach seems to be more complex than the thermodynamic approach, or the direct approaches which consist of postulating, a priori, the constitutive equations in terms of internal variables. However, I think that microscopic approaches are of interest because of the physical meaning of the internal parameters which are geometrically well understood. Moreover, the relation between the microscopic scale and the macroscopic scale is also well understood, after the works of Hill (1966), Mandel and his school (1966), Kröner (1958) and Hutchinson (1970).

With regard to the present discussion, I mention the recent work of Dangvan and Cordier (1980) who made use of the microscopic–macroscopic relationships and of a criterion of fracture along slip planes for their model of damage.

References

Bui, H. D., Ehrlacher, A. and Nguyen, Q. S. (1979), *C. R. Acad. Sci. Paris*, B, **289**, 211.
Bui, H. D., Ehrlacher, A. and Nguyen, Q. S. (1980), *J. Mécanique*, to be published.
Bui, H. D. and Ehrlacher, A. (1980), *C. R. Acad. Sci. Paris*, B, **290**, 273.
Bui, H. D. (1980), *C. R. Acad. Sci. Paris*, B, **290**, 345.
Bui, H. D., Zaoui, A. and Zarka, J. (1972), *Foundation of Plasticity*, A. Sawczuk (ed.), 51.
Cherepanov, G. P. (1969), *Int. J. Solids Struct.*, **5**, 863.
Dafalias, Y. F. (1977), *Int. J. Non-Linear Mech.*, **12**, 327.
Dangvan, K. and Cordier, G. (1980), *Symp. IUTAM Senlis, France*.
Dragon, A. and Mroz, Z. (1979), *Int. J. Rock Mech.*, **16**, 253.
Eshelby, J. D. (1968), *Solid State Phys.*, II, Academic Press, New York, 79.
Fuller, K. N. G., Fox. P. G. and Field, J. E. (1975), *Proc. Roy. Soc.*, **A341**, 537.
Gurtin, M. E. (1979), *Int. J. Solids Struct.*, **15**, 553.
Gurson, A. L. (1977), ASME, *Engng. Mat. Tech.*, **99**, 2.
Hill, R. (1966), *J. Mech. Phys. Solids*, **14**, 95.
Hutchinson, J. W. (1970), *Proc. Roy. Soc.* **319**, 247.
Kfouri, A. P. and Rice, J. R. (1977), *Fracture ICF4*, **1**, 43.
Kröner, E. (1958), *Z. Physik*, 151.
Lemaitre, J., Cordebois, J. P. and Dufailly, J. (1979), *C. R. Acad. Sci. Paris*, B, 5.
Mandel, J. (1966), *Proc. 11th Int. Cong. Appl. Mech.*, Munich, 502.
Mroz, Z. (1977), Private communication.
Nguyen, Q. S. (1980), *J. Mécanique*, to be published.
Nguyen, Q. S. and Bui, H. D. (1974), *J. Mécanique*, **13**, 321.
Rice, J. R. (1978), *J. Mech. Phys. Solids*, **26**, 61.
Rice, J. R. and Levy, N. (1969), *Phys. Strength Plasticity*, Cambridge, 277.
Sidoroff, F. (1975), *Arch. Mech.* **27**, 807.
Weichert, R. and Schönert, K. (1978), *J. Mech. Phys. Solids*, **26**, 151.

SESSION 7

Chairman: A. Sawczuk
               Polish Academy of Sciences, Institute of Fundamental
               Technological Research, Warsaw, Poland

Authors: T. Yokobori, A. T. Yokobori, Jr., H. Sakata, I. Maekawa
               A. Dragon
               H. Lippmann

Discussion: Z. Mróz

Replies: T. Yokobori
               H. Lippmann

S. Nemat-Nasser, Editor
THREE-DIMENSIONAL CONSTITUTIVE RELATIONS AND DUCTILE FRACTURE
North-Holland Publishing Company (1981) 365–386

CONSTITUTIVE EQUATIONS AND GLOBAL CRITERIA FOR DUCTILE FRACTURE

Takeo YOKOBORI, A. Toshimitsu YOKOBORI, Jr.,
Hiroshi SAKATA and Ichiro MAEKAWA

Tohoku University, Sendai, Japan

Constitutive equations and global criteria for fracture which involve large scale yielding, as in ductile fracture, are presented. Unstable and stable ductile fracture are treated. The conclusions obtained are as follows: (1) For unstable ductile fracture, involving large scale yielding and void formation, the global criterion for fracture is obtained by the inclusion of a singularity at the tip of the plastic region; (2) For stable ductile fracture at high temperature, the following specific results are obtained: (a) The global criterion, as parametrically represented for crack growth in creep, fatigue and creep-fatigue interaction, is obtained for the Region II of crack growth curve of stainless steel: (b) The parameter in the criterion thus obtained has both physical and practical significance just like the Larson-Miller parameter for creep fracture time in the case of an unnotched or uncracked specimen: (c) A fracture parameter, P, is introduced which is related to the macroscopic local stress at the crack tip.

0. Notation

a	half length of the crack
a_0	the length of initial notch
a^*	actual crack length
a_{eff}	effective crack length taking into account the initial notch
A_p, A_f, L_1, L_2, L_3 and M	constant
G	modulus of rigidity
J'	modified J-integral
k	Boltzmann's constant
$\dot{l}$	elongation rate
L	potential energy of applied load
n	the number of nucleated voids
P	proposed fracture parameter
s	the plastic zone length
T	absolute temperature
t_h	holding time
W	half width of the specimen
W_e	strain energy

W_p plastic work accompanying the crack extension
W_v energy dissipated by the void nucleation
$\dot{\varepsilon}$ creep strain rate
γ specific surface energy
γ_e equivalent surface energy
λ strain hardening exponent
σ_g gross section stress
σ_1 local stress near the crack tip
σ_Y yield stress
ρ the radius of the crack tip
ν Poisson's ratio

1. Introduction

For ductile fracture accompanying large scale yielding, Yokobori and Ichikawa (1966) showed that the energy change of the system accompanying the crack extension in the Dugdale–Bilby–Cottrell–Swinden (DBCS) Model (Dugdale, 1960; Bilby, Cottrell and Swinden, 1963) is always positive, and thus crack growth cannot be unstable. This is inconsistent with the real case where unstable ductile fracture occurs. This paradox has been solved by Yokobori *et al.* (1968; Yokobori, 1968) by proposing a new model with the singularity at the tip of the plastic region. The model is different from the DBCS model in that the singularity exists at the outer tip of the plastic zone. Moreover, the model is a continuum model and not a microscopic one. Using this model a global criterion for the fracture accompanying large scale yielding was proposed (Yokobori *et al.*, 1968; Yokobori, 1968). In the first half of this paper, the theory is extended by taking into account the effect of voids which accompany the extension of the crack tip.

In the latter half of this paper, as examples of stable ductile fracture accompanying large scale yielding, new constitutive equations and global criteria are derived for creep (Yokobori *et al.*, 1980b), fatigue and creep-fatigue interaction (Yokobori *et al.*, 1980c) for the Region II of crack growth curve of stainless steel at high temperatures; these are based on the experimental (Yokobori and Sakata, 1980) and analytical studies of Yokobori *et al.* (1980a). The parameter *P*, which appears in the proposed criterion has both physical and practical significance, just like the Larson-Miller parameter for creep fracture time in the case of an unnotched or uncracked specimen. The parameter *P* relates to the macroscopic local stress at the crack tip.

2. Unstable ductile fracture

Yokobori and Ichikawa (1966) showed that the energy change of the system accompanying the crack extension in the Dugdale–Bilby–Cottrell–Swinden (DBCS) Model (Dugdale, 1960; Bilby *et al.*, 1963) (Fig. 1) is always positive, and thus crack growth cannot be unstable. This is inconsistent with the real case where unstable ductile fracture occurs. This paradox has been resolved by Yokobori *et al.* (1968; Yokobori, 1968) by proposing a new model which takes into account the cohesive force at the actual crack tip. The model is shown in Fig. 2, and is different from the DBCS model in that a singularity exists at the outer tip of the plastic zone. Moreover, the model is a continuum model and not a microscopic one. Using this model a global criterion for the fracture accompanying large scale yielding has been proposed (Yokobori *et al.*, 1968; Yokobori, 1968). In this section, the theory is extended by taking into account the effect of voids which accompany the crack extension.

It may be assumed that voids are produced in the plastic zone (Fig. 3). Some energy is needed to nucleate these voids. Therefore, this energy should be included in the energy change of the system as follows:

$$\delta E = \delta W_e + \delta L + \delta W_p + 4\gamma \delta a + \delta W_v, \tag{1}$$

where δW_e, δL, δW_p, and δW_v are the change of the strain energy, the potential energy of the external load, the plastic work accompanying the

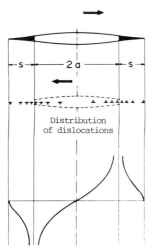

Fig. 1. Dugdale–Bilby–Cottrell–Swinden model: A model of a slit crack of length $2a$ with plastic zone of length s.

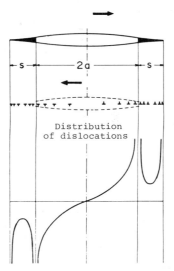

Fig. 2.   Yokobori–Kamei–Ichikawa model: A model of a slit crack of length $2a$ with plastic zone of length s.

crack extension, and the energy dissipated by the void nucleation, respectively. $a=$ half length of the crack and $\gamma=$ specific surface energy. In this case, Eq. (1) may be regarded as a constitutive equation. δW_e, δL and δW_p are given by Yokobori *et al.* (1968) and Yokobori (1968), as follows:

$$\delta W_e = \frac{8(1-\nu)\sigma_Y^2}{\pi G}\left[a\log\frac{a+s}{a}+\left(\frac{\pi}{2}\frac{\sigma}{\sigma_Y}-\cos^{-1}\frac{a}{a+s}\right)\sqrt{(a+s)^2-a^2}\right.$$
$$\left.+\frac{a+s}{2}\left(\frac{\pi}{2}\frac{\sigma}{\sigma_Y}-\cos^{-1}\frac{a}{a+s}\right)^2\left(\frac{\partial(a+s)}{\partial a}\right)\right]\delta a, \qquad (2)$$

$$\delta L = -\frac{4(1-\nu)\sigma_Y\sigma}{G}\left[\sqrt{(a+s)^2-a^2}\right.$$
$$\left.+(a+s)\left(\frac{\pi\sigma}{2\sigma_Y}-\cos^{-1}\frac{a}{a+s}\right)\left(\frac{\partial(a+s)}{\partial a}\right)\right]\delta a, \qquad (3)$$

$$\delta W_p = \frac{8(1-\nu)\sigma_Y^2}{\pi G}\left[\sqrt{(a+s)^2-a^2}\cos^{-1}\frac{a}{a+s}-a\log\frac{a+s}{a}+(a+s)\right.$$
$$\left.\times\left(\frac{\pi\sigma}{2\sigma_Y}-\cos^{-1}\frac{a}{a+s}\right)\left(\cos^{-1}\frac{a}{a+s}\right)\left(\frac{\partial(a+s)}{\partial a}\right)\right]\delta a. $$
$$(4)$$

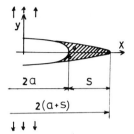

Fig. 3.   A model of a slit crack of length $2a$ with plastic zone in which voids are nucleated.

Denote the surface energy per void by γ'. Then δW_v is,

$$\delta W_v = \gamma'(\partial n/\partial a)\delta a, \tag{5}$$

where n is the number of nucleated voids. Let γ_e be the equivalent surface energy,

$$4\gamma_e = 4\gamma + \gamma'\frac{\partial n}{\partial a}. \tag{6}$$

Let us assume the similarity of crack extension, $\partial a/a = \partial(a+s)/(a+s)$, that is[1], $\partial(a+s)/\partial a = (a+s)/a$. Furthermore, from the relation $a/(a+s) = (\pi/2)1/f(\sigma/\sigma_Y)$ (Hult and McClintock, 1957), and by substituting Eqs. (2)–(5) into Eq. (1), we arrive at

$$\delta E = \left[-\frac{4(1-\nu)\sigma_Y^2 a}{\pi G}\{F(\sigma/\sigma_Y)\}^2 + 4\gamma_e \right]\delta_a, \tag{7}$$

where

$$F(\sigma/\sigma_Y) = f(\sigma/\sigma_Y)\left[\frac{\sigma}{\sigma_Y} - \frac{2}{\pi}\cos^{-1}\left(\frac{\pi}{2}\frac{1}{f(\sigma/\sigma_Y)} \right) \right], \tag{8}$$

and

$$f(\sigma/\sigma_Y) = \frac{1+(\sigma/\sigma_Y)^2}{1-(\sigma/\sigma_Y)^2}E\left(\frac{2(\sigma/\sigma_Y)}{1+(\sigma/\sigma_Y)^2} \right); \tag{}$$

here E is the complete elliptic integral of the second kind; σ_Y = yield stress and ν = Poisson's ratio. The criterion for unstable fracture of this type is,

$$\delta E = 0. \tag{9}$$

Thus we have the following criterion:

$$F(\sigma_c/\sigma_Y) = \sqrt{\frac{\pi G\gamma_e}{(1-\nu)\sigma_Y^2 a}} \tag{10}$$

---

[1]This condition is obtained as follows: $s = \phi(a, \sigma, \sigma_Y, E, \nu)$. By dimensional analysis, we have $s = a\psi(\sigma, \sigma_Y, E, \nu)$. Thus, $ds = da\psi(\sigma, \sigma_Y, E, \nu)$, and $s/a = ds/da$.

which has the same form as in the previous paper except that γ is replaced by γ_e. $F(\sigma/\sigma_Y)$ calculated from Eq. (8), is shown against σ/σ_Y in Fig. 4. It can be seen that $F(\sigma/\sigma_Y)$ takes a maximum value, F_{max}, at $\sigma/\sigma_Y \cong 0.6$. Then using Eq. (7) it follows that Eq. (9) has real roots when $\sqrt{\pi G \gamma_e / [(1-\nu)\sigma_Y^2 a]} \leqslant F_{max}$. In this case, as σ is increased monotonically from zero, the length of the plastic zone increases until σ reaches the value σ_c corresponding to the intersection point U, and then under the stress σ_c the crack extends. On the contrary, when $\sqrt{\pi G \gamma_e / [(1-\nu)\sigma_Y^2 a]} > F_{max}$, again using Eq. (7), it follows that Eq. (9) does not have real roots, and the length of the plastic zone increases with increasing σ until σ reaches σ_Y, that is, general yielding will occur. Thus, it is concluded that unstable fracture occurs provided that

$$\sigma_Y^2 a \geqslant \frac{\pi G \gamma_e}{(1-\nu)F_{max}^2}. \tag{11}$$

Assuming $G = 10^2$ GPa, $\gamma = 1$ dJ/m and $\nu = 1/3$ as reasonable values for steel, the condition for unstable fracture is

$$\sigma_Y^2 a \geqslant \beta 96.2 \, (\mathrm{MPa})^2 \mathrm{m}, \tag{12}$$

or

$$\frac{\sigma_Y^2}{\beta} a \geqslant 96.2 \, (\mathrm{MPa})^2 \mathrm{m}, \tag{13}$$

where $\beta = \gamma'/\gamma$. Since $\partial n/\partial a$ is considered to be positive, $\beta > 1$. This predicts that the critical crack length for unstable fracture is larger by the factor of β than that for the case without void nucleation.

On the other hand, for fracture occurrence a second condition should be satisfied (Yokobori et al., 1977): the breaking of the atomic bonds. This may be controlled by limiting local stress involving microscopic local stress. This microscopic local stress will be given, say, by the dynamically piling-up

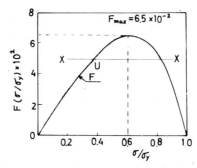

Fig. 4. The value of $F(\sigma/\sigma_Y)$ versus σ/σ_Y calculated by Eq. (8).

dislocations (Yokobori and Yokobori, 1980). Coupling this microscopic stress with the macroscopic one, a second condition is obtained. In this manner the overall or global fracture criterion is given by satisfying the two conditions. This is a subject for a future study.

3. Stable ductile fracture

3.1. *Parametric representation of global criterion for stable crack growth*

As examples of stable ductile fracture accompanying large scale yielding, let us consider in this section constitutive equations and global criteria for stable crack growth under creep, fatigue and creep–fatigue interaction at high temperatures.

The experimental studies on high temperature crack growth behavior of 304 stainless steel with notch have been carried out using a high temperature microscope. During creep, fatigue and creep–fatigue interaction tests at high temperatures, without interrupting the tests, in a vacuum of 10^{-5} mmHg, crack length measurements have been performed (Yokobori and Sakata, 1980). The load waves were controlled as shown in Figs. 5a–c for the fatigue, creep and creep–fatigue interaction tests, respectively. Furthermore, the geometrical change of notch shape from the instant of load application has been observed during the tests without interrupting the tests (Yokobori *et al.*, 1980a). Correspondingly, the effective crack length, a_{eff}, has been calculated by the finite element method, estimating the local stress distribution near the tip of the crack which has been initiated from an initial notch root. The specimens are provided with 30° V-type double notch with depth of 0.25 mm, and the specimen width and thickness are 4 and 1 mm, respectively. The ratio of $\alpha\sqrt{a_{\text{eff}}}\,\sigma_{\text{g}}$ (obtained by using FEM method) to the stress intensity factor $K_{\text{I}}(\equiv\alpha\sqrt{a}\,\sigma_{\text{g}})$ can be approximately expressed by the following equation:

$$\alpha\sqrt{a_{\text{eff}}}\,\sigma_{\text{g}}/K_{\text{I}}=1-0.4\exp\{-47.3a^*/(W-a_0)\} \qquad (14)^2$$

or the effective crack length, a_{eff}, is given by (Yokobori *et al.*, 1980a),

$$a_{\text{eff}}=a\left[1-0.4\exp\{-47.3a^*/(W-a_0)\}\right]^2 \qquad (\text{see Fig. 6}),$$
$$(15)$$

[2] The geometrical change of the notch shape during the loading process is nearly completed by the time the crack initiates at the notch root. The shape of the notch at the instant of the crack initiation is almost independent of the experimental conditions such as temperature, gross section stress and holding time (Yokobori *et al.*, 1980a).

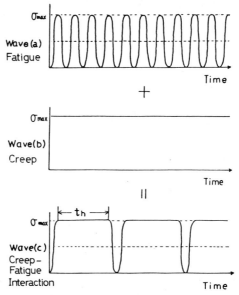

Fig. 5.   Loading wave: (a) fatigue test; (b) creep test; (c) creep–fatigue interaction test.

where

$$\alpha = 1.98 + 0.36\left(\frac{a}{W}\right) - 2.12\left(\frac{a}{W}\right)^2 + 3.42\left(\frac{a}{W}\right)^3, \qquad a \leqslant 0.7W.$$

(16)

Thus the experimental data on the crack growth rate are analyzed in terms of the parameters $\alpha\sqrt{a_{\text{eff}}}\,\sigma_g$, where σ_g is the gross section stress, and absolute temperature T.

A typical example of the crack growth rate, $\mathrm{d}a/\mathrm{d}t$, obtained from the experimental data under high temperature creep, fatigue and creep–fatigue interaction condition is shown against $\alpha\sqrt{a_{\text{eff}}}\,\sigma_g$ (Yokobori et al., 1980a) in Fig. 7. It can be seen from Fig. 7 that by using the plot of the logarithm of $\mathrm{d}a/\mathrm{d}t$ vs logarithm of $\alpha\sqrt{a_{\text{eff}}}\,\sigma_g$, the curve is distinctly divided into three regions, denoted by I, II and III. By comparing this curve with the curve of the logarithm of $\mathrm{d}a/\mathrm{d}t$ vs logarithm K_I, it can be seen that by taking $\alpha\sqrt{a_{\text{eff}}}\,\sigma_g$ as abscissa, Region II covers a much wider range, as in Fig. 7; here a typical example of the relation between crack growth rate, $\mathrm{d}a/\mathrm{d}t$, and the usual parameter K_I is shown by the dotted line.

Based on the series of experimental (Yokobori and Sakata, 1980) and analytical data (Yokobori et al., 1980a), the following new mathematical equation has been derived for the prediction of high temperature crack

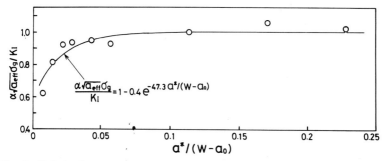

Fig. 6. Relation between $\alpha\sqrt{a_{\text{eff}}}\,\sigma_g/K_I$ and nondimensional crack length, $a^*/(W-a_0)$.

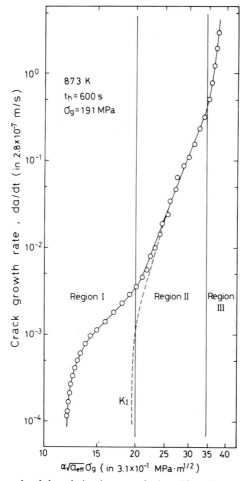

Fig. 7. Typical example of the relation between the logarithm of crack growth rate and the logarithm of fracture mechanics parameter $\alpha\sqrt{a_{\text{eff}}}\,\sigma_g$.

growth rate in Region II (SI units are used):

$$\log(\mathrm{d}a/\mathrm{d}t) = -8.67 + (8.48 \times 10^3/T) \log\left(\alpha\sqrt{a_{\mathrm{eff}}}\,\sigma_{\mathrm{g}}/1.44 \times 10^2\right)$$
$$+ 5.64 \log \sigma_{\mathrm{g}} \quad \text{for creep,} \tag{17}$$

$$\log(\mathrm{d}a/\mathrm{d}t) = -7.32 + (8.28 \times 10^3/T) \log\left(\alpha\sqrt{a_{\mathrm{eff}}}\,\sigma_{\mathrm{g}}/1.43 \times 10^2\right)$$
$$+ 4.89 \log \sigma_{\mathrm{g}} \quad \text{for creep–fatigue interaction } (t_{\mathrm{h}} = 600\text{s}), \tag{18}$$

$$\log(\mathrm{d}a/\mathrm{d}t) = 0.35 + \left(\frac{3.47 \times 10^3}{T} + 5.03\right) \log\left(\alpha\sqrt{a_{\mathrm{eff}}}\,\sigma_{\mathrm{g}}/1.77 \times 10^3\right)$$
$$+ 5.57 \log \sigma_{\mathrm{g}} \quad \text{for creep–fatigue interaction } (t_{\mathrm{h}} = 60 \text{ s}), \tag{19}$$

$$\log(\mathrm{d}a/\mathrm{d}t) = 5.80 + \left(\frac{8.79 \times 10^2}{T} + 3.71\right) \log\left(\alpha\sqrt{a_{\mathrm{eff}}}\,\sigma_{\mathrm{g}}/6.79 \times 10^3\right)$$
$$\text{for fatigue.} \tag{20}$$

Let P denote the significant part in Eqs. (17), (18), (19) and (20), as follows:

$$P = (8.48 \times 10^3/T) \log\left(\alpha\sqrt{a_{\mathrm{eff}}}\,\sigma_{\mathrm{g}}/1.44 \times 10^2\right) + 5.64 \log \sigma_{\mathrm{g}}$$
$$\text{for creep,} \tag{21}$$

$$P = (8.28 \times 10^3/T) \log\left(\alpha\sqrt{a_{\mathrm{eff}}}\,\sigma_{\mathrm{g}}/1.43 \times 10^2\right) + 4.89 \log \sigma_{\mathrm{g}}$$
$$\text{for creep–fatigue interaction } (t_{\mathrm{h}} = 600\text{s}), \tag{22}$$

$$P = \left(\frac{3.47 \times 10^3}{T} + 5.03\right) \log\left(\alpha\sqrt{a_{\mathrm{eff}}}\,\sigma_{\mathrm{g}}/1.77 \times 10^3\right) + 5.57 \log \sigma_{\mathrm{g}}$$
$$\text{for creep–fatigue interaction } (t_{\mathrm{h}} = 60 \text{ s}), \tag{23}$$

$$P = \left(\frac{8.79 \times 10^2}{T} + 3.71\right) \log\left(\alpha\sqrt{a_{\mathrm{eff}}}\,\sigma_{\mathrm{g}}/6.79 \times 10^3\right)$$
$$\text{for fatigue.} \tag{24}$$

Taking the parameter P as abscissa, and plotting the experimental data of crack growth rate as ordinate, with a_{eff}, σ_{g}, and temperature as parameters, we arrive at Figs. 8–11, where the solid lines represent Eqs. (17) to (20). It can be seen from these figures that the high temperature crack growth rate of 304 stainless steel in Region II is very well characterized by the proposed new parameter. That is, Eqs. (21)–(24) predict fairly accurately the experimental data on crack growth rate.

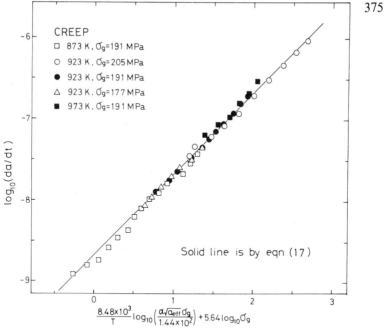

Fig. 8. The presentation of the creep crack growth rate by a new proposed parameter; experimental data from Yokobori and Sakata (1980).

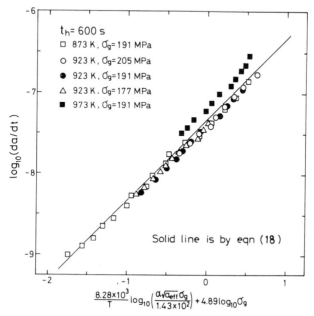

Fig. 9. The presentation of the creep–fatigue interaction crack growth rate by a new proposed parameter; $t_h = 600$ s; experimental data from Yokobori and Sakata (1980).

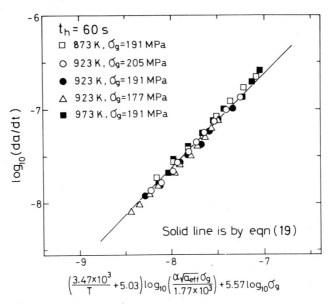

Fig. 10. The presentation of the creep–fatigue interaction crack growth rate by a new proposed parameter; $t_h = 60$ s; experimental data from Yokobori and Sakata (1980).

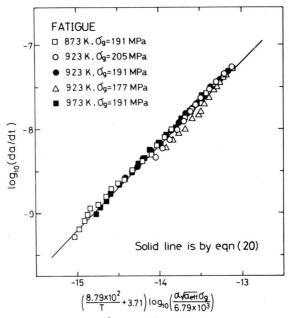

Fig. 11. The presentation of the fatigue crack growth rate by a new proposed parameter; experimental data from Yokobori and Sakata (1980).

3.2. Physical meaning of parameter P

We observe that Eqs. (17)–(20) have the form associated with the thermal activation process theory. Indeed, it has been found that the activation energy obtained from experimental data is nearly equal to self-diffusion energy for the case of creep and creep–fatigue interaction ($t_h = 600$ s), and is about the same as the energy required for the dislocation movement in the case of fatigue. These show that whatever the micro-mechanism of the rate-determining process may be, the constitutive equations and the global criteria for the stable crack growth with large scale yielding of this type at high temperatures are solely determined by the corresponding parameter P, suggesting that P is a physically based parameter.

It is also interesting and practically significant to note that high temperature crack growth rate in Region II for $t_h \geqslant 600$ s condition is very well characterized by the parameter proposed for creep. That is, Eq. (17) predicts fairly accurately the crack growth rate under creep and creep–fatigue interaction for $t_h \geqslant 600$ s condition at high temperatures, as shown in Fig. 12.

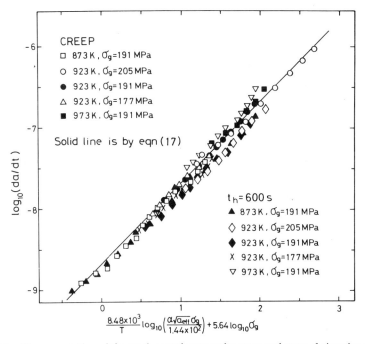

Fig. 12. The presentation of the crack growth rate under creep and creep–fatigue interaction ($t_h = 600$ s) at high temperature by a new proposed parameter; experimental data from Yokobori and Sakata (1980).

T. Yokobori et al.

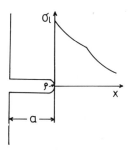

Fig. 13.   Elastic-plastic stress distribution near the flat surfaced notch with cycloidal tip in a work-hardening materials accompanying large scale yielding.

It can be seen from the analysis mentioned in Subsections 3.1 and 3.2 that the new parameter P for predicting crack growth rate in cracked specimen has both practical and fundamental significance almost similar to the Larson–Miller parameter (Larson and Miller, 1952) for predicting the creep fracture time in unnotched or uncracked specimens. An interpretation of the physical meaning of the Larson–Miller parameter has been attempted in literature (Yokobori, 1965). For an understanding of the connection between ours (cracked specimen) and Larson–Miller's (uncracked specimen), it is useful to note that da/dt in our case corresponds to $1/t$ in their case, where t is time to fracture.

3.3.   Mechanical meaning of parameter P

The local stress distribution near the flat surface notch tip with cycloidal tip for Mode III under large scale yielding condition has been given by

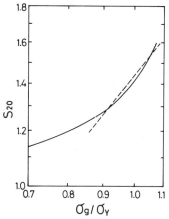

Fig. 14.   The sum of the series given by Eq. (26) versus σ_g'/σ_Y; the dotted line is given by Eq. (28).

Yokobori and Konosu (1976) (Fig. 13). For Mode I also the local stress, σ_1, may be expressed by a similar expression except for a numerical factor. Then we may express σ_1 as,

$$\sigma_{1(x)} = \sigma_Y \left[\sqrt{\frac{(1+\lambda)a}{\rho + (1+\lambda)^2 x} \frac{\sigma_g}{\sigma_Y}} \right]^{\frac{2\lambda}{1+\lambda}} \left[f\left(\frac{\sigma_g}{\sigma_Y}, \lambda \right) \right]^{\frac{\lambda}{1+\lambda}}, \qquad (25)$$

where a = the notch length, σ_g = the gross section stress, σ_Y = yield stress, λ = strain hardening exponent, ρ = crack tip radius, x = the distance from the notch in the crack direction, and $f(\sigma_g/\sigma_Y, \lambda)$ is given by a series as follows:

$$f\left(\frac{\sigma_g}{\sigma_Y}, \lambda \right) = 1 + A_1 \left(\frac{\sigma_g}{\sigma_Y} \right)^2 + A_2 \left(\frac{\sigma_g}{\sigma_Y} \right)^4 + A_3 \left(\frac{\sigma_g}{\sigma_Y} \right)^6$$

$$+ A_4 \left(\frac{\sigma_g}{\sigma_Y} \right)^8 + A_5 \left(\frac{\sigma_g}{\sigma_Y} \right)^{10} + \cdots A_i \left(\frac{\sigma_g}{\sigma_Y} \right)^{2i} + \cdots, \qquad (26)$$

$$A_1 = \frac{1}{2} C_1, \qquad A_2 = \frac{1}{4} C_1^2, \qquad A_3 = \frac{1}{8} \left(\frac{3}{4} C_2 + C_1^3 \right),$$

$$A_4 = \frac{1}{16} \left(\frac{3}{2} C_1 C_2 + C_1^4 \right), \qquad A_5 = \frac{1}{32} \left(\frac{5}{4} C_3 + \frac{9}{4} C_1^2 C_2 + C_1^5 \right), \cdots,$$

$$A_{10} = \frac{7}{512} \left(\frac{63}{64} C_1 C_5 + \frac{15}{32} C_2 C_4 + \frac{25}{32} C_1^3 C_4 + \frac{45}{224} C_3^2 + \frac{225}{224} C_1^2 C_2 C_3 \right.$$

$$\left. + \frac{15}{28} C_1^5 C_3 + \frac{3}{16} C_1 C_2^3 + \frac{75}{112} C_1^4 C_2^2 + \frac{3}{7} C_1^7 C_2 + \frac{1}{14} C_1^{10} \right),$$

$$A_{21} = \cdots, \cdots$$

$$i = 1, 2, \cdots,$$

$$C_k = \frac{2k - 1 - \mu_k}{2k - 1 + \mu_k},$$

$$\mu_k = \left\{ \frac{(1-\lambda)^2}{4} + (2k-1)^2 \lambda \right\}^{1/2} - \frac{1-\lambda}{2}$$

$$k = 1, 2, 3, \cdots.$$

Let us consider in more detail, $f(\sigma_g/\sigma_Y, \lambda)$ given by Eq. (26) which is an increasing function with respect to σ_g/σ_Y.

Denoting the sum of the series given by Eq. (26) by S_{2i}, and summing it up to $i = 10$ for $0.9 \geqslant \sigma_g/\sigma_Y$, it follows that $f(\sigma/\sigma_g, \lambda)$ converges, and that $|S_{2(i+1)} - S_{2i}|/S_{2(i+1)} \leqslant 0.0005$. Also, for the case of $1.05 \geqslant \sigma_g/\sigma_Y \geqslant 0.9$, the series converges and $|S_{2(i+1)} - S_{2i}|/S_{2(i+1)} \leqslant 0.01$. The logarithm of the sum, S_{20}, calculated up to $i = 10$ in Eq. (26), is shown by the solid line against the logarithm of σ_g/σ_Y in Fig. 14. It can be seen from Fig. 14 that for the range of $1.05 \geqslant \sigma_g / \sigma_Y \geqslant 0.9$ which concerns our experiments, S_{20} is fairly well approximated by the straight line as shown by the dotted line. That is, it can be seen that $f(\sigma_g / \sigma_Y, \lambda)$ can be fairly well approximated by,

$$f\left(\frac{\sigma_g}{\sigma_Y}, \lambda\right) \simeq 1.44\left(\frac{\sigma_g}{\sigma_Y}\right)^{1.23} \tag{28}$$

Then, substituting Eq. (28) and $\lambda = 0.35$ into Eq. (25), we obtain

$$\sigma_1 = \sigma_Y\left(\sqrt{\frac{(1+\lambda)a}{\rho}} \frac{\sigma_g}{\sigma_Y}\right)^{0.52}\left(\frac{\sigma_g}{\sigma_Y}\right)^{0.32}, \tag{29}$$

where $\rho \gg x$, that is, at the crack tip.

On the other hand, for instance, for the temperature range from 600 to 700°C the crack growth rate for this materials is written from Eq. (17) as

$$\frac{da}{dt} = M\left(\alpha\sqrt{a_{\text{eff}}}\,\sigma_g\right)^{9.20}\sigma_g^{5.64}, \tag{30}$$

where M = constant value.

Assuming the values of σ_g / σ_Y and λ in Eq. (29) correspond to the averaged ones for the range of the temperature from 600 to 700°C, and comparing Eq. (30) with Eq. (29), we see that as far as both gross stress, σ_g, and the crack length, a (or a_{eff}), are concerned,

$$\frac{da}{dt} \propto \left\{\left(\sqrt{a_{\text{eff}}}\,\sigma_g\right)^{0.52}\sigma_g^{0.32}\right\}^{17.7},$$

or

$$\frac{da}{dt} \propto \sigma_1^{17.7}. \tag{31}$$

A similar relation may be obtained for the case of creep–fatigue interaction for $t_h = 600$s and 60s. This suggests that the global criterion for crack growth with large scale yielding at high temperature creep and creep–fatigue interaction condition may be expressed by local stress σ_1 at the crack tip as the rate controlling fracture mechanical parameter; at least, the global criterion may include σ_1.

3.4.   Comparison with the modified J integral

Some investigators (Landes and Begley, 1976; Kubo et al., 1979) have suggested that the high temperature crack growth rate, da/dt, is characterized by the J integral, modified by using the creep strain rate for the strain; the so-called modified J integral, J'. Based on an assumed model,

Kubo *et al.* (1979) considered the following formula:

$$\frac{da}{dt} = L_1 J', \tag{32}$$

where L_1 is a constant independent of the applied stress. They also used

$$J' = L_2 \sigma_g \dot{l}, \tag{33}$$

where $\dot{l}$ = elongation rate. Thus they concluded that Eq. (32) is in agreement with experimental data.

Let us plot our experimental data, the logarithm of da/dt as ordinate against the logarithm of J' as abscissa; Figs. 15–18. From these figures, it is seen that da/dt is not characterized by the single parameter, J', independent of the applied stress, σ_g, and temperature, T. Moreover, by comparing each of Figs. 8–11 with each of Figs. 15–18, respectively, it is seen that parametric representation by P is in much better agreement with the experimental data than representation by the modified J integral, J'.

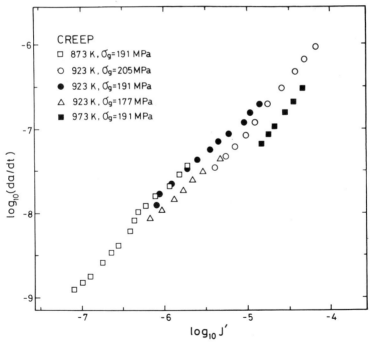

Fig. 15. The presentation of the creep crack growth rate by the modified J-integral, J'; experimental data from Yokobori and Sakata (1980).

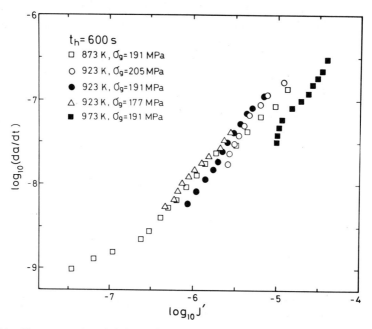

Fig. 16. The presentation of the creep–fatigue interaction crack growth rate by the modified *J*-integral, *J'*; $t_h = 600$ s; experimental data from Yokobori and Sakata (1980).

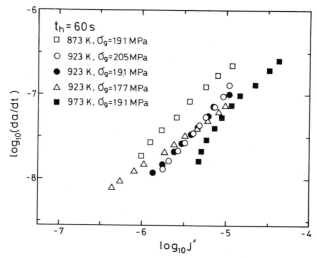

Fig. 17. The presentation of the creep–fatigue interaction crack growth rate by the modified *J*-integral, *J'*, $t_h = 60$ s; experimental data from Yokobori and Sakata (1980).

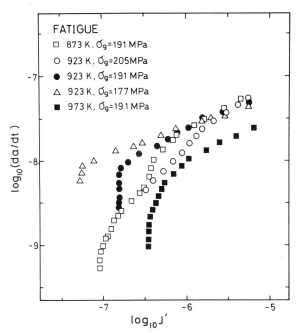

Fig. 18. The presentation of the fatigue crack growth rate by modified *J*-integral, *J'*; experimental data from Yokobori and Sakata (1980).

It can be shown why the J' integral cannot characterize da/dt, at least as a single parameter alone, as follows: Creep strain rate $\dot{\varepsilon}$ may be expressed in general form as,

$$\dot{\varepsilon} = A_\mathrm{p} \exp\left(-\frac{\Delta H_\mathrm{p} - \phi_\mathrm{p}(\sigma_\mathrm{g})}{kT}\right),\tag{34}$$

where ΔH_p = activation energy for creep flow, $\phi_\mathrm{p}(\sigma_\mathrm{g})$ = activation energy decrease by applied stress, and A_p = constant. On the other hand, da/dt may also be expressed in the form of thermal activation process. That is,

$$\frac{da}{dt} = A_\mathrm{f} \exp\left(-\frac{\Delta H_\mathrm{f} - \phi_\mathrm{f}(\sigma_\mathrm{g})}{kT}\right),\tag{35}$$

where ΔH_f = activation energy for crack extension, $\phi_\mathrm{f}(\sigma_\mathrm{g})$ = activation energy decrease by the applied stress. Using Eq. (34), Eq. (35) is rewritten as,

$$\frac{da}{dt} = \frac{A_\mathrm{f}}{A_\mathrm{p}}\dot{\varepsilon}\exp\left(-\frac{\Delta H_0 - \phi_0(\sigma_\mathrm{g})}{kT}\right),\tag{36}$$

where $\Delta H_0 = \Delta H_{\mathrm{f}} - \Delta H_{\mathrm{p}}$, $\phi_0(\sigma_{\mathrm{g}}) = \phi_{\mathrm{f}}(\sigma_{\mathrm{g}}) - \phi_{\mathrm{p}}(\sigma_{\mathrm{g}})$. Or using Eq. (33),

$$\frac{\mathrm{d}a}{\mathrm{d}t} = \frac{A_{\mathrm{f}}}{L_3 A_{\mathrm{p}}} J' \frac{1}{\sigma_{\mathrm{g}}} \exp\left(-\frac{\Delta H_0 - \phi_0(\sigma)}{kT}\right), \tag{37}$$

where $L_3 = $ constant. Equation (37) indicates that $\mathrm{d}a/\mathrm{d}t$ is not characterized by the J' integral alone, and, in turn, factors such as $(1/\sigma_{\mathrm{g}})\exp(-(\Delta H_0 - \phi_0(\sigma)/kT))$ should be multiplied. The factor of this kind inevitably includes not only σ_{g}, but also $\phi_{\mathrm{p}}(\sigma_{\mathrm{g}})$ and $\phi_{\mathrm{f}}(\sigma_{\mathrm{g}})$ which are ambiguous functions and difficult for mathematical formulation.

3.5. Discussion

As is seen by inspecting Eq. (26), this series diverges as $\sigma_{\mathrm{g}}/\sigma_{\mathrm{Y}}$ becomes large. Thus for the case of large scale yielding corresponding to the large values of $\sigma_{\mathrm{g}}/\sigma_{\mathrm{Y}}$, the crack tip stress distribution cannot be described by Eq. (25), and the stress distribution would become more and more even: thus the global criterion for crack growth would not be strongly affected by the macroscopic local stress distribution at the crack tip, but rather it would probably be determined by the stress, averaged over a larger domain from the crack tip.

The experimental data (Yokobori and Sakata, 1980) used, concern specimens with the same shape and size, and, thus, the size effect has not been included. However, according to our analysis on the stress distribution near the crack tip under large scale yielding, as above, it is expected that the specimen shape and size effect will be included in the proportionality term, α, in Eqs. (21)–(24). The study of the general formulation of the term α in Eqs. (21)–(24), is a subject for a future study.

4. Conclusions

The conclusions obtained are as follows:

(1) For unstable ductile fracture involving large scale yielding and void formation, the global criterion for fracture is obtained by the inclusion of a singularity at the tip of the plastic region.

(2) For stable ductile fracture at high temperature, the following specific results are obtained:

(a) The global criterion, as parametrically represented for crack growth in creep, fatigue and creep–fatigue interaction, is obtained for the Region II of crack growth curve of stainless steel;

(b) The parameter in the criterion thus obtained has both physical and practical significance just like the Larson–Miller parameter for creep fracture time in the case of an unnotched or uncracked specimen;

(c) A fracture parameter, P, is introduced which is related to the macroscopic local stress at the crack tip.

Acknowledgments

The authors acknowledge Professor Y. Yamada, University of Tokyo for useful discussions concerning the calculation of the effective crack length by FEM. Thanks are also made to Professor M. Ichikawa, Tokyo Electro-Communication University for discussion on unstable ductile fracture. Part of the work was financially sponsored by the Mitsubishi Foundation.

References

Bilby, B. A., Cottrell, A. H. and Swinden, K. H. (1963), "The Spread of Plastic Yield from a Notch," *Proc. Roy. Soc.*, Ser. A **272**, 304.

Dugdale, D. S. (1960), "Yielding of Steel Sheets Containing Slits," *J. Mech. Phys. Solids.*, **8**, 100.

Hult, J. A. H. and McClintock, F. A. (1957), "Elastic-Plastic Stress and Strain Distribution Around Sharp Notches Under Repeated Shear," *9th Int. Cong. Appl. Mech.*, **8**, 51.

Kubo, S., Ohji, K. and Ogura, K. (1979), "An Analysis of Creep Crack Propagation on the Basis of the Plastic Singular Stress Field," *Eng. Frac. Mech.*, **11**, 315.

Landes, J. D. and Begley, J. A. (1976), "A Fracture Mechanics Approach to Creep Crack Growth," *ASTM STP 590*, 128.

Larson, F. R. and Miller, J. (1952), "A Time-Temperature Relationship for Rupture and Creep Stresses," *Trans. ASME*, **74**, 765.

Yokobori, T. (1955), *Zairyo Kyodo Gaku*, Gihodo, Tokyo, 143. English edition, Yokobori, T. (1965), *Strength, Fracture and Fatigue of Materials*, Noordoff, Groningen, The Netherlands, 186.

Yokobori, T. (1968), "Criteria for Nearly Brittle Fracture," *Int. J. Fract. Mech.* **4**, 179.

Yokobori, T. and Ichikawa, M. (1966), "The Energy Principle as Applied to the Elastic-Plastic Crack Model for Fracture Criterion," *Rep. Res. Inst. Str. Fract. Mat.*, Tohoku Univ., Sendai, Japan **2**, 21.

Yokobori, T., Kamei, A. and Ichikawa, M. (1968), "A Criterion for Unstable Elastic-Plastic Fracture Based on Energy Balance Considerations," *Rep. Res. Inst. Str. Fract. Mat.*, Tohoku Univ., Sendai, Japan, **4**, 1.

Yokobori, T. and Konosu, S. (1976), "The Elastic-Plastic Stress Distribution near the Flat Surfaced Notch with Cylindrical Tip in a Work-Hardening Materials," *Ing-Arch* **45**, 243.

Yokobori, T., Konosu, S. and Yokobori, A. T. Jr., (1977), "Micro-and-Macro Fracture Mechanical Approach to Fracture and Fatigue Crack Propagation," *Fracture, ICF4*, Univ. of Waterloo, Waterloo, **1**, 165.

Yokobori, T. and Sakata, H. (1980), "Studies on Crack Growth Rate under High Temperature Creep, Fatigue and Creep–Fatigue Interaction—I On the Experimental Studies on Crack Growth Rate as Affected by $\sqrt{a_{\text{eff}}}\,\sigma_g$, σ_g and Temperature," *J. Eng. Fracture Mech.*, **13**, 509.

Yokobori, T., Sakata, H., and Yokobori, A. T. Jr. (1980a), "Studies on Crack Growth Rate under High Temperature Creep, Fatigue and Creep–Fatigue Interaction—II On the Role of Fracture Mechanics Parameter $\sqrt{a_{\text{eff}}}\,\sigma_g$," *J. Eng. Fracture Mech.*, **13**, 523.

Yokobori, T., Sakata, H. and Yokobori, A. T. Jr. (1980b), "A New Parameter for Prediction of Creep Crack Growth Rate at High Temperature," *J. Engng. Fract. Mech.*, **13**, 533.

Yokobori, T. and Yokobori, A. T. Jr. (1980), "Physical and Phenomenological Model with Nonlinearity in Ductile Fracture and Fatigue Crack Growth," *Proc. IUTAM Symposium*, CETIM, Springer Verlag, Berlin.

S. Nemat-Nasser, Editor
THREE-DIMENSIONAL CONSTITUTIVE RELATIONS AND DUCTILE FRACTURE
North-Holland Publishing Company (1981) 387

LIMITATION TO DUCTILITY SET BY INCLUSION-INDUCED DISTURBANCE OF YIELDING*

A. DRAGON

Institute of Fundamental Technological Research, Polish Academy of Sciences, Warsaw, Poland

A plasticity model is elaborated to describe advanced stages of yielding accompanied by damage under general stress. It fits for the behavior met in singular localized zones like, e.g., the blunted-crack tip region, neck-center, etc. Our particular interest was in modelling of the disturbance of simple yielding pattern on the macroscale due to averaged misfit effect of inclusions with respect to "ductile" matrix. The latter misfit is commonly known to incite the process of ductile fracturing in metals.

An internal damage-variable plasticity conception has been found to form a useful framework for simulation of process of disturbed yielding. The simple scalar damage-variable is employed to control void-induced dilatational softening. Furthermore, the more precise tensorial-variable concept is proposed. It is shown that such a variable may represent some directional effects of void-growth under advanced plastic straining of the "unit-cell".

Two stages of disturbed yielding, namely the "composite" stage and porous-metal yielding are distinguished as governed by evolution of internal variable simulating respective phases of damage, i.e., (i) separation of inclusion, (ii) further void-growth and coalescence. An idea of so called internal irreversibility condition coupled with increment of damage variable is employed. Such a condition is assumed to interact with plastic potential at the separation stage causing the departure from normality rule for matrix-inclusion assemblage (normality was presumed for initial nondisturbed yielding with inclusion presence not yet manifesting itself). In particular, a vertex is formed in the stress-space on the yield surface by intersection with internal irreversibility surface, the latter acting as another plastic quasi-potential. The uniqueness of plastic strain-rate thus determined is discussed.

Quantitative analysis is pursued involving particular representations for the yield and internal irreversibility criteria and hardening functions for succeeding stages of plastic deformation. The damage-induced softening is considered to overbalance the hardening mechanisms, the state of mutual hardening–softening balance being postulated to form a threshold condition (a necessary one) for transition from damage to macrocrack growth. This is correlated with singular transition of the evolution equation for damage variable.

The damage-variable formulation of plasticity seems a promising one for modelling of some ductile fracture phenomena regarded as process involving disturbed plastic deformation. Its power can be improved provided the mathematical formulae involving the notion and evolution of damage are sufficiently accurate and well-founded experimentally thus approximating decisive features of interaction between internal deterioration and inelastic straining in metal.

*Full text of presented lecture not included in this book.

S. Nemat-Nasser, Editor
THREE-DIMENSIONAL CONSTITUTIVE RELATIONS AND DUCTILE FRACTURE
North-Holland Publishing Company (1981) 389–404

DUCTILITY CAUSED BY PROGRESSIVE FORMATION OF SHEAR CRACKS

Horst LIPPMANN

Lehrstuhl A für Mechanik, Technische Universität, Pf. 202420, D-8000 München 2, Fed. Rep. Germany

Rheological models are developed in order to explain the elastic-plastic coupling, i.e. the interrelation between the elastic and crack-plastic ("clastic") parts of deformation, for rock-like material under compression. Special emphasis is given to the "anomalous" coupling for which the elastic strain *decreases* with growing stress or total strain.

1. Introduction

Figure 1 shows experimental stress–strain (σ–ε) diagrams taken from uniaxial compression tests ($\sigma \geqslant 0$, $\varepsilon \geqslant 0$). They start with a reversible elastic branch 00' ($0 \leqslant \varepsilon \leqslant \varepsilon_0$) followed by a zone of transition 0'M'R' ($\varepsilon_0 < \varepsilon < \varepsilon_R$) after which, a totally broken granular state is reached ($\varepsilon_R \leqslant \varepsilon$). Materials with such a property will in the context of this paper, be called "rock-like".

In the zone of transition 0'M'R', subsequent and growing arrays of internal shear cracks are formed giving rise to friction between their faces. The corresponding quasi-ductile dissipative deformation is often referred to as "crack-plastic", "clastic", or simply "plastic". Behind M' the increase of cracks leads to a strain softening so-called "post-failure" zone with "failure" at M'.

The superimposed elastic strain ε^E is measured after unloading, e.g. along A'A, B'B, C'C, or M'M, to deliver ε_A^E, ε_B^E, etc. The thus obtained relationship $\varepsilon^E = \varepsilon^E(\varepsilon)$, in which the total strain ε might be substituted by the plastic one ε^P via

$$\varepsilon = \varepsilon^E + \varepsilon^P,$$

expresses a coupling effect between both parts of deformation. It is in the prefailure zone of transition, $\varepsilon_0 < \varepsilon < \varepsilon_M$, generally assumed as monotonically increasing ("normal" coupling) because, tensile (!) cracks do weaken the material (Budiansky and O'Connel, 1976). Corresponding constitutive laws for the clastic regime first published by Dougill (1976), were generalized by Dragon (1976), Dragon and Mróz (1979), or by Bažant and Kim (1979) so as to cover compression and shear as well.

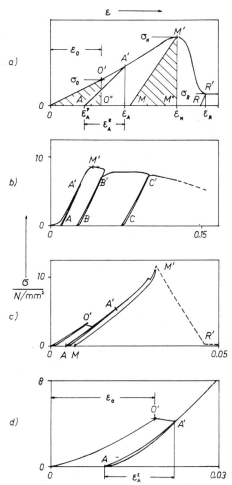

Fig. 1. Stress–strain diagrams for uniaxial compression with intermediate unloading, after Burgert (1979). (——) measured, (– – –) completed from similar tests. (a) General. (b) Plaster of Paris ($\dot{\varepsilon}=0.4$ min^{-1}). (c) Araldite, 3% hardener ($\dot{\varepsilon}=0.53$ min^{-1}). (d) Araldite, 5% hardener ($\dot{\varepsilon}=0.53$ min^{-1}).

They can, however, not explain an "anomalous" coupling for which the function $\varepsilon^E = \varepsilon^E(\varepsilon)$ in the transitional prefailure regime, contains monotonically decreasing arcs. This effect had already been observed from experiments; however, no special emphasis was attributed to it (Hueckel, 1976). In numerous papers by Hueckel and coworkers (Hueckel, 1976; Maier and Hueckel, 1979), the coupling functions (defined similar to $\varepsilon^E(\varepsilon)$) were left unspecified.

Recently it was found by Lippmann (1979) that for theoretical reasons the coal in its underground strata has to show an anomalous elastic-plastic coupling just in order to be prone to translatory rock-bursts; these are most dangerous events in coal mining, a mechanical explanation of which has been given by Lippmann (1978/1979). After a long search for model materials which could simulate rock bursts in the laboratory only one material was found, namely, the photoelastic polymer "araldite" (CIBA, Wehr, Germany) if the percentage of its hardener was chosen between 3 and 10. It is exactly this material which shows in Fig. 1d the effect of anomalous coupling, i.e. $\varepsilon_A^E < \varepsilon_0^E$, where $\varepsilon_0^E = \varepsilon_0$ denotes the initial elastic strain. So it seems worth while to study the mechanical background of that effect by setting up an adequate material model.

2. Discrete rheological model

A model for linear elastic-plastic material, described, e.g., by Persoz (1969), is generalized according to Fig. 2 to consist of a parallel arrangement of "elements", each of which contains a generally nonlinear spring and a crack constituent. The spring stress $\sigma^j = \sigma^j(\varepsilon)$ is assumed continuously monotonically increasing, while the crack term behaves rigid until $|\sigma^j|$ reaches a certain threshold $\sigma_b^j > 0$ at which fracture takes place. After this, slipping friction $|\sigma^j| = \sigma_g^j$ occurs with another threshold given, $\sigma_g^j > 0$. We assume that,

$$0 < \zeta^j < 1; \qquad \zeta^j = \sigma_g^j / \sigma_b^j. \tag{1}$$

Then under loading conditions, the sawtooth-like stress-strain diagram of

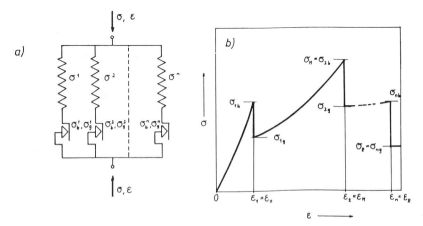

Fig. 2. (a) Discrete rheological model. (b) Corresponding stress–strain diagram.

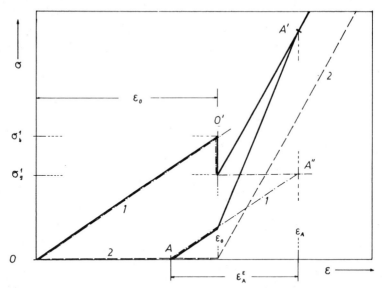

Fig. 3. Stress–strain diagram with unloading A′A for a two-element model. (– · – · –) element characteristics 1; (– – –) element characteristics 2.

Fig. 2b may arise in which the jump discontinuities represent the transition from sticking to slipping friction in the crack terms.

Let us now consider in Fig. 3, a simple two-element model with one linear spring characteristic $\sigma^1 - \sigma^1(\varepsilon)$, while the second one $\sigma^2(\varepsilon)$ represents a strongly concave curve to be idealized in a way such that σ^2 disappears from 0 to ε_0, but increases for $\varepsilon > \varepsilon_0$ much stronger than $\sigma^1(\varepsilon)$ does. The limit of fracture σ_b^1, and the slipping threshold σ_g^1 of spring 1 are finite while σ_g^2, σ_b^2 shall be large enough so that they do not influence the resulting σ-ε-diagram.

When the model is loaded along 00′A′ (Fig. 3), then the stress of element 1 moves until A″. Therefore the stress during unloading, starting from A′, follows a curve which is the sum of the broken lines 1 and 2, and ends up at A. One recognizes, using elementary geometry, that in this example the elastic strain

$$\varepsilon_A^E = \left(\sigma_g^1 / \sigma_b^1 \right) \varepsilon_0 - \zeta^1 \varepsilon_0$$

becomes smaller than ε_0 so that in fact an anomalous coupling exists. Note that there is a step in Fig. 1c,d similar to the sawtooth in Fig. 3, though less steep. This could mean that the true internal mechanism of deformation in araldite is actually comparable to the model. In this event, spring 1 might be interpreted as to represent the elasticity of the matrix of pure material, while spring 2 results from the parallel shear displacements between adjacent chain molecules which are, as a consequence of the hardening constituent, transversely interconnected (meshed).

3.   Stochastic model

3.1.   *Loading*

In order to flatten the sawteeth in Fig. 2b, or even to obtain smooth stress-strain diagrams, a stochastic distribution of dimensionless spring characteristics shall be considered (Fig. 4), while for reasons of simplicity, the parameter ζ^j introduced by Eq. (1) is assumed to be an independent random variable so that it could be replaced by its mean value:

$$\tau^j = \tau^j(\varepsilon) = \sigma^j(\varepsilon)/\sigma_b^j, \qquad \zeta = \sigma_g^j/\sigma_b^j = \text{const.} \tag{2}$$

The functions $\tau^j(\varepsilon)$ are like $\sigma^j(\varepsilon)$ continuously monotonically increasing for $-\infty < \varepsilon < \infty$, but form a stochastic process otherwise. For each fixed deformation ε they are weighted by the distribution function $F(\tau, \varepsilon)$ and the distribution density $f(\tau, \varepsilon)$ to fulfill,

$$F(-\infty, \varepsilon) = 0, \qquad F(+\infty, \varepsilon) = 1, \qquad f(\tau, \varepsilon) = \partial F(\tau, \varepsilon)/\partial \tau \geqslant 0. \tag{3}$$

At the initial stage $\varepsilon = 0$ we also count the individual deformations of all springs to be zero while, of course, prestresses must be allowed for. They can not exceed the dimensionsless limit of fracture, 1, so that $-1 \leqslant \tau^j(0) \leqslant 1$ holds, i.e.

$$F(\tau, 0) = 1, \qquad f(\tau, 0) = 0 \quad \text{if } \tau \geqslant 1,$$

$$F(\tau, 0) = 0, \qquad f(\tau, 0) = 0 \quad \text{if } \tau \leqslant -1. \tag{4}$$

Moreover, all crack terms are assumed to be still solid at $\varepsilon = 0$. Then at any loading stage $\varepsilon \geqslant 0$, the elements with $\tau^j(\varepsilon) < 1$ behave elastically with still unbroken crack terms, while for $\tau^j(\varepsilon) \geqslant 1$, the dimensionless slipping stress ζ acts. So adding up the weighted stresses of all elements, introducing the

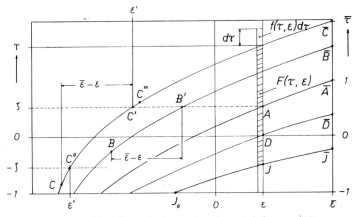

Fig. 4.   Field of dimensionless spring characteristics $\tau = \tau^j(\varepsilon)$.

dimension factor S which might be interpreted as their total number, and regarding (3), the compressional stress-strain dependence $\sigma=\sigma(\varepsilon)$ is found according to

$$\sigma(\varepsilon)=S\left\{\int_{-\infty}^{1} \tau f(\tau, \varepsilon)\,d\tau+\zeta[1-F(1, \varepsilon)]\right\}. \tag{5}$$

As an example we consider a band with thickness T of uniformly distributed parallel spring characteristics, the middle one of which is defined by means of the ascending function $g(\varepsilon)$. We assume

$$dg(\varepsilon)/d\varepsilon>0, \qquad g(0)=0, \qquad 0<T<2. \tag{6}$$

Then the approach

$$f(\tau, \varepsilon)=0, \qquad F(\tau, \varepsilon)=0 \quad \text{if } \tau<g(\varepsilon)-\frac{T}{2},$$

$$f(\tau, \varepsilon)=\frac{1}{T}, \qquad F(\tau, \varepsilon)=\frac{1}{T}\left[\varepsilon-g(\varepsilon)+\frac{T}{2}\right]$$

$$\text{if } g(\varepsilon)-\frac{T}{2}\leqslant\tau\leqslant g(\varepsilon)+\frac{T}{2}, \tag{7}$$

$$f(\tau, \varepsilon)=0, \qquad F(\tau, \varepsilon)=1 \quad \text{if } \tau>g(\varepsilon)+\frac{T}{2},$$

obeys the conditions (3) and (4), while (5) leads to the diagrams of Fig. 5 in which,

$$\frac{\sigma(\varepsilon)}{S}=g(\varepsilon) \quad \text{if } 0\leqslant\varepsilon\leqslant\varepsilon_0 \text{ (elastic)},$$

$$\frac{\sigma(\varepsilon)}{S}=\frac{1}{2T}\left\{1-\left[g(\varepsilon)-\frac{T}{2}\right]^2+2\zeta\left(\left[g(\varepsilon)+\frac{T}{2}\right]-1\right)\right\}$$

$$\text{if } \varepsilon_0\leqslant\varepsilon\leqslant\varepsilon_R \text{ (transitional)}, \tag{8}$$

$$\frac{\sigma(\varepsilon)}{S}=\zeta \qquad \text{if } \varepsilon_R\leqslant\varepsilon_R \text{ (granular)}.$$

The limit deformations ε_0 or ε_R of the elastic or the transitional state, respectively, follow from the equations

$$g(\varepsilon_0)=1-\frac{T}{2}, \qquad g(\varepsilon_R)=1+\frac{T}{2}. \tag{9}$$

Obviously, Fig. 5 represents well known forms of stress-strain diagrams for rock-like material.

As a second example we try to generalize the discrete model defined by the two spring characteristics in Fig. 3. Now the linear one will be split off into a fan while the second one becomes a family of steeply ascending

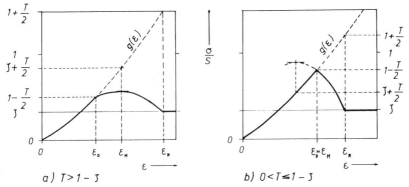

a) $T > 1 - \mathfrak{J}$ b) $0 < T \leqslant 1 - \mathfrak{J}$

Fig. 5. Examples of stress–strain loading diagrams: (a) $T=1$; (b) $T=2/5$.

parallel straight lines which coincide mutually on their horizontal part along the ε-axis (Fig. 6). So we obtain the two families τ_1^α, τ_2^k according to

$$\tau_1^\alpha(\varepsilon) = \alpha\varepsilon, \qquad \frac{1}{b} \leqslant \alpha \leqslant \frac{1}{a};$$

$$\tau_2^k(\varepsilon) = \max\left(0, \frac{\varepsilon - k}{\beta}\right), \qquad a \leqslant k \leqslant b, \tag{10a}$$

in which the given constants a, b, β as well as another one γ, to be introduced below, are to obey

$$0 < a < b, \qquad \beta = b - a, \qquad 0 \leqslant \gamma \leqslant 1. \tag{10b}$$

The characteristics shall be uniformly distributed inside both families, while

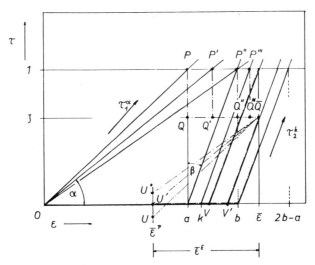

Fig. 6. Two families τ_1^α and τ_2^k, of spring characteristics for loading (——); element characteristics after fracture, and for unloading (– · – · –).

the families as such are weighted by γ or $(1-\gamma)$, respectively:

$$F(\tau, \varepsilon) = F_1(\tau, \varepsilon) + F_2(\tau, \varepsilon),$$

$$f(\tau, \varepsilon) = f_1(\tau, \varepsilon) + f_2(\tau, \varepsilon), \tag{11}$$

$$F_1(\tau, \varepsilon) = \gamma \begin{cases} 1 & \text{if } \tau > \varepsilon/a; \\ \dfrac{\tau - (\varepsilon/b)}{(\varepsilon/a) - (\varepsilon/b)} & \text{if } \varepsilon/b < \tau \leqslant \varepsilon/a, \\ 0 & \text{if } \tau \leqslant \varepsilon/b; \end{cases} \tag{12a}$$

$$F_2(\tau, \varepsilon) = \begin{cases} 1-\gamma & \text{if } \tau > \max\!\left(0, \dfrac{\varepsilon - a}{\beta}\right), \\ (1-\gamma)\left\{ 1 - \max\!\left(0, \dfrac{\varepsilon - a}{\beta}\right) + \tau \right\} \\ \qquad \text{if } \max\!\left(0, \dfrac{\varepsilon - b}{\beta}\right) < \tau \leqslant \max\!\left(0, \dfrac{\varepsilon - a}{\beta}\right), \\ 0 \qquad \text{if } \tau \leqslant \max\!\left(0, \dfrac{\varepsilon - b}{\beta}\right). \end{cases} \tag{12b}$$

Differentiation with respect to τ yields

$$f_1(\tau, \varepsilon) = \gamma \begin{cases} 0 & \text{if } \tau > \varepsilon/a \text{ or } \tau > \varepsilon/b, \\ \dfrac{1}{\varepsilon\left(\dfrac{1}{a} - \dfrac{1}{b}\right)} & \text{if } \varepsilon/b \leqslant \tau \leqslant \varepsilon/a; \end{cases} \tag{13a}$$

$$f_2(\tau, \varepsilon) = (1-\gamma) \begin{cases} 0 \;\; \text{if } \tau > \max\!\left(0, \dfrac{\varepsilon - a}{\beta}\right) \text{ or } \tau < \max\!\left(0, \dfrac{\varepsilon - b}{\beta}\right), \\ \left[1 - \max\!\left(0, \dfrac{\varepsilon - a}{\beta}\right) \right] \delta(\tau) + 1 \\ \qquad \text{if } \max\!\left(0, \dfrac{\varepsilon - b}{\beta}\right) \leqslant \tau \leqslant \max\!\left(0, \dfrac{\varepsilon - a}{\beta}\right) \end{cases} \tag{13b}$$

in which $\delta(\tau)$ denotes Dirac's δ-function. Then by virtue of (5), the

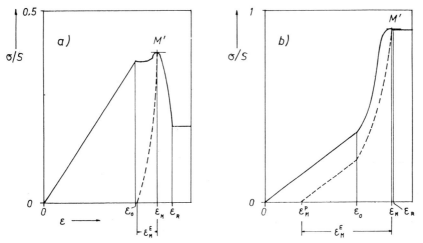

Fig. 7. Examples for stress–strain diagrams with unloading from $\dot{M}'$ (– – –) with $b/a=1.2$, $\gamma=0.4$. (a) $\varepsilon_M^E/\varepsilon_0=0.219$ if $\zeta=0.2$, (b) $\varepsilon_M^E/\varepsilon_0=0.982$ if $\zeta=0.9$.

following equations for the stress–strain diagram in Fig. 7 are obtained,

$$\frac{\sigma(\varepsilon)}{S}=\frac{\gamma}{2}\left(\frac{1}{a}+\frac{1}{b}\right)\varepsilon \quad \text{if } 0\leqslant\varepsilon<a \text{ (elastic)},$$

$$\frac{\sigma(\varepsilon)}{S}=\frac{1}{2(b-a)}\left[\gamma a\left((1-2\zeta)\frac{b}{\varepsilon}-\frac{\varepsilon}{b}+2\zeta\frac{b}{a}\right)+\frac{1-\gamma}{b-a}b^2\left(\frac{\varepsilon}{b}-\frac{a}{b}\right)^2\right]$$

$$\text{if } a\leqslant\varepsilon<b \text{ (transitional)},$$

$$\frac{\sigma(\varepsilon)}{S}=\frac{1}{2}(1-\gamma)\left[1-\left(\frac{\varepsilon-b}{b-a}\right)^2\right]+\zeta\left(1-(1-\gamma)\frac{2b-a-\varepsilon}{b-a}\right)$$

$$\text{if } b\leqslant\varepsilon<2b-a, \text{ (transitional)},$$

$$\frac{\sigma(\varepsilon)}{S}=\zeta \text{ if } 2b-a\leqslant\varepsilon \text{ (granular)}.$$

Note that the diagram in Fig. 7a shows a step similar to Fig. 1c, d. Here, as in Fig. 7b, the maximum (failure) point M' is seen to be situated in the interval $b\leqslant\varepsilon<2b-a$. Then the condition $d(\sigma/S)/d\varepsilon=0$ delivers, because of $\sigma_M=\sigma(\varepsilon_M)$,

$$\varepsilon_M=b+\zeta(b-a), \qquad \frac{\sigma_M}{S}=\zeta+\frac{1}{2}(1-\gamma)(1-\zeta)^2. \tag{14}$$

3.2. Ordered spring characteristics: unloading and reloading

After having loaded until $\bar{\varepsilon}>0$ we now shall study the unloading process, at the beginning of which the individual dimensionless stress points $\bar{\tau}$ of the elements are distributed on the right hand vertical line in Fig. 4. Unfortunately, the information given by the functions F and f is not sufficient to solve the problem so that we need an additional assumption, e.g., the following one:

Assumption. *There is exactly one dimensionless, continuous, strictly monotonically increasing spring characteristic,*

$$\tau = \tau^0(\varepsilon; \bar{\tau}, \bar{\varepsilon}), \qquad -\infty < \varepsilon < \infty, \tag{15}$$

through each point $(\bar{\varepsilon}, \bar{\tau})$ of the right hand vertical line in Fig. 4.

So there exists just one family of spring characteristics the field of which shall be called "ordered".

Now we unload down to ε under the conditions,

$$0 \leqslant \varepsilon \leqslant \bar{\varepsilon}, \qquad 0 \leqslant \bar{\varepsilon} - \varepsilon \leqslant \bar{\varepsilon}, \tag{16}$$

and denote by $\bar{\sigma}(\varepsilon, \bar{\varepsilon})$ the total stress during unloading. Then the following formula will be shown to hold,

$$\frac{\bar{\sigma}(\varepsilon; \bar{\varepsilon})}{S} = \int_{-\infty}^{1} \tau^0(\varepsilon; \bar{\tau}, \bar{\varepsilon}) f(\bar{\tau}, \bar{\varepsilon}) \, d\bar{\tau}$$

$$+ \int^{K} \tau^0(\varepsilon' - [\bar{\varepsilon} - \varepsilon]; \bar{\tau}, \bar{\varepsilon}) f(\bar{\tau}, \bar{\varepsilon}) \, d\bar{\tau}$$

$$- \zeta \int^{K'} f(\bar{\tau}, \bar{\varepsilon}) \, d\bar{\tau}. \tag{17}$$

Its first integral sums up the contributions of all elements which are still elastic at $\bar{\varepsilon}$, i.e. $\bar{\tau} \leqslant 1$ according to the points $\overline{A}$, $\overline{D}$, $\overline{J}$ in Fig. 4. They also remain elastic, because of (16), during unloading. The last two integrals of Eq. (17) refer to points like $\overline{B}$, $\overline{C}$ in Fig. 4. The crack terms corresponding to them in the elements (Fig. 2) are broken so that their unloading starts from the dimensionless level ζ of slipping friction (cf. Eq. (2)), i.e. from the points B' or C' in Fig. 4, respectively. For them, the individual strains ε' of the elements may be found using Eq. (15), by solving the equation

$$\tau^0(\varepsilon'; \bar{\tau}, \bar{\varepsilon}) = \zeta. \tag{18}$$

During unloading, ε' reduces additionally by $\bar{\varepsilon} - \varepsilon$. If the stress point remains in the elastic domain $-\zeta < \tau < \zeta$ (B in Fig. 4) then it contributes to the second integral in Eq. (17) while otherwise, to the third one (point C″ in Fig. 4 which can not move until C). In order to distinguish between these two possibilities, the interval $\bar{\tau} \geq 1$ on the vertical axis at $\bar{\varepsilon}$ is subdivided into two subsets which consequently form the domains of integration in Eq. (17):

$$K = \{\bar{\tau} \mid \bar{\tau} > 1, \tau^0(\varepsilon' - [\bar{\varepsilon} - \varepsilon]; \bar{\tau}, \bar{\varepsilon}) \geq -\zeta\},$$

$$K' = \{\bar{\tau} \mid \bar{\tau} > 1, \tau^0(\varepsilon' - [\bar{\varepsilon} - \varepsilon]; \bar{\tau}, \bar{\varepsilon}) < -\zeta\}. \tag{19}$$

After unloading until ε^*, we reload up to ε observing the conditions

$$0 \leq \varepsilon^* \leq \varepsilon \leq \bar{\varepsilon}. \tag{20}$$

The expression for the reloading stress $\sigma^*(\varepsilon; \bar{\varepsilon}, \varepsilon^*)$ can be derived following similar considerations as above to yield,

$$\begin{aligned}
\frac{\sigma^*(\varepsilon; \bar{\varepsilon}, \varepsilon^*)}{S} &= \int_{-\infty}^{1} \tau^0(\varepsilon; \bar{\tau}, \bar{\varepsilon}) f(\bar{\tau}, \bar{\varepsilon}) \, d\bar{\tau} \\
&+ \int^{K} \tau^0(\varepsilon' - [\bar{\varepsilon} - \varepsilon]; \bar{\tau}, \bar{\varepsilon}) f(\bar{\tau}, \bar{\varepsilon}) \, d\bar{\tau} \\
&+ \int^{K^*} \tau^0(\varepsilon'' + [\varepsilon - \varepsilon^*]; \bar{\tau}, \bar{\varepsilon}) f(\bar{\tau}, \bar{\varepsilon}) \, d\bar{\tau} \\
&+ \zeta \int^{\bar{K}} f(\bar{\tau}, \bar{\varepsilon}) \, d\bar{\tau}.
\end{aligned} \tag{21}$$

Here ε'' (belonging to C″ in Fig. 4), and the subsets K^*, $\bar{K}$ of the $\bar{\tau}$-axis at $\bar{\varepsilon}$ are defined by means of,

$$\tau^0(\varepsilon''; \bar{\tau}, \bar{\varepsilon}) = -\zeta,$$

$$K^* = \{\bar{\tau} \mid \bar{\tau} > 1, \tau^0(\varepsilon' - [\bar{\varepsilon} - \varepsilon^*]; \bar{\tau}, \bar{\varepsilon}) < -\zeta,$$

$$\tau^0(\varepsilon'' + [\varepsilon - \varepsilon^*]; \bar{\tau}, \bar{\varepsilon}) \leq \zeta\},$$

$$\bar{K} = \{\bar{\tau} \mid \bar{\tau} > 1, \tau^0(\varepsilon' - [\bar{\varepsilon} - \varepsilon^*]; \bar{\tau}, \bar{\varepsilon}) < -\zeta,$$

$$\tau^0(\varepsilon'' + [\varepsilon - \varepsilon^*]; \bar{\tau}, \bar{\varepsilon}) > \zeta\}. \tag{22}$$

It may be demonstrated that, because of (16) and (20), the relations

$$\sigma(\varepsilon) \geq \sigma^*(\varepsilon; \bar{\varepsilon}, \varepsilon^*) \geq \bar{\sigma}(\varepsilon; \bar{\varepsilon}), \qquad \sigma(\bar{\varepsilon}) = \bar{\sigma}(\bar{\varepsilon}, \bar{\varepsilon}) = \sigma^*(\bar{\varepsilon}; \bar{\varepsilon}, \varepsilon^*)$$

hold, so that unloading and subsequent reloading generate loops in the stress-strain diagram which are situated below the loading curve, and intersect with it at the point $\bar{\varepsilon}$, $\sigma(\bar{\varepsilon})$.

We also mention that the definition (15) of "ordered" spring characteristics may be generalized in a way that $m \geqslant 1$ families $\tau = \tau_j^0(\varepsilon; \bar{\tau}, \bar{\varepsilon})$ through the points $\bar{\tau}, \bar{\varepsilon}$ exist, $j = 1, \ldots, m$. Then, similar to (11), the functions F and f are composed of m distribution functions F_j, or densities f_j respectively. The values $\varepsilon' = \varepsilon_j'$, $K = K_j$, $K' = K_j'$, $\varepsilon'' = \varepsilon_j''$, $K^* = K_j^*$, $\bar{K} = \bar{K}_j$ have to be formed for each family in a separate manner. Also, the right hand sides of Eqs. (17) and (21) must be written down separately, and added up in order to give $\bar{\sigma}$ or σ^*, respectively.

We now proceed to the example given in Fig. 5, and assume that during unloading no reverse slipping in the crack terms occurs, i.e.

$$K' = \emptyset. \tag{23}$$

The family (15) of characteristics parallel to $g(\varepsilon)$ may be defined according to

$$\tau^0(\varepsilon; \bar{\tau}, \bar{\varepsilon}) = g - \bar{g} + \bar{\tau}; \qquad g = g(\varepsilon), \quad \bar{g} = g(\bar{\varepsilon}). \tag{24}$$

Then we specify $g(\varepsilon)$ to be an exponential function

$$g(\varepsilon) = \Omega[e^{\omega \varepsilon} - 1]; \qquad \omega = \text{const}, \quad \Omega = \text{const}, \quad \omega \Omega > 0. \tag{25}$$

This gives us after some lengthy but elementary calculations,

$$\frac{\bar{\sigma}(\varepsilon; \bar{\varepsilon})}{S} = g(\varepsilon \quad \bar{\varepsilon}^{\mathrm{P}}),$$

in which $\bar{\varepsilon}^{\mathrm{P}}$ denotes the plastic part of deformation for which $\bar{\varepsilon} = \bar{\varepsilon}^{\mathrm{E}} + \bar{\varepsilon}^{\mathrm{P}}$ holds. So the resulting elastic unloading curve, not drawn in Fig. 5, becomes parallel to the initial elastic loading curve. Therefore the elastic-plastic coupling is normal. In order to fulfill the assumption (23) it can be shown that the following condition has to be observed,

$$\frac{1}{\Omega T}\left[\zeta\left(\bar{g} + \frac{T}{2} - 1\right) + \frac{1}{2}\left(\bar{g} - \frac{T}{2} + 1\right)\left(-\bar{g} + \frac{T}{2} + 1\right)\right](\Omega + \bar{g} + \zeta - 1) \leqslant 2\zeta.$$

As a second example we unload from the failure point M′ in Fig. 7. It corresponds, because of (14), to the vertical line at $\bar{\varepsilon}$ in Fig. 6. Therefore one may easily see that all unloading characteristics either pass through $\overline{Q}$, or through the point $\varepsilon = a$ on the horizontal axis so that the total unloading stress $\bar{\sigma}$ becomes zero at the point U′, at which the average value of the element stresses τ equals 0. For this consideration to be valid, the tangent of the center line passing through U′ of the fan (see Fig. 6) amounts to $\alpha' = \frac{1}{2}((1/b) + (1/a))$, and it must not exceed the corresponding tangent $1/\beta$

of the second family τ_2^k. This means, because of (10b) and using $b^2 - a^2 \leqslant 2ab$, that,

$$b \leqslant a(1 + \sqrt{2})$$

has to be fulfilled. Then the elastic part of deformation $\bar{\varepsilon}^E$ is in Fig. 6 seen to obey $\zeta/\bar{\varepsilon}^E = \alpha'$, i.e.

$$\bar{\varepsilon}^E = \frac{2\zeta}{\dfrac{1}{b} + \dfrac{1}{a}}.$$

It always leads, for the data given in Fig. 7, to an anomalous coupling $\varepsilon_M^E/\varepsilon_0 < 1$.

3.3. Unloading for stochastic spring characteristics

In order to replace the condition of ordered spring characteristics by a truly stochastic one, we introduce for any three pairs of stress and strain the triple distribution function,

$$\overline{\Pi}(\tau, \varepsilon; \tau', \varepsilon'; \bar{\tau}, \bar{\varepsilon}) = \Pr\{\tau^j(\varepsilon) > \tau, \tau^j(\varepsilon') > \tau', \tau^j(\bar{\varepsilon}) > \bar{\tau}\}. \quad (26)$$

In Eq. (26), Pr means "probability" of the event specified between { }, and $\tau^j(\varepsilon)$ or $\varepsilon^j(\tau)$ denotes the entity of spring characteristics.

The function $\overline{\Pi}$ cannot be chosen arbitrarily. So it follows from the definition that,

$$\overline{\Pi}(\tau, \varepsilon; \tau', \varepsilon'; \bar{\tau}, \bar{\varepsilon}) = \overline{\Pi}(\tau', \varepsilon'; \bar{\tau}, \bar{\varepsilon}; \tau, \varepsilon) = \overline{\Pi}(\bar{\tau}, \bar{\varepsilon}; \tau, \varepsilon; \tau', \varepsilon')$$

$$= \overline{\Pi}(\tau', \varepsilon'; \tau, \varepsilon; \bar{\tau}, \bar{\varepsilon}) = \overline{\Pi}(\tau, \varepsilon; \bar{\tau}, \bar{\varepsilon}; \tau', \varepsilon')$$

$$= \overline{\Pi}(\bar{\tau}, \bar{\varepsilon}; \tau', \varepsilon'; \tau, \varepsilon), \quad (27)$$

and also that,

$$\frac{\partial}{\partial \tau} \overline{\Pi}(\tau, \varepsilon; \tau', \varepsilon'; \bar{\tau}, \bar{\varepsilon}) \leqslant 0;$$

$$\overline{\Pi}(-\infty, \varepsilon; -\infty, \varepsilon'; -\infty, \bar{\varepsilon}) = 1, \qquad \overline{\Pi}(\infty, \varepsilon; \tau', \varepsilon'; \bar{\tau}, \bar{\varepsilon}) = 0,$$

$$\quad (28)$$

provided all spring characteristics exist for $-\infty < \varepsilon < \infty$. This also means that the functions

$$\overline{X}(\tau, \varepsilon; \tau', \varepsilon') = \overline{\Pi}(\tau, \varepsilon; \tau'\varepsilon'; -\infty, \bar{\varepsilon}),$$

$$\overline{F}(\tau, \varepsilon) = \overline{\Pi}(\tau, \varepsilon; -\infty, \varepsilon'; -\infty, \bar{\varepsilon}), \quad (29)$$

must not depend on $\bar{\varepsilon}$ or ε', respectively. The former consideration leading to Eq. (4) means now that,

$$\overline{\Pi}(\tau,0;-\infty,\varepsilon';-\infty,\bar{\varepsilon})=\begin{cases}0 & \text{if } \tau>1 \\ 1 & \text{if } \tau<1.\end{cases} \quad (30)$$

As we assume again that all spring characteristics are continuous and strictly monotonic, the condition $\tau^j(\varepsilon')>\tau'$, for instance, implies $\tau^j(\bar{\varepsilon})>\bar{\tau}$ provided that $\bar{\varepsilon}\geqslant\varepsilon'$ and $\bar{\tau}\leqslant\tau'$ holds, i.e.,

$$\overline{\Pi}(\tau,\varepsilon;\tau',\varepsilon';\bar{\tau},\bar{\varepsilon})=\overline{X}(\tau,\varepsilon;\tau',\varepsilon')$$

$$\text{if } \bar{\varepsilon}\geqslant\varepsilon' \quad \text{and} \quad \bar{\tau}\leqslant\tau';$$

$$\overline{\Pi}(\tau,\varepsilon;\tau',\varepsilon';\bar{\tau},\bar{\varepsilon})=\overline{F}(\tau,\varepsilon) \quad (31)$$

$$\text{if } \bar{\varepsilon}\geqslant\varepsilon, \varepsilon'\geqslant\varepsilon \quad \text{and} \quad \bar{\tau}\leqslant\tau, \tau'\leqslant\tau.$$

As another consequence of monotony the weighted numer of spring characteristics obeying $\tau(\varepsilon)>\tau$ increases with ε, i.e.

$$\frac{\partial}{\partial\varepsilon}\overline{\Pi}(\tau,\varepsilon;\tau',\varepsilon';\bar{\tau},\bar{\varepsilon})\geqslant 0. \quad (32)$$

Now we define the distribution functions,

$$F(\tau,\varepsilon)=1-\overline{F}(\tau,\varepsilon),$$

$$\Pi(\tau,\varepsilon;\tau',\varepsilon';\bar{\tau},\bar{\varepsilon})=\Pr\{\tau^j(\varepsilon)\leqslant\tau, \tau^j(\varepsilon')>\tau', \tau^j(\bar{\varepsilon})>\bar{\tau}\}$$

$$=\overline{X}(\tau',\varepsilon';\bar{\tau},\bar{\varepsilon})-\overline{\Pi}(\tau,\varepsilon;\tau',\varepsilon';\bar{\tau},\bar{\varepsilon}), \quad (33)$$

$$X(\tau,\varepsilon;\tau',\varepsilon')=\Pr\{\tau^j(\varepsilon)\leqslant\tau, \tau^j(\varepsilon')\leqslant\tau'\}$$

$$=F(\tau',\varepsilon')-\Pr\{\tau^j(\varepsilon)>\tau, \tau^j(\varepsilon')\leqslant\tau'\}$$

$$=F(\tau',\varepsilon')-\overline{F}(\tau,\varepsilon)+\overline{X}(\tau,\varepsilon;\tau',\varepsilon'),$$

of which F was already introduced in Eq. (3). The densities,

$$\xi(\tau,\varepsilon;\bar{\tau},\bar{\varepsilon})=\frac{\partial}{\partial\tau}X(\tau,\varepsilon;\bar{\tau},\bar{\varepsilon}),$$

$$s(\tau,\varepsilon;\tau',\varepsilon';\bar{\tau},\bar{\varepsilon})=\frac{\partial}{\partial\varepsilon'}\Pi(\tau,\varepsilon;\tau',\varepsilon';\bar{\tau},\bar{\varepsilon}), \quad (34)$$

$$t(\tau,\varepsilon;\tau',\varepsilon';\bar{\tau},\bar{\varepsilon})=\frac{\partial}{\partial\tau}s(\tau,\varepsilon;\tau',\varepsilon';\bar{\tau},\bar{\varepsilon})$$

$$=\frac{\partial^2}{\partial\tau\partial\varepsilon'}\Pi(\tau,\varepsilon;\tau',\varepsilon';\bar{\tau},\bar{\varepsilon}),$$

are formed provided the differentiation may be carried out also in the sense of Dirac's δ-function.

Let us return to the proper unloading problem. Obviously the three integrals in the following expression have the same physical background as the three integrals in Eq. (17):

$$
\frac{\bar{\sigma}(\varepsilon, \bar{\varepsilon})}{S} = \int_{-\infty}^{1} \tau \Pr\{\tau < \tau^j(\varepsilon) \leqslant \tau + d\tau, \tau^j(\bar{\varepsilon}) \leqslant 1\}
$$

$$
+ \int_{\tau=-\zeta}^{\infty} \int_{\varepsilon'=-\infty}^{\bar{\varepsilon}} \tau \Pr\{\tau < \tau^j(\varepsilon' - [\bar{\varepsilon} - \varepsilon]) \leqslant \tau + d\tau,
$$

$$
\varepsilon' \leqslant \varepsilon^j(\zeta) < \varepsilon' + d\varepsilon', \tau^j(\bar{\varepsilon}) > 1\}
$$

$$
- \zeta \int_{\tau=-\infty}^{-\zeta} \int_{\varepsilon'=-\infty}^{\bar{\varepsilon}} \Pr\{\tau < \tau^j(\varepsilon' - [\bar{\varepsilon} - \varepsilon]) \leqslant \tau + d\tau,
$$

$$
\varepsilon' \leqslant \varepsilon^j(\zeta) < \varepsilon' + d\varepsilon', \tau^j(\bar{\varepsilon}) > 1\}.
$$

Because of monotony, the conditions $\tau^j(\varepsilon') > \tau'$ and $\varepsilon^j(\tau') < \varepsilon'$ are equivalent, so that using Eqs. (33) and (34), the unloading stress can be expressed as

$$
\frac{\bar{\sigma}(\varepsilon, \bar{\varepsilon})}{S} = \int_{-\infty}^{1} \tau \xi(\tau, \varepsilon; 1, \bar{\varepsilon}) d\tau
$$

$$
+ \int_{-\infty}^{\bar{\varepsilon}} d\varepsilon' \int_{-\zeta}^{\infty} \tau t(\tau, \varepsilon' - [\bar{\varepsilon} - \varepsilon]; \zeta, \varepsilon'; 1, \bar{\varepsilon}) d\tau \qquad (35)
$$

$$
- \zeta \int_{-\infty}^{\bar{\varepsilon}} s(-\zeta, \varepsilon' - [\bar{\varepsilon} - \varepsilon]; \zeta, \varepsilon'; 1, \bar{\varepsilon}) d\varepsilon'.
$$

It becomes identical to Eq. (17) if for an ordered field, according to (15), the triple density $\overline{\overline{\Pi}}$ has the following form:

$$
\overline{\overline{\Pi}}(\tau, \varepsilon; \tau', \varepsilon'; \tilde{\tau}, \bar{\varepsilon}) = \int_{\tilde{\tau}}^{\infty} H(\tau^0(\varepsilon; \tilde{\tau}, \bar{\varepsilon}) - \tau) H(\tau^0(\varepsilon'; \tilde{\tau}, \bar{\varepsilon}) - \tau')
$$

$$
\times f(\tilde{\tau}, \bar{\varepsilon}) d\tilde{\tau}.
$$

Here f denotes the distribution density already introduced in Eq. (3), and $H(x)$ ($=1$ if $x > 0$; $=0$ if $x \leqslant 0$) is the Heaviside function. Further independent examples for $\overline{\overline{\Pi}}$ have not yet been found because of the heavy restrictions imposed by virtue of the Eqs. (27–32).

Acknowledgments

The investigation was financially supported by the Bergbau-Forschung GmbH, Essen, upon a grant from the land Nordrhein-Westfalen (Fed. Rep. Germany). The author is indebted to Mr. W. Burgert, Munich, and to Dr. G. Bräuner, Essen, for many helpful discussions.

References

Bažant, Z. P. and Kim, S. S. (1979), "Plastic-Fracturing Theory for Concrete," *J. Eng. Div. ASCE*, **105**, 407.
Budiansky, B. and O'Connel, R. J. (1976), "Elastic Moduli of Cracked Solids," *Int. J. Solids Struct.*, **12**, 81.
Burgert, W. (1979), Personal communication (part of Ph.D. thesis under preparation).
Dougill, J. W. (1976), "On Stable Progressively Fracturing Solids," *ZAMP*, **27**, 423.
Dragon, A. (1976), "On Phenomenological Description of Rock-like Materials With Account For Kinetics of Brittle Fracture," *Arch. Mech.*, **28**, 13.
Dragon, A. and Mróz, Z. (1979), "A Continuum Model for Plastic-Brittle Behaviour of Rock and Concrete," *Int. J. Eng. Sci.*, **17**, 121.
Hueckel, T. (1976), "The Flow Law For the Granular Solids and Rocks With Variable Unloading Rule," *Problèmes de Rhéologie de Mécanique des Sols*, Symp. Franco-Polonais, Nice, 1974, PWN, Warszawa, 203.
Lippmann, H. (1978/1979), "The Mechanics of Translatory Rock Bursting," *Advances in Analysis of Geotechnical Instabilities*, J. C. Thompson (ed.), University of Waterloo Press, Waterloo, Ontario, 25.
Lippmann, H. (1979), *Materialverhalten gebirgsschlaggefährdeter Flöze*, internal report given to the Bergbau-Forschung GmbH, Essen.
Maier, G. and Hueckel, T. (1979), "Nonassociated and Coupled Flow Rules of Elastoplasticity for Rock-Like Materials," *Int. J. Rock. Mech. Min. Sci.*, **16**, 77.
Persoz, B. (1969), "Modèles Non-Linéaire," *La Rhéologie*, B. Persoz (ed.), Masson, Paris, 45.

S. Nemat-Nasser, Editor
THREE-DIMENSIONAL CONSTITUTIVE RELATIONS AND DUCTILE FRACTURE
North-Holland Publishing Company (1981) 405–409

DISCUSSION ON SESSION 7

Z. MRÓZ

Institute of Fundamental Technological Research, Warsaw, Poland

The papers presented during this Session are concerned with crack initiation, stable and unstable ductile crack propagation, and progressive failure due to brittle rupture of structure subelements. Let me discuss each paper separately.

Remarks on *A. Dragon: Limitation to Ductility Set by Inclusion Induced Disturbance of Yielding*

Let us briefly recapitulate the basic assumptions of this work. It is assumed that the initially homogeneous and incompressible plastic flaw is perturbed by the formation of voids due to decohesion at the inclusion–matrix interfaces. The author introduces initial porosity, β_i, inclusion concentration parameters β_k of particles of different ranges of diameters and the total parameter β_t represents the porosity and the fraction of fractured inclusions. The yield condition is assumed to depend on kinematic and isotropic hardening parameters α and $\mathcal{K}$, as well as on a set of β_i; thus $f(\sigma, \alpha, \mathcal{K}, \beta_i) = 0$. Besides the yield condition, there exists a set of conditions $f_k(\sigma, \beta_k) = 0$ which represent the mechanisms of void nucleation for particular sets of inclusions. After reaching a critical value of porosity, $\beta = \beta_{cr}$, only the mechanism of yielding of a porous material occurs with the associated dilatational hardening and softening.

The work in the present form provides the general structure of constitutive relations which account for the void nucleation process, and introduces the inclusion concentration as an essential material parameter. However, it may be more effective to separate the void nucleation process from void growth due to viscous or plastic flow. Here, I present an alternative and somewhat different version of the model which could be easily applied in numerical simulation of initiation of fracture.

Consider the yield condition of the form,

$$f(\sigma, \mathcal{K}, \beta') = \frac{\sigma_{kk} - I^0(\beta')}{a^2(\mathcal{K}, \beta')} + S_{ij}S_{ij} - K^2(\mathcal{K}, \beta') = 0, \tag{1}$$

where β' denotes the actual porosity, and $\mathcal{K}$ is the scalar hardening

405

parameter. The three material functions, $I^0(\beta')$, $a(\mathcal{H}, \beta')$ and $K(\mathcal{H}, \beta')$, allow the incorporation of experimental data for a large class of materials. The yield surface is a rotational ellipsoid whose center and diameter vary with $\mathcal{H}$ and β'. Assume further that there exists *a fracture surface*,

$$f_c(\sigma, \beta_0, \beta') = \tfrac{1}{2} S_{ij} S_{ij} - m(\beta')[J(\beta') - \sigma_{kk}] = 0, \qquad (2)$$

where β_0 denotes the initial volume fraction of inclusions, and $m(\beta')$ and $J(\beta')$ are functions depending on varying porosity. Thus for the stress states corresponding to the yield surface (1), only plastic flow occurs, whereas for stress states corresponding to the fracture surface (2), the brittle fracture develops. For stress states satisfying $f=0$ and $f_c=0$, both plastic flow and brittle fracture develop simultaneously. The brittle fracture corresponds to the mechanism of decohesion at interfaces and the *fracture tensor* r affects the yield condition and elastic stiffness, but does not provide contribution to plastic strain.

The plastic flow rule and the fracture growth rule are assumed in the associated forms,

$$\dot{\varepsilon}^P = \frac{1}{K} n \dot{\sigma}_n, \qquad (3)$$

$$\dot{r} = \frac{1}{K_c} n_c \dot{\sigma}_{nc}, \qquad (4)$$

where

$$\dot{\sigma}_n = \dot{\sigma} \cdot n = \dot{\sigma} \cdot \frac{\partial f}{\partial \sigma} \left(\frac{\partial f}{\partial \sigma} \cdot \frac{\partial f}{\partial \sigma} \right)^{-1/2},$$

$$\dot{\sigma}_{nc} = \dot{\sigma} \cdot n_c = \dot{\sigma} \cdot \frac{\partial f_c}{\partial \sigma} \left(\frac{\partial f_c}{\partial \sigma} \cdot \frac{\partial f_c}{\partial \sigma} \right)^{-1/2}, \qquad (5)$$

n and n_c are the normalized gradient tensors, $\partial f / \partial \sigma$ and $\partial f_c / \partial \sigma$. Further, we can write the evolution rules for hardening parameters in the form,

$$\mathcal{H} = (\dot{e}^P \cdot \dot{e}^P)^{1/2}, \qquad (6)$$

$$\beta' = a \operatorname{tr} \dot{\varepsilon}^P + b \operatorname{tr} \dot{r}, \qquad (7)$$

where a and b are material parameters. Here $\dot{e}^P$ and S are the deviatoric plastic strain rate and stress, whereas tr denotes the trace of the tensor. Equation (7) indicates that the growth of porosity is due to volume increase of existing voids and due to the decohesion process on inclusions. The assumption that the specific elastic strain energy, $U = U(\varepsilon^e, r)$, depends on the fracture tensor completes our description. The form of this function constitutes a separate problem.

It is seen that the system of relations (1)–(7) is coupled and differs from that discussed in Dragon's paper by postulating the evolution rule (7) and the fracture rule (4). The hardening and fracture moduli K and K_c occurring in (3) and (4) can be found by satisfying the consistency conditions, namely, Plastic flow:

$$f=\dot{f}=0; \quad f_c<0 \quad \text{or} \quad f_c=0, \dot{f}_c<0, \dot{r}=0;$$

Brittle decohesion:

$$f_c=\dot{f}_c=0; \quad f<0 \quad \text{or} \quad f=0, \dot{f}<0, \dot{\varepsilon}^\text{p}=0;$$

Combined mechanism:

$$f=f_c=\dot{f}=\dot{f}_c=0.$$

Initially, the yield surface is represented by the von Mises cylinder whereas the fracture surface is the rotational paraboloid, intersecting the hydrostatic tension axis at the point $\sigma_{kk}=J(\beta_0)$. For stress states close to hydrostatic tension, the fracture surface $f_c=0$ is first reached and the decohesion mechanism operates, thus affecting the shape of the yield surface which becomes a rotational ellipsoid with changing semiaxes due to growing porosity β'. On the other hand, for stress states with predominant deviatoric part, the plastic flow occurs first, thus inducing hardening, and the fracture surface is reached subsequently. The incremental analysis can thus provide the value of the maximal stress attained in the deformation process and the unstable behavior can be treated by following the evolution of hardening and softening.

Remarks on *H. Lippmann: Ductility Caused by Progressive Formation of Shear Cracks*

The present paper is concerned with the analysis of a model composed of a set of parallel spring and friction elements with maximal and residual values of the friction force. Assuming nonlinear spring characteristics, the author demonstrates existence of the "stiffening" effect due to residual stresses locked in particular nonlinear springs. It is speculated that the phenomenon of rock bursting is closely associated with this anomalous elastic-plastic coupling.

The concept of a composite model with breaking or softening subelements seems more proper in study of progressing facturing and of inelastic response than a single macrocrack approach, since cracks are densely distributed within the representative macroelement. However, the loss of stability is associated with the elastic and softening response and it is not clear why the author tries to relate the effect of elastic coupling (variation of elastic stiffness with irreversible strain) to the mode of unstable and uncontrollable rupture.

To discuss this problem in more detail, consider a certain state of loading of a body within which there are two domains:

—domain V_1 where softening occurs after reaching critical state level, so that $d\sigma \cdot d\varepsilon < 0$ in V_1;

—and domain V_2 which is elastic, so that $d\sigma \cdot d\varepsilon > 0$ in V_2.

Denoting the potential energy of the body by $\Pi(u) = W(u) - \Pi_e(u)$, where u denotes the displacement field and W, Π_e are the elastic strain energy of the body and the potential energy of surface tractions. The necessary stability condition then requires that,

$$\frac{\partial^2 \Pi}{\partial u_i \partial u_k} \delta u_i \delta u_k = \int \delta\boldsymbol{\sigma} \cdot \delta\boldsymbol{\varepsilon} \, dV_2 + \int \delta\boldsymbol{\sigma} \cdot \delta\boldsymbol{\varepsilon} \, dV_1 - \int \delta T^0 \cdot \delta u^0 \, dS > 0, \quad (1)$$

where $\delta u, \delta\varepsilon$ are the kinematically admissible variations of the displacement and strain fields, $\delta\sigma$ is the variation of stress field related by constitutive relations to $\delta\varepsilon$. When surface tractions are specified on S_T, then $\delta T^0 = 0$ and for specified displacements on S_u there is $\delta u^0 = 0$. In this case, the inequality (1) becomes,

$$\int \delta\boldsymbol{\sigma} \cdot \delta\boldsymbol{\varepsilon} \, dV_1 + \int \delta\boldsymbol{\sigma} \cdot \delta\boldsymbol{\varepsilon} \, dV_2 > 0. \quad (2)$$

If plastic strain increment occurs, then $\delta\boldsymbol{\sigma} \cdot \delta\boldsymbol{\varepsilon} = \delta\boldsymbol{\sigma} \cdot \delta\boldsymbol{\varepsilon}^e + \delta\boldsymbol{\sigma} \cdot \delta\boldsymbol{\varepsilon}^p$ and the condition (2) applies. Since one term of (2) is positive, the other negative, the onset of instability occurs when (2) is violated and this depends on the rate of softening and on the elastic stiffness at the limit point. The instability point also depends on a class of admissible displacement variations and therefore on the kind of surface tractions or displacements. Thus, for a uniaxial loading of the speciment composed of stable and unstable elements, the uncontrollable dynamic process under applied stress commences at the point of maximal stress where $d\sigma/d\varepsilon = 0$, whereas for a strain controlled program, the instability occurs when $d\varepsilon/d\sigma = 0$, that is when the softening stress–strain curve possesses a vertical tangent. The brittle characteristic employed by the author is a limiting case of the softening behavior when the tangent softening modulus tends to infinity.

The elegant analysis presented by Professor Lippmann clarifies the elastic coupling effect and lays a foundation for the modelling of a fracturing process through the composite model. On the other hand, the rock bursts and catastrophic rupture phenomena are described by the stability or instability conditions such as (1) or (2) where these coupling phenomena seem to be of secondary importance.

Remarks on T. Yokobori et al.: Constitutive Equations and Global Criteria for Ductile Fracture

In the first part of this work, the authors extend the previously discussed model of crack growth by accounting for void nucleation at the crack tip. In other works, they introduce a damage zone by including the additional energy dissipation due to void nucleation and derive the respective stability condition.

A similar problem was recently studied by H. D. Bui (*C. R. Acad. Sci.*, **290B** (1980) 345–348) who provided an exact solution for the damage and plastic zones. The damage is accounted for by the limit strain condition. In the present paper, however, the voids are not assumed to change the plastic properties and their effect on crack behavior is not clear. Thus it seems natural to also include the term that accounts for change of dissipation due to porosity of plastic zone.

The second part of the work is concerned with the stable crack growth during creep. The empirical parameter P is found which enables one to correlate all experimental data through a simple analytical expression. It is stated that the J-integral does not provide a sufficiently good correlation with experiment as regards to the growth rate of crack.

The analysis seems to be based on steady-state solution for creep and it is not clear how the transient term affects both the stress distribution and the propagation rate. This question was also discussed by Professor Leckie, in his paper, who provided some arguments for using the steady-state solution.

The problem of crack propagation under creep condition also poses a question of accounting for material degradation in the zone adjacent to the crack tip. For instance, the Kachanov theory of creep damage could be combined with the crack propagation rule, which would provide a better understanding of the crack–damage interaction.

S. Nemat-Nasser, Editor
THREE-DIMENSIONAL CONSTITUTIVE RELATIONS AND DUCTILE FRACTURE
North-Holland Publishing Company (1981) 411

REPLY BY T. YOKOBORI

Concerning the question about the first part of our paper, we have neglected the change in dissipation caused by the porosity in the plastic zone. It is desirable to take this into account. Nevertheless, we believe that the main conclusions may not be affected much.

The first question to the second part of this paper seems to be of a rather general nature. The purpose of our studies in the second part of the paper is to extract rate-determining factors for high temperature crack growth rate, and to formulate the crack growth equation in terms of independent variables, as described in the text. This is important from the standpoint of both mechanics and practical application. Also for this reason we focused on Region II in the $\ln da/dt$ versus $\ln \alpha\sqrt{a_{eff}}\,\sigma_g$-plot. Of course, the linear relation between $\ln da/dt$ and $\ln \alpha\sqrt{a_{eff}}\,\sigma_g$ does not mean steady state, because if da/dt is plotted against time, da/dt does not show constant value with respect to time. Therefore, it is natural to expect that the stress distribution changes gradually with time. This will mean that the value of power coefficients m^* and n^* in our proposed formula, $da/dt = M(\alpha\sqrt{a_{eff}}\,\sigma_g)^{m^*}\sigma_g^{n^*}$ (see Eq. (30)), may probably change with time in Region II. Nevertheless, the fact that the proposed equation is in very good agreement with experimental data may mean in turn that m^* and n^* are hardly affected by changes of this kind.

The second question about the second part of our paper seems to have an answer similar to that of the first question about the second part of the paper. We observed the void formation during high temperature crack growth, and, in that sense the material degradation occurs in the zone adjacent to the crack tip. As we described above, the change in degradation with time may not have much effect on the values of m^* and n^*. Even in uncracked materials, degradation must occur with time, and, nevertheless, Larson–Miller's formula is in good agreement with the experimental data on creep rupture in uncracked materials. This also may justify our formula. In addition to this, we would like to remark on the suggestion of combining Kachanov theory of creep damage. The concept of damage in Kachanov's theory is represented mathematically as the change of the true exerted stress, and the damage in this sense will be much affected by the reduction of the cross sectional area of the specimen in the tertiary stage of creep test. As mentioned above, we have considered Region II and not Region III, the final stage from the design standpoint. Still, it is to be noted that the first question may also be raised to the Kachanov theory, in which the power coefficient and other parameters are assumed constant while the damage is changing.

S. Nemat-Nasser, Editor
THREE-DIMENSIONAL CONSTITUTIVE RELATIONS AND DUCTILE FRACTURE
North-Holland Publishing Company (1981) 413

REPLY BY H. LIPPMANN

The intention of my paper is, to give by means of simple rheological models an intuitive understanding of several experimentally observed effects which arise during the uniaxial compression test of rock-like materials. Though model experiments show that one of those effects, i.e. the anomalous elastic-plastic coupling, might have to do with so-called rock-bursts in coal mines (cp. the short remark at the end of the Introduction), this problem is not at all treated in the present contribution.

Despite that, the discussion by Professor Mróz deals in the first place with his interesting ideas regarding rock-bursting instabilities. Obviously both of us are following up quite different approaches. My approach has nothing to do with the minimum of the elastic potential energy, and even more so as dissipation prevails, but is based on the concept of load carrying capacity under conditions of internal and external friction. Details may be found in the original literature (cp. Lippmann 1978, 1979) but I am sure that Professor Mróz and I will continue our dialogue at occasions, being more closely related to mining mechanics than this Symposium is.

SESSION 8

Chairman: D. Radenkovic
           Ecole Polytechnique, Palaiseau, France

Authors: Y. C. Gao, K. C. Hwang

S. Nemat-Nasser, Editor
THREE-DIMENSIONAL CONSTITUTIVE RELATIONS AND DUCTILE FRACTURE
North-Holland Publishing Company (1981) 417–434

ELASTIC-PLASTIC FIELDS IN STEADY CRACK GROWTH

Yu-chen GAO

Harbin Shipbuilding Engineering Institute, Harbin, China

Keh-chih HWANG

Tsinghua University, Peking, China

By use of Von Mises yield criterion and the associated flow rule for elastic perfectly-plastic material, the history-dependent yield condition and constitutive relations for the case of plane strain are derived. The concept of strain-jump is introduced and two cases of strong and weak discontinuities across boundaries of neighboring domains are distinguished. Four contiguity conditions for neighboring domains are set up and given in simple form. For incompressible material ($\nu = \frac{1}{2}$) it is shown that the mixed-mode (I and II) crack cannot grow steadily, and the asymptotic near-tip solution for pure Mode I is obtained. The method can be extended to the strain-hardening material, and some results for power-law hardening material are presented.

0. Introduction

It is well-known that the near-tip stress and strain fields for a crack in growth are quite different from those for a stationary crack, and their analysis is essential to the criteria of ductile fracture. Rice (1968, 1978) has given the theoretical analysis of deformations at a growing crack tip, without regard to the existence of unloading zone behind the tip. Chitaley and McClintock (1971) have successfully analyzed the elastic-plastic fields in steady growth of Mode III crack for an elastic perfectly-plastic material. To avoid considering each previous increment of crack growth, a steady-state was studied. It was shown by the authors that the crack tip is surrounded by the primary plastic zone, the unloading "wake" region and the secondary plastic zone of reloading in the reverse sense. The authors attempted to confirm that not only φ, $\partial\varphi/\partial n$ (φ being the stress function for anti-plane shear), but also $\nabla^2\varphi$ were continuous across the boundary Γ_2 between the primary plastic zone and the wake region. But the proof given by them was inadequate. Slepjan (1974)[1] pointed out the continuity across Γ_2 of $\nabla^2\varphi$ is

---

[1] The authors of this paper are greatly indebted to Dr. Q. S. Nguyen for his kindness in informing us of the works of Slepjan (1973, 1974).

true only asymptotically as $r \to 0$. By use of Tresca–St. Venant's yield criterion and the associated flow rule, Slepjan (1974) worked out the asymptotic solutions for the fields near the crack tips of pure Mode I and pure Mode II in the case of plane strain.

In this paper are studied the elastic-plastic fields for mixed-Mode (I and II) crack growing steadily in an elastic perfectly-plastic medium. By use of Von Mises yield criterion and the associated flow rule, the history-dependent yield condition and constitutive relations for the case of plane strain are derived. In the formulation of contiguity conditions, i.e. the conditions to be satisfied at the boundary of neighboring domains, the concept of strain-jump is introduced and two cases of strong and weak discontinuities are distinguished. Four contiguity conditions are set up and given in simple form. The Airy stress function is expanded as a power series of r, and as the first-order near-tip asymptotic solution, only the first term is retained. For incompressible material ($\nu = \frac{1}{2}$) it is shown that the mixed-Mode (I and II) crack cannot grow steadily, and the complete solution for pure Mode I is obtained. This solution exhibits logarithmic singularity of near-tip strains. Similar to the Mode III results of Chitaley and McClintock (1971) there exist in the upper (or lower) half plane two plastic domains with an intermediate unloading zone. The primary plastic domain is a sector with $|\theta| < 112.1°$, while the secondary plastic domain is a sector adjacent to the free crack surface with $162.1° < |\theta| < 180°$. Finally the explicit expressions for the crack opening displacements are given.

The method used in this paper can be extended to strain-hardening material, and some results for power-law hardening material are presented.

1. Basic equations for plane strain

In Fig. 1 are shown the various domains in a steadily growing crack problem.

1.1. Stress function and stresses

In terms of the stress function, φ, the stress components are

$$\sigma_x = \frac{\partial^2 \varphi}{\partial y^2}, \qquad \sigma_y = \frac{\partial^2 \varphi}{\partial x^2}, \qquad \tau_{xy} = -\frac{\partial^2 \varphi}{\partial x \partial y}. \tag{1}$$

The strain components are split into the elastic and plastic parts,

$$\overset{*}{\varepsilon}_x = \overset{*}{\varepsilon}{}^e_x + \overset{*}{\varepsilon}{}^P_x, \qquad \overset{*}{\varepsilon}_y = \overset{*}{\varepsilon}{}^e_y + \overset{*}{\varepsilon}{}^P_y,$$

$$\overset{*}{\varepsilon}_{xy} = \overset{*}{\varepsilon}{}^e_{xy} + \overset{*}{\varepsilon}{}^P_{xy}, \qquad \overset{*}{\varepsilon}_z = \overset{*}{\varepsilon}{}^e_z + \overset{*}{\varepsilon}{}^P_z, \tag{2}$$

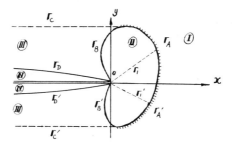

Fig. 1. I—Elastic domain; II—Primary plastic domain, loading; III—Wake, plastic unloading; IV—Secondary plastic domain, reverse loading.

where the elastic part is related to the stress by Hooke's law,

$$\overset{*}{\varepsilon}{}^e_x = \frac{\sigma_x}{E} - \frac{\nu}{E}(\sigma_y + \sigma_z), \qquad \overset{*}{\varepsilon}{}^e_y = \frac{\sigma_y}{E} - \frac{\nu}{E}(\sigma_x + \sigma_z),$$

$$\overset{*}{\varepsilon}{}^e_{xy} = \frac{1+\nu}{E}\tau_{xy}, \qquad \overset{*}{\varepsilon}{}^e_z = \frac{\sigma_z}{E} - \frac{\nu}{E}(\sigma_x + \sigma_y). \tag{3}$$

ν and E being the Poisson and the Young moduli, respectively.

The Reuss relations for perfect plasticity are used to define the plastic strain increments,

$$\mathrm{d}\overset{*}{\varepsilon}{}^p_x = \frac{\lambda}{3}(2\sigma_x - \sigma_y - \sigma_z)\,\mathrm{d}a, \qquad \mathrm{d}\overset{*}{\varepsilon}{}^p_y = \frac{\lambda}{3}(2\sigma_y - \sigma_x - \sigma_z)\,\mathrm{d}a,$$

$$\mathrm{d}\overset{*}{\varepsilon}{}^p_{xy} = \lambda\tau_{xy}\,\mathrm{d}a, \qquad \mathrm{d}\overset{*}{\varepsilon}{}^p_z = \frac{\lambda}{3}(2\sigma_z - \sigma_x - \sigma_y)\,\mathrm{d}a, \tag{4}$$

where $2a$ denotes the crack length, and $\lambda > 0$. We confine attention to plane strain problems. Then σ_z is determined from the condition

$$\mathrm{d}\varepsilon_z = \mathrm{d}\overset{*}{\varepsilon}{}^e_z + \mathrm{d}\overset{*}{\varepsilon}{}^p_z = 0. \tag{5}$$

For steady crack growth, if we observe the field in a reference system moving with the crack tip, the distribution of stresses and strains remain time-invariant. In this paper the crack is assumed to grow steadily along the x-direction; then for any quantity given in Cartesian coordinates,

$$\frac{\partial}{\partial a} = -\frac{\partial}{\partial x}. \tag{6}$$

From (3)–(5) we obtain

$$\sigma_z = \nu(\sigma_x + \sigma_y) + (\tfrac{1}{2} - \nu)\sigma_p, \tag{7}$$

where

$$\sigma_p = \frac{2E}{3} \exp\left(-\frac{2E\Lambda}{3}\right) \int_x^{x_A} \lambda(\sigma_x + \sigma_y) \exp\left(\frac{2E\Lambda}{3}\right) dx, \qquad (8)$$

$$\Lambda = \int_x^{x_A} \lambda \, dx, \qquad (9)$$

and x_A denotes the value of x at the plastic boundary, Γ_A, Fig. 1.

Substituting (7) into (3) and (4), and using (8), we can rewrite (2) in the form

$$\varepsilon_x = \varepsilon_x^e + \varepsilon_x^p, \qquad \varepsilon_y = \varepsilon_y^e + \varepsilon_y^p, \qquad \varepsilon_{xy} = \varepsilon_{xy}^e + \varepsilon_{xy}^p, \qquad (10)$$

where

$$\varepsilon_x^e = \frac{1-\nu^2}{E}\sigma_x - \frac{\nu}{E}(1+\nu)\sigma_y, \qquad \varepsilon_y^e = \frac{1-\nu^2}{E}\sigma_y - \frac{\nu(1+\nu)}{E}\sigma_x,$$

$$\varepsilon_{xy}^e = \frac{1+\nu}{E}\tau_{xy}, \qquad (11)$$

$$\varepsilon_x^p = \frac{1}{2}\int_x^{x_A}\lambda(\sigma_x - \sigma_y)\,dx + \left(\frac{1}{2}-\nu\right)^2\frac{\sigma_p}{E},$$

$$\varepsilon_{xy}^p = \int_x^{x_A}\lambda\tau_{xy}\,dx, \qquad \varepsilon_y^p = \frac{1}{2}\int_x^{x_A}\lambda(\sigma_y - \sigma_x)\,dx + \left(\frac{1}{2}-\nu\right)^2\frac{\sigma_p}{E}. \qquad (12)$$

1.2. Yield condition

From Von Mises yielding condition and (7), we obtain

$$\tfrac{1}{3}\sigma^2 \equiv \tfrac{1}{4}(\sigma_x - \sigma_y)^2 + \tau_{xy}^2 + \tfrac{1}{3}\left(\tfrac{1}{2}-\nu\right)^2(\sigma_x + \sigma_y - \sigma_p)^2 = k^2, \qquad (13)$$

in which σ is the equivalent stress and k is the yield shear stress.

1.3. Compatibility equation

For plane strain problems, we have,

$$\frac{1-\nu^2}{E}\Delta\Delta\varphi + \frac{\partial^2\varepsilon_x^p}{\partial y^2} + \frac{\partial^2\varepsilon_y^p}{\partial x^2} - 2\frac{\partial^2\varepsilon_{xy}^p}{\partial x\partial y} = 0. \qquad (14)$$

In the initial elastic domain (I, Fig. 1), $\varepsilon_x^p = \varepsilon_y^p = \varepsilon_{xy}^p = 0$, while in the wake of plastic unloading (III, Fig. 1), ε_x^p, ε_y^p and ε_{xy}^p are functions of y only.

2. Contiguity conditions

2.1. General relations

We assume that the boundary, Γ, between neighboring domains is an arbitrary curve. We take the parallel-curves family of Γ and their straight normals to be the coordinate lines, and represent the coordinate parameters by s and n (Fig. 2).

Let $[\psi]_\Gamma$ denote the jump of ψ across Γ, i.e. $[\psi]_\Gamma = \psi|_{n=+0} - \psi|_{n=-0}$. Then from the continuity of tractions, $[\sigma_n]_\Gamma = [\tau_{ns}]_\Gamma = 0$, we obtain

$$[\varphi]_\Gamma = \left[\frac{\partial \varphi}{\partial n}\right]_\Gamma = 0. \tag{15}$$

It is easy to prove that the contiguity condition of displacements for both sides of Γ can be written as

$$[\varepsilon_s]_\Gamma = 0,$$
$$\left[\frac{\partial \varepsilon_s}{\partial n}\right]_\Gamma - 2\frac{\mathrm{d}}{\mathrm{d}s}[\varepsilon_{ns}]_\Gamma - \frac{\mathrm{d}\vartheta}{\mathrm{d}s}[\varepsilon_n]_\Gamma = 0, \tag{16}$$

In which ϑ is the angle from the x-axis to the normals of Γ. Equations (15) and (16) are the four contiguity conditions that must be satisfied at boundaries of neighboring domains.

If the stresses or strains suffer jump discontinuities across Γ, it is called strong discontinuity. If the stresses or strains themselves are continuous, but their derivatives along the normal direction n suffer jump discontinuities across Γ, it is called weak discontinuity.

2.2. Strong discontinuity

Assume the yield condition (13) is satisfied at both sides of the line of strong discontinuity, Γ. Then $[\sigma_s]_\Gamma$ can be expressed in terms of $[\sigma_p]_\Gamma$,

$$[\sigma_s]_\Gamma = \psi([\sigma_p]_\Gamma). \tag{17}$$

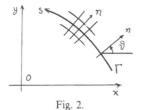

Fig. 2.

By use of (10)–(12) and the first equation in (16), we obtain

$$[\varepsilon_n]_\Gamma = \frac{2}{E}(\tfrac{1}{2}-\nu)^2[\sigma_p]_\Gamma + \frac{2}{E}(1+\nu)(\tfrac{1}{2}-\nu)\psi([\sigma_p]_\Gamma). \tag{18}$$

Substitution of (18) into the second of (16) gives

$$\left[\frac{\partial\varepsilon_s}{\partial n}\right]_\Gamma - 2\frac{d}{ds}[\varepsilon_{ns}]_\Gamma - \frac{d\vartheta}{ds}\left\{ \frac{2}{E}(\tfrac{1}{2}-\nu)^2[\sigma_p]_\Gamma \right.$$
$$\left. + \frac{2}{E}(1+\nu)(\tfrac{1}{2}-\nu)\psi([\sigma_p]_\Gamma) \right\} = 0. \tag{19}$$

2.3. Strong discontinuity without stress-jumps

We can prove that, in the elastic-plastic evolutionary problem, the stress-jumps cannot exist inside the plastic domain unless it is an elastic-core. As for the strain-jumps, they can only exist in the case of $\nu=\tfrac{1}{2}$. For $\nu=\tfrac{1}{2}$, it follows from (13) and (18) that,

$$[\sigma_s]_\Gamma = 0, \qquad [\varepsilon_n]_\Gamma = 0. \tag{20}$$

Then (19) becomes

$$2\frac{d}{ds}[\varepsilon_{ns}]_\Gamma - \left[\frac{\partial\varepsilon_s}{\partial n}\right]_\Gamma = 0. \tag{21}$$

By use of (10)–(12) and the first equation in (16), (21) can be transformed into

$$\frac{1-\nu^2}{E}\left[\frac{\partial^3\varphi}{\partial n^3}\right]_\Gamma = \frac{[\lambda]_\Gamma}{\cos\vartheta}(\sigma_s-\sigma_n)/2$$
$$-2\frac{d\vartheta}{ds}\tan\vartheta\cdot[\varepsilon^p_{ns}]_\Gamma + 2\frac{d}{ds}[\varepsilon^p_{ns}]_\Gamma, \tag{22}$$

where $\nu=\tfrac{1}{2}$. Besides, the first equation in (20) leads to

$$\left[\frac{\partial^2\varphi}{\partial n^2}\right]_\Gamma = 0. \tag{23}$$

Thus, only ε^p_{ns} and $\partial^3\varphi/\partial n^3$ are discontinuous across Γ, with their jump discontinuities satisfying (22). It follows from discontinuity of ε^p_{ns} and $\partial^3\varphi/\partial n^3$ that Γ must be the slip line, i.e.

$$\sigma_s - \sigma_n = 0 \quad \text{on } \Gamma. \tag{24}$$

Finally, from (24), Eq. (22) leads to

$$\frac{1-\nu^2}{E}\left[\frac{\partial^3\varphi}{\partial n^3}\right]_\Gamma = 2\frac{d}{ds}[\varepsilon^p_{ns}]_\Gamma - 2\frac{d\vartheta}{ds}\tan\vartheta\cdot[\varepsilon^p_{ns}]_\Gamma, \qquad \left(\nu=\frac{1}{2}\right). \tag{25}$$

Equations (15), (23) and (25) are the four contiguity conditions for the case of strong discontinuity which can occur only for $\nu = \frac{1}{2}$ across the slip line.

2.4. Weak discontinuity

For weak discontinuity the second equation of (16) becomes

$$\frac{1-\nu^2}{E}\left[\frac{\partial^3 \varphi}{\partial n^3}\right]_\Gamma = \frac{[\lambda]_\Gamma}{\cos \vartheta}\left\{\frac{1}{2}(\sigma_s - \sigma_n) + \frac{2}{3}\left(\frac{1}{2}-\nu\right)^2 (\Delta\varphi - \sigma_p)\right\}.$$

(26)

Equations (15), (23) and (26) are the four contiguity conditions across the weak discontinuity line which is usually the boundary between the plastic domain and the elastic or plastic unloading domain.

2.5. The fifth contiguity condition

Expressing the compatibility equation, (14), in the n, s-coordinates and using the four contiguity conditions, we obtain

$$\frac{1-\nu^2}{E}\left[\frac{\partial^4 \varphi}{\partial n^4}\right]_\Gamma - \frac{1-\nu^2}{E}\tan\vartheta \cdot \frac{d}{ds}\left[\frac{\partial^3 \varphi}{\partial n^3}\right]_\Gamma$$

$$-\frac{1}{\cos\vartheta}\left[\frac{\partial}{\partial n}\left(\lambda\left\{\frac{1}{2}(\sigma_s - \sigma_n) + \frac{2}{3}\left(\frac{1}{2}-\nu\right)^2(\Delta\varphi - \sigma_p)\right\}\right)\right]_\Gamma$$

$$+\frac{d\vartheta}{ds}\frac{[\lambda]_\Gamma}{\cos\vartheta}\left\{\frac{1}{2}(\sigma_n - \sigma_s) + \frac{2}{3}\left(\frac{1}{2}-\nu\right)^2(\Delta\varphi - \sigma_p)\right\}$$

$$+\frac{2}{\cos\vartheta}\frac{d}{ds}[\sigma_{ns}\lambda]_\Gamma - 2\frac{d^2\vartheta}{ds^2}[\varepsilon_{ns}^p]_\Gamma = 0,$$

(27)

from which the jump discontinuity, $[\partial^4\varphi/\partial n^4]_\Gamma$, can be found. We shall call (27) the fifth contiguity condition.

2.6. Theorem for the unloading boundary

For the unloading boundary, i.e. the boundary between the primary plastic domain and the unloading wake region, an unloading condition should be supplemented. We can prove the following theorem:

Theorem. *Let Γ be an unloading boundary, not coincident with the slip line* (24). *Then we have*

$$\left.\frac{\partial\sigma}{\partial x}\right|_{\Gamma(e)} = \lambda|_{\Gamma(p)} = 0,$$

(28)

where σ is the equivalent stress, as defined in (13),

$$\sigma=\sqrt{3}\left[\tfrac{1}{4}(\sigma_s-\sigma_n)^2+\tau_{ns}^2+\tfrac{1}{3}(\tfrac{1}{2}-\nu)^2(\Delta\varphi-\sigma_p)^2\right]^{1/2}, \tag{29}$$

and the subscripts $\Gamma(e)$ and $\Gamma(p)$ are used to denote the values on the boundary Γ at the sides of unloading and loading domains, respectively.

Proof. Differentiating (29) with respect to x, we obtain

$$2\sigma\frac{\partial\sigma}{\partial x}=3\left\{\frac{1}{2}(\sigma_s-\sigma_n)\left(\frac{\partial\sigma_s}{\partial x}-\frac{\partial\sigma_n}{\partial x}\right)+2\tau_{ns}\frac{\partial\tau_{ns}}{\partial x}\right.$$
$$\left.+\frac{2}{3}\left(\frac{1}{2}-\nu\right)^2(\Delta\varphi-\sigma_p)\left(\frac{\partial\Delta\varphi}{\partial x}-\frac{\partial\sigma_p}{\partial x}\right)\right\}. \tag{30}$$

It follows from (8) and (9) that

$$\frac{\partial\sigma_p}{\partial x}=-\frac{2E\lambda}{3}(\Delta\varphi-\sigma_p). \tag{31}$$

From (15), (23), (26) and (31) it follows that,

$$\left[2\sigma\frac{\partial\sigma}{\partial x}\right]_\Gamma=2\sigma\left[\frac{\partial\sigma}{\partial x}\right]_\Gamma$$
$$=\frac{3E}{1-\nu^2}\left\{\left(\frac{1}{2}(\sigma_s-\sigma_n)+\frac{2}{3}\left(\frac{1}{2}-\nu\right)^2(\Delta\varphi-\sigma_p)\right)^2\right.$$
$$\left.+\frac{4}{9}(1-\nu^2)\left(\frac{1}{2}-\nu\right)^2(\Delta\varphi-\sigma_p)^2\right\}[\lambda]_\Gamma. \tag{32}$$

However, since

$$\frac{\partial\sigma}{\partial x}\bigg|_{\Gamma(p)}=0, \qquad \lambda|_{\Gamma(e)}=0, \tag{33}$$

we obtain

$$2\sigma\frac{\partial\sigma}{\partial x}\bigg|_{\Gamma(e)}=-\frac{3E}{1-\nu^2}\left\{\left(\frac{1}{2}(\sigma_s-\sigma_n)+\frac{2}{3}\left(\frac{1}{2}-\nu\right)^2(\Delta\varphi-\sigma_p)\right)^2\right.$$
$$\left.+\frac{4}{9}(1-\nu^2)\left(\frac{1}{2}-\nu\right)^2(\Delta\varphi-\sigma_p)^2\right\}\lambda|_{\Gamma(p)}. \tag{34}$$

Besides, the flow rule and the unloading condition in the wake region demand that,

$$\lambda|_{\Gamma(p)}\geqslant 0 \quad\text{and}\quad \frac{\partial\sigma}{\partial x}\bigg|_{\Gamma(e)}\geqslant 0. \tag{35}$$

Comparing (34) and (35), we arrive at,

$$\frac{\partial \sigma}{\partial x}\bigg|_{\Gamma(e)} = 0 \quad \text{and} \quad \lambda|_{\Gamma(p)} = 0$$

which complete the proof.

3. Conditions on the crack surfaces and at infinity

Let ρ be the dimensional scale of the plastic domain. For the case of small-scale yielding, φ is required to yield the elastic solution,

$$\varphi = r^{3/2} \left\{ \frac{K_I}{\sqrt{2\pi}} \left(\cos \frac{\theta}{2} + \frac{1}{3} \cos \frac{3\theta}{2} \right) - \frac{K_{II}}{\sqrt{2\pi}} \left(\sin \frac{\theta}{2} + \sin \frac{3\theta}{2} \right) \right\},$$

(36)

for $r \gg \rho$ and $r \ll a$. At $r \gg a$, we have $\sigma_x = \sigma$, $\tau_{xy} = \tau$. In (36) K_I and K_{II} are the stress-intensity factors, i.e.

$$K_I = \sigma\sqrt{\pi a}, \qquad K_{II} = \tau\sqrt{\pi a},$$

(37)

and r and θ are the polar coordinates with origin at the crack tip. The boundary conditions on the crack surfaces are

$$\sigma_\theta = \tau_{r\theta} = 0 \quad \text{for} \quad \theta = \pm\pi.$$

(38)

4. The asymptotic stress solution for the case of $\nu = \frac{1}{2}$ (mixed-Mode I and II)

4.1. Power series expansion

To study the asymptotic behavior near the crack tip, we assume

$$\varphi = \sum_{n=0}^{\infty} r^{n+2} f_n(\theta).$$

(39)

For the case of $\nu = \frac{1}{2}$, the yield condition, (13), reduces to

$$\frac{1}{4} \left(\frac{\partial^2 \varphi}{\partial y^2} - \frac{\partial^2 \varphi}{\partial x^2} \right)^2 + \left(\frac{\partial^2 \varphi}{\partial x \partial y} \right)^2 = k^2.$$

(40)

Substituting (39) into (40) and equating the coefficients of like terms r^n, we obtain equations for $f_0(\theta)$ and $f_m(\theta)$, $m \geq 1$,

$$\frac{1}{4}[f_0''(\theta)]^2 + [f_0'(\theta)]^2 = k^2,$$

$$\frac{1}{4}\sum_{i=0}^{m}[i(i+2)f_i(\theta) - f_i''(\theta)][(m-i)(m+2-i)f_{m-i}(\theta) - f_{m-i}''(\theta)]$$

$$+ \sum_{i=0}^{m}(i+1)(m-i+1)f_i'(\theta)f_{m-i}'(\theta) = 0. \tag{42}$$

The stress function in the plastic unloading domain satisfies compatibility, (14),

$$\frac{1-\nu^2}{E}\Delta\Delta\varphi + \frac{d^2}{dy^2}\varepsilon_x^p(y) = 0. \tag{43}$$

Let

$$\frac{d^2}{dy^2}\varepsilon_x^p(y) = \sum_{n=0}^{\infty} D_n y^{n-2}. \tag{44}$$

Then from substitution of (44) and (39) into (43) it follows that,

$$\left(\frac{d^2}{d\theta^2} + n^2\right)\left[\frac{d^2}{d\theta^2} + (n+2)^2\right]f_n(\theta) + \frac{ED_n}{1-\nu^2}(\sin\theta)^{n-2} = 0. \tag{45}$$

4.2. The first approximation (n=0)

For the plastic domains, solution of (41) gives

$$f_0(\theta) = \frac{k}{2}[b_1 \pm \cos 2(\theta - \theta_0)], \tag{46}$$

or

$$f_0(\theta) = \frac{k}{2}(b_2 \pm 2\theta). \tag{47}$$

For the plastic unloading domain, the general solution of (45) for $n=0$ is

$$f_0(\theta) = C_1 + C_2\theta + C_3 \cos 2\theta + C_4 \sin 2\theta + f_0^*(\theta), \tag{48}$$

where

$$f_0^*(\theta) = \frac{-ED_0}{4(1-\nu^2)}\left[(\cos 2\theta - 1)\ln\sin\theta + \left(\theta + \frac{1}{2}\cot\theta\right)\sin 2\theta\right]:$$

$$\tag{49}$$

As shown in Fig. 3, the near-tip region consists of the various plastic domains (①, ②, ②′ in primary ⑪ and ④, ④′ in secondary ⑭)) and the plastic unloading domains (③, ③′ in wake ⑪). In Fig. 3 the nets of slip lines are also shown. The boundary conditions are (15) and (23) for boundaries between neighboring domains and (38) for the crack surfaces $\theta = \pm \pi$. For the first approximation ($n = 0$) of (39), these conditions can be reduced to

$$[f_0(\theta)]_\Gamma = [f_0'(\theta)]_\Gamma = [f_0''(\theta)]_\Gamma = 0, \qquad f_0(\pm\pi) = f_0'(\pm\pi) = 0. \tag{50}$$

After a lengthy calculation using (46)–(50), we obtain the expressions $f_0(\theta)$ for the various domains in the upper half plane,

$$f_0(\theta) = \begin{cases} \dfrac{k}{2}[b + \cos 2(\theta - \theta_0)] & \text{for } -\dfrac{\pi}{4} + \theta_0 \leqslant \theta \leqslant \dfrac{\pi}{4} + \theta. \\[2mm] \dfrac{k}{2}\Big[b + \dfrac{\pi}{2} - 2(\theta - \theta_0)\Big] & \text{for } \dfrac{\pi}{4} + \theta_0 \leqslant \theta \leqslant \pi - \beta, \\[2mm] C_1 + C_2\theta + C_3 \cos 2\theta \\ \qquad + C_4 \sin 2\theta + f_0^*(\theta) & \text{for } \pi - \beta \leqslant \theta \leqslant \pi - \gamma, \\[2mm] \dfrac{k}{2}(1 - \cos 2\theta) & \text{for } \pi - \gamma \leqslant \theta \leqslant \pi, \end{cases} \tag{51}$$

where

$$C_1 = -(\pi - \gamma)C_2 + \frac{k}{2} - \frac{ED_0}{4(1 - \nu^2)} \ln \sin \gamma,$$

$$b = \frac{3\pi}{2} - 2\theta_0 - 2\beta + 1 - \frac{2}{k}\Big[C_2(\beta - \gamma) + \frac{ED_0}{4(1 - \nu^2)} \ln \frac{\sin \gamma}{\sin \beta}\Big], \tag{52}$$

$$C_3 = -\frac{1}{2}(C_2 \sin 2\gamma + k) + \frac{ED_0}{4(1 - \nu^2)}[\cos^2 \gamma + \ln \sin \gamma],$$

$$C_4 = -\frac{C_2}{2} \cos 2\gamma + \frac{ED_0}{4(1 - \nu^2)}\Big(\pi - \gamma - \frac{1}{2} \sin 2\gamma\Big), \tag{53}$$

$$C_2 = k\Big\{\sin \beta\,(1 - \sin 2\beta) + \Big[\cos\Big(\frac{\pi}{4} + \theta_0\Big) + \cos \beta\Big] \\ \qquad \times \Big(\cos 2\gamma - \cos 2\beta + 2 \ln \frac{\sin \gamma}{\sin \beta}\Big)\Big\} \\ \qquad \times \Big(\cos \beta - \cos(2\gamma - \beta) - 2 \cos \beta \ln \frac{\sin \gamma}{\sin \beta}\Big)^{-1},$$

$$D_0 = -\frac{4(1 - \nu^2)}{E \sin \beta}\Big[k \cos\Big(\frac{\pi}{4} + \theta_0\Big) + (k + C_2) \cos \beta\Big]. \tag{54}$$

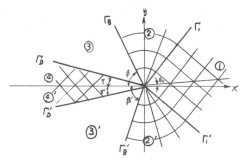

Fig. 3.   The plastic domains of the near-tip crack region.

Here the expression for D_0 in (54) is obtained by strain analysis, i.e., by calculating $\varepsilon_x^p(y)$ in the wake domain ③ (Fig. 3) and substituting this expression into (44). Besides, the angles β and γ are related as

$$\sin(\beta-\gamma)(\sin\gamma-\cos\gamma)-2\cos\left(\frac{\pi}{4}+\theta_0\right)\sin^2(\beta-\gamma)$$

$$-\left[\cos\left(\frac{\pi}{4}+\theta_0\right)(\cos 2\gamma-\cos 2\beta)+\cos\beta\cos 2\gamma\right]\ln\frac{\sin\gamma}{\sin\beta}$$

$$-\left[\cos\left(\frac{\pi}{4}+\theta_0\right)(\sin 2\beta-\sin 2\gamma)+\cos\beta(1-\sin 2\gamma)\right](\beta-\gamma)$$

$$=0. \tag{55}$$

The corresponding expressions for the lower half plane are obtained from (51)–(55) by replacing θ, θ_0 with $-\theta$, $-\theta_0$, and the set of constants b, C_1, C_2, C_3, C_4, D_0, β, γ with another set denoted by primes $b',\ldots,\gamma'$. Since the domain ① is common for the upper and lower half planes, we must have $b=b'$, which leads to a relation between the constants,

$$k(2\theta_0+\beta-\beta')+C_2(\beta-\gamma)-C_2'(\beta'-\gamma')+\frac{ED_0}{4(1-\nu^2)}\ln\frac{\sin\gamma}{\sin\beta}$$

$$-\frac{ED_0'}{4(1-\nu^2)}\ln\frac{\sin\gamma'}{\sin\beta'}=0. \tag{56}$$

5. The asymptotic strain solution for the case of $\nu=\frac{1}{2}$ (mixed-Mode I and II)

For the case of $\nu=\frac{1}{2}$, Eq. (12) becomes

$$\varepsilon_x^P = -\varepsilon_y^P = \frac{1}{2}\int_x^{x_A}\lambda(\sigma_x-\sigma_y)\,\mathrm{d}x,$$

$$\varepsilon_{xy}^P = \int_x^{x_A}\lambda\tau_{xy}\,\mathrm{d}x. \tag{57}$$

At the boundary between neighboring domains, the contiguity conditions (25) or (26) must be satisfied. For the first approximation, they can be written as

$$[\varepsilon_{r\theta}^P]_\Gamma = \frac{1-\nu^2}{2E}\int_{r_0}^r\frac{1}{r}[f_0'''(\theta)]_\Gamma\,\mathrm{d}r \tag{58}$$

for strong discontinuity, and as

$$[\lambda]_\Gamma = -\frac{1-\nu^2}{E}\frac{2\cos\vartheta}{r}\frac{1}{f_0''(\theta)}[f_0'''(\theta)]_\Gamma \tag{59}$$

for weak discontinuity. Within the various domains of Fig. 3, the compatibility (14) must be satisfied, with the fifth contiguity condition as its consequence. Besides, the strains and their jump discontinuities must damp out as the distance r increases. Considering all these requirements, we obtain finally the strain distribution in all the domains of Fig. 3.

5.1. In domain ①

Here,

$$\lambda=\lambda_1(y')+\lambda_2(x'), \tag{60}$$

where λ_1 and λ_2 are arbitrary functions, and

$$x'=x\sin\left(\frac{\pi}{4}+\theta_0\right)-y\cos\left(\frac{\pi}{4}+\theta_0\right),$$

$$y'=x\cos\left(\frac{\pi}{4}+\theta_0\right)+y\sin\left(\frac{\pi}{4}+\theta_0\right).$$

The functions λ_1 and λ_2 are determined from the contiguity condition on Γ_A (Fig. 1). It can be expected that the strains do not have any singularities in domain ①,

$$\varepsilon_r\sim\varepsilon_\theta\sim\varepsilon_{r\theta}\sim0(1). \tag{61}$$

5.2. Jump discontinuities of plastic strains across Γ_1

These are,

$$\Delta_{\Gamma_1} \varepsilon_y^P = -\Delta_{\Gamma_1} \varepsilon_x^P = \frac{2k(1-\nu^2)}{E} \sin 2\left(\frac{\pi}{4}+\theta_0\right) \ln\frac{A}{r},$$

$$\Delta_{\Gamma_1} \varepsilon_{xy}^P = \frac{2k(1-\nu^2)}{E} \cos 2\left(\frac{\pi}{4}+\theta_0\right) \ln\frac{A}{r} \tag{62}$$

Here A is a constant of integration, which is related to the length of the discontinuity line, and may be taken as the dimensional scale of the plastic domain.

5.3. In domain ②

From the compatibility condition in domain ②, we obtain

$$\lambda = \frac{g(\theta)}{r} - \frac{2(1-\nu^2)}{E}\cos\left(\frac{\pi}{4}+\theta_0\right)\frac{\ln r}{r}, \tag{63}$$

where $g(\theta)$ is an arbitrary function, depending on boundary conditions. Hence,

$$\varepsilon_x^P = -\varepsilon_y^P = -\frac{4k(1-\nu^2)}{E}\cos\left(\frac{\pi}{4}+\theta_0\right)\sin\theta \ln\frac{A}{r}+0(1),$$

$$\varepsilon_{xy}^P = \frac{2k(1-\nu^2)}{E}\left[\cos\left(\frac{\pi}{4}+\theta_0\right)\left(\ln\frac{\tan(\theta/2)}{\tan[(\pi/8)+(\theta_0/2)]}\right.\right.$$

$$\left.\left. +2\cos\theta\right)-1\right]\ln\frac{A}{r}+0(1). \tag{64}$$

5.4. Jump discontinuities of plastic strains across Γ_B

These are

$$\Delta_{\Gamma_B} \varepsilon_y^P = -\Delta_{\Gamma_D} \varepsilon_x^P = \frac{4(1-\nu^2)}{E}\cot\beta$$

$$\times\left\{C_2+k+k\cos\left(\frac{\pi}{4}+\theta_0\right)\cos\beta\right\}\ln\frac{A}{r},$$

$$\Delta_{\Gamma_B} \varepsilon_{xy}^P = -\frac{2(1-\nu^2)}{E}\frac{\cos 2\beta}{\sin^2\beta}$$

$$\times\left\{C_2+k+k\cos\left(\frac{\pi}{4}+\theta_0\right)\cos\beta\right\}\ln\frac{A}{r}. \tag{65}$$

5.5. In domain ③

Here $\lambda=0$, and ε_x^P, ε_y^P, ε_{xy}^P are functions of y only.

5.6. In domain ④

We have

$$\varepsilon_x^P = -\varepsilon_y^P = k\Lambda^* + \varepsilon_x^P(y) + 0(1),$$
$$\varepsilon_{xy}^P = \varepsilon_{xy}^P(y) + 0(1), \tag{66}$$

where $\varepsilon_x^P(y)$ and $\varepsilon_{xy}^P(y)$ are the values of plastic strains at corresponding point on Γ_D with the same ordinate y.

$$\Lambda^* = p \ln\left(\frac{(\cot \gamma - 1)y}{-(x+y)}\right) + q \ln\left(\frac{(\cot \gamma + 1)y}{-x+y}\right), \tag{67}$$

where

$$p = \frac{1}{2k}\left[D_0 - \frac{4C_2}{E}(1-\nu^2)\right],$$

$$q = \frac{1}{2k}\left[D_0 + \frac{4C_2}{E}(1-\nu^2)\right]. \tag{68}$$

Now we turn to the determination of the angles in Fig. 3. From (65) can be obtained the jump discontinuity of $\varepsilon_{r\theta}^P$ across Γ_B,

$$\underset{\Gamma_B}{\Delta} \varepsilon_{r\theta}^P = -\frac{2(1-\nu^2)}{E \sin^2 \beta}\left\{C_2 + k + k \cos\left(\frac{\pi}{4} + \theta_0\right)\cos \beta\right\}\ln \frac{A}{r}.$$

Since $\tau_{r\theta} = k > 0$ on Γ_B, $\Delta_{\Gamma_B}\varepsilon_{r\theta}^P$ must be non-negative and damping with increasing r. This demands

$$C_2 + k + k \cos\left(\frac{\pi}{4} + \theta_0\right) \cos \beta \leq 0. \tag{69}$$

The unloading condition of the domain ③ (Fig. 3) follows from Eq. (41) as follows:

$$\tfrac{1}{4}\left[f_0''(\theta)\right]^2 + \left[f_0'(\theta)\right]^2 \leq k^2, \qquad \pi - \beta \leq \theta \leq \pi - \gamma. \tag{70}$$

It can be proved that (70) holds true when

$$C_2 \cos(\beta - \gamma) + k \cos \gamma\left[\cos\left(\frac{\pi}{4} + \theta_0\right) + \cos \beta\right] \geq 0,$$
$$C_2 + k + k \cos\left(\frac{\pi}{4} + \theta_0\right) \cos \beta \geq 0. \tag{71}$$

From (69) and the second of (71) it follows that,

$$C_2 + k + k \cos\left(\frac{\pi}{4} + \theta_0\right) \cos \beta = 0, \tag{72}$$

in which C_2 can be substituted from the first equation in (54). Simultaneously solving (72), (55) and an equation corresponding to (55) for the lower half plane, we obtain

$$\theta_0 = 0, \qquad \beta = 67.9°, \qquad \gamma = 17.9°, \tag{73}$$

It can be verified numerically that the unloading condition (70) is actually satisfied. The angles β and γ coincide with the results obtained by Slepjan (1974).

Equation (73) corresponds to the steady growth of Mode I crack, which is the solution given by Gao (1980). The above analysis leads us to the conclusion that there is not any solution corresponding to $\theta_0 \neq 0$, and therefore the mixed-Mode crack (Modes I and II) cannot grow steadily.

6. Crack surface displacements

The fields of displacements can be obtained through integrating the expressions for strains. Here we give only the displacements at the crack surface, $\theta = \pi - 0$,

$$u_x|_{\theta=\pi-0} = -D_0 x \ln(-x) + 0(1),$$

$$u_y|_{\theta=\pi-0} = -\frac{4(1-\nu^2)}{E} C_2 x \ln(-x) + 0(1), \tag{74}$$

in which D_0 and C_2 are constants as given in (54). Equations (74) hold true when $|x|$ is sufficiently small.

7. About the Mode III crack

Slepjan (1974) pointed out that the continuity of $\nabla^2 \varphi$ across the unloading boundary Γ_2, as assumed by Chitaley and McClintock (1971), is true only asymptotically as $r \to 0$. We shall here give a proof for the continuity of $\nabla^2 \varphi$ across Γ_2 not only near the tip but also far away from it. The method used for proof is very similar to that for (28).

By a procedure similar to the derivation of contiguity conditions in this paper, we can obtain for the steady growth of Mode III crack,

$$[\varphi]_\Gamma = \left[\frac{\partial \varphi}{\partial n}\right]_\Gamma = 0, \tag{75}$$

$$\frac{1+\nu}{E}\left[\frac{\partial^2 \varphi}{\partial n^2}\right]_\Gamma - \frac{\partial \varphi}{\partial n}\frac{[\lambda]_\Gamma}{\cos \vartheta} = 0, \tag{76}$$

where φ is the stress function for anti-plane shear, and ϑ is the angle shown in Fig. 2. Here (76) is for weak discontinuity. Let $\tau=(\tau_s^2+\tau_n^2)^{1/2}$, and use (75), (76), to obtain

$$\left[2\tau\frac{\partial\tau}{\partial x}\right]_\Gamma=2\tau\left[\frac{\partial\tau}{\partial x}\right]_\Gamma=\frac{2E}{1+\nu}\left(\frac{\partial\varphi}{\partial n}\right)^2[\lambda]_\Gamma. \tag{77}$$

However

$$\frac{\partial\tau}{\partial x}\bigg|_{\Gamma(\mathrm{p})}=0,\qquad\lambda\big|_{\Gamma(\mathrm{e})}=0, \tag{78}$$

and then

$$2\tau\frac{\partial\tau}{\partial x}\bigg|_{\Gamma(\mathrm{e})}=-\frac{2E}{1+\nu}\left(\frac{\partial\varphi}{\partial n}\right)^2\lambda\bigg|_{\Gamma(\mathrm{p})}. \tag{79}$$

The flow rule and the unloading condition for the wake region demand, respectively, that

$$\lambda\big|_{\Gamma(\mathrm{p})}\geqslant0, \tag{80}$$

$$\frac{\partial\tau}{\partial x}\bigg|_{\Gamma(\mathrm{e})}\geqslant0. \tag{81}$$

Assume $\partial\varphi/\partial n\neq0$, as is true at the unloading boundary Γ_2 far away from the tip. Then (79)–(81) lead to,

$$\lambda\big|_{\Gamma(\mathrm{p})}=\frac{\partial\tau}{\partial x}\bigg|_{\Gamma(\mathrm{e})}=0. \tag{82}$$

From (82) and (76) we have $[\partial^2\varphi/\partial n^2]_\Gamma=0$ which leads to the continuity of $\nabla^2\varphi$, i.e.

$$[\nabla^2\varphi]_\Gamma=0.$$

8. Strain-hardening material

The method developed in this paper for the perfectly-plastic medium can be extended to strain-hardening materials. Referring to another paper[2] for the details, we will limit ourselves to the presentation of some final results for power-law hardening material. Assuming the hardening exponent $n>1$, the stress function φ is found to be of the form

$$\varphi=r^2\left(\ln\frac{A}{r}\right)^{1/(n-1)}\left\{f(\theta)+\left(\ln\frac{A}{r}\right)^{-1}f_1(\theta)+\left(\ln\frac{A}{r}\right)^{-2}f_2(\theta)+\cdots\right\}. \tag{83}$$

[2]Gao, Y. C. and Hwang, K. C., *Elastic-Plastic Field in Steady Crack Growth in a Strain-Hardening Material*, a paper to be submitted to ICF5, 1981.

Following the method used in this paper, we can obtain the fifth-order differential equation for $f(\theta)$ as the first approximation of equation of compatibility. We have for a plastic domain,

$$\left(\sin\theta\, g^{(n-3)/2}g'f''\right)'' + \sin\theta\, g^{(n-3)/2}g'f'' = 0, \tag{84}$$

where

$$g = g(\theta) = \frac{1}{4}(f'')^2 + (f')^2, \quad (\ )' \equiv \frac{d}{d\theta}, \tag{85}$$

and for an elastic domain

$$((f''' + 4f')\sin\theta)'' + (f''' + 4f')\sin\theta = 0. \tag{86}$$

Using (84), (86) and five contiguity conditions, we can obtain $f(\theta)$ expressed in the same form as (51) for perfectly-plastic material.

References

Chitaley, A. D. and McClintock, F. A. (1971), "Elastic-Plastic Mechanics of Steady Crack Growth under Anti-Plane Shear," *J. Mech. Phys. Solids*, **19**, 147.

Gao, Y. C. (1980), "Elastic-Plastic Field at the Tip of a Crack Growing Steadily in a Perfectly Plastic Medium," *Acta Mechanica Sinica*, No. 1, 48.

Rice, J. R. (1968), *Fracture: An Advanced Treatise*, H. Liebowitz (ed.), Vol. 2, 191.

Rice, J. R. and Sorensen, E. P. (1978), "Continuing Crack-Tip Deformation for Plane-Strain Crack Growth in Elastic-Plastic Solids," *J. Mech. Phys. Solids*, **26**, 163.

Slepjan, L. I. (1973), "Deformatsija u Kraja Rastushchei Treshchiny," *Meh. Tver. Tela*, **8**, 139.

Slepjan, L. I. (1974), "Rastushchaja Treshchina pri Ploskoi Deformatsii Uprugo-Plasticheskovo Tela," *Meh. Tver. Tela*, **9**, 57.

LIST OF PARTICIPANTS

Austria
H.P. Stüwe, Österreichische Akademie der Wissenschaften, Erich-Schmid-Institut für Festkörperphysik, A-8700 Leoben

Bulgaria
A. Baltov, Institut de Mécanique et Biomécanique, Bul. Acad. Sci., P.O. Box 373, 1090 Sofia

Canada
R.N. Dubey, University of Waterloo, Faculty of Engineering, Department of Mechanical Engineering, Waterloo, Ontario
K. Neale, Département de Génie Civil, Université de Sherbrooke, Sherbrooke, Québec

China (P.R)
Hwang Keh-chih, Qinghua (Tsinghua) University, Beijing (Peking)
Li Hao, Huazhong Institute of Technology, Wuhan

Denmark
J. Christoffersen, Technical University of Denmark, Department of Solid Mechanics, Building 404, DK-2800 Lyngby
V. Tvergaard, Technical University of Denmark, Department of Solid Mechanics, Building 404, DK-2800 Lyngby

France
M. Abouaf, Ecole des Mines, Sophia Antipolis, 06560 Valbonne
M. Amestoy, Laboratoire de Mécanique des Solides, Ecole Polytechnique, 91120 Palaiseau
M. Bathias, Université de Compiègne, Laboratoire de Mécanique, B.P. 233, 60206 Compiègne
G. Baylac, EDF/SEPTEN, Tour EDF/GDF, 92000 Paris la Defense
S. Bhandari, NOVATOME, 20, rue Edouard Herriot, 92350 Le Plessis Robinson
M. Bossut, Manufacture Française des Pneumatiques Michelin, 63000 Clermont Ferrand

R. Boutin, Centre de Recherches Aluminium Péchiney, B.P. 24, 38340 Voreppe

H.D. Bui, Laboratoire de Mécanique des Solides, Ecole Polytechnique, 91120 Palaiseau

Th. Charras, CEA-DENT, B.P. 2, 91190 Gif sur Yvette

K. Dang Van, Laboratoire de Mécanique des Solides, Ecole Polytechnique, 91120 Palaiseau

A. Ehrlacher, Laboratoire de Mécanique des Solides, Ecole Polytechnique, 91120 Palaiseau

Y. d'Escatha, Service de l'Industrie et des Mines de Bourgogne Franche Comté, 3, rue Devosge, 21000 Dijon

Cl. Faidy, EDF/SEPTEN, Tour EDF/GDF, 92000 Paris la Defense

D. Fougeres, C.I.S.I., 6, rue de la Maison Rouge, 94120 Fontenay sous Bois

J. de Fouquet, E.N.S.M.A., rue Guillaume VII, 86034 Poitiers Cedex

D. François, U.T.C. Département de Génie Mécanique, Université de Technologie de Compiègne, B.P. 233, 60206 Compiègne

J. Gerald, Laboratoire de Mécanique des Solides, Ecole Polytechnique, 91120 Palaiseau

P. Germain, Département de Mécanique, Ecole Polytechnique, 91120 Palaiseau

P. Habib, Laboratoire de Mécanique des Solides, Ecole Polytechnique, 91120 Palaiseau

J.P. Hutin, E.D.F., 3, rue de Messine, 75008 Paris

R. Labbens, CREUSOT-LOIRE, 15, rue Pasquier, 75383 Paris Cedex 08

A. Le Douaron, R.N.U.R. Direction des Laboratoires–Service 0852, 8 et 10, avenue Emile Zola, 92109 Boulogne Billancourt

J. Lemaitre, Université Paris VI, Laboratoire de Mécanique et de Technologie ENSET, 61, avenue du Président Wilson, 94230 Cachan

M. Mandel, Laboratoire de Mécanique des Solides, Ecole Polytechnique, 91120 Palaiseau

D. Miannay, C.E.A.–Sce Métallurgie/M.S., B.P. 561, 92542 Montrouge Cedex

F. Montheillet, Ecole des Mines, Sophia Antipolis, 06560 Valbonne

F. Mudry, Centre des Matériaux de l'E.N.S.M.P., B.P. 87, 91003 Evry

Nguyen Q.S., Laboratoire de Mécanique des Solides, Ecole Polytechnique, 91120 Palaiseau

M. Noel, EDF/SEPTEN, Tour EDF/GDF, 92000 Paris la Defense

A. Pineau, Centre des Matériaux E.M.P., B.P. 87, 91003 Evry Cedex

M. Predeleanu, Université Paris VI, Laboratoire de Mécanique et de Technologie ENSET, 61, avenue du Président Wilson, 94230 Cachan

D. Radenkovic, Laboratoire de Mécanique des Solides, Ecole Polytechnique, 91120 Palaiseau

G. Regazzoni, Ecole des Mines, Sophia Antipolis, 06560 Valbonne

R. Roche, C.E.A., DEMT/CEN. SACLAY, B.P. 2, 91190 Gif sur Yvette

G. Rousselier, Electricité de France, Dpt. Etude des Matériaux, Moret-sur-Loing

G. Sanz, Institut de Recherches de la Sidérurgie Française (IRSID), 185, rue Président Roosevelt, 78105 St-Germain-en-Laye Cedex

Jh. Schmitt, Laboratoire de Physique et Technologie des Matériaux, Faculté des Sciences-Ile de Saulez, 57000 Metz

R. Sutterlin, 3, rue Méchain, 75014 Paris

V. Tuong, Laboratoire de Rhéologie, ISMCM, 3, rue Fernand Hainaut, 93407 Saint Ouen

A. Zaoui, L.P.M.T.M., Avenue J.B. Clément, 93430 Villetaneuse

J. Zarka, Laboratoire de Mécanique des Solides, Ecole Polytechnique, 91120 Palaiseau

Germany (F.R.)

O. Bruhns, Ruhr-Universität Bochum, Institut für Mechanik, Pf. 10.21.48, D-4630 Bochum 1

E. Kröner, Institut für Theoret. und Angew. Physik der Universität Stuttgart, Pfaffenwaldring 57/VI, 7000 Stuttgart 80

Th. Lehmann, Ruhr-Universität Bochum, Institut für Mechanik, Pf. 10.21.48, D-4630 Bochum 1

H. Lippmann, Lehrstuhl A für Mechanik, Technische Universität, Pf. 202420, D-8000 München

Italy

M. Como, Istituto di Technica delle Costruzioni, Facoltà di Ingegneria, Piazzale Tecchio, Napoli

A. Giarda, Via G. Fauser, 4, 28100 Novara

A. Grimaldi, Dipartimento di Strutture, Università della Calabria, Cosenza

J.C. Grossetie, Euratom, C.C.R., ISPRA, 21020 Ispra (Varese)

F. Maceri, Dipartimento di Strutture, Università della Calabria, Cosenza

Japan

T. Imura, Dept. of Metallurgy, Faculty of Engineering, Nagoya University, Furo-Cho, Chikusa-ku, Nagoya 464

T. Inoue, Dept. of Mechanical Engineering, Kyoto University, Yoshida-honmachi, Sakyo-ku 606, Kyoto

J. Suhara, Dept. of Naval Architecture, Faculty of Engineering, Kyushu University, Hakozaki, Higashiku, Fukuoka, 812

M. Tanaka, Dept. of Mechanical Engineering, Osaka University, Yamada-kami, Osaka

Y. Yamada, Institute of Industrial Science, University of Tokyo, 22-1 Roppongi 7 Chome, Minato-ku, Tokyo 106

T. Yokobori, Dept. of Mechanical Engineering II, Tohoku University, Sendai

Poland

A. Dragon, Polish Academy of Sciences, Institute of Fundamental Technological Research, Swietokrzyska 21, 00-049 Warsaw

Z. Mróz, Polish Academy of Sciences, Institute of Fundamental Technological Research, Swietokrzyska 21, 00-049 Warsaw

A. Sawczuk, Polish Academy of Sciences, Institute of Fundamental Technological Research, Swietokrzyska 21, 00-049 Warsaw

The Netherlands

A.U. de Koning, Mauritsstraat 15, 8019 XR Zwolle

Sweden

H. Broberg, National Testing Institute, Division of Solid Mechanics, Box 857, 50115 Borås

B. Storåkers, Royal Institute of Technology, Strength of Materials and Solid Mechanics, S-100 44 Stockholm

United Kingdom

B.A. Bilby, University of Sheffield, Dept. of Materials, Mappin Street, Sheffield S1 SJD

F.M. Burdekin, University of Manchester Institute of Science and Technology, Dept. of Civil and Structural Engineering, P.O. Box 88, Manchester M60 1QD

I. Collins, University of Manchester Institute of Science and Technology, P.O. Box 88, Manchester M60 1QD

M.J. Cowling, Dept. of Mechanical Engineering, Glasgow University, Glasgow G12 8QQ

B.F. Dyson, National Physical Laboratory, Division of Materials Applications, Teddington, Middlesex TW11 0LW

J. Gittus, U.K.A.E.A., SNL Salwick, Preston PR4 0RR

J. Hancock, Dept. of Mechanical Engineering, Glasgow University, Glasgow G12 8QQ

P. Kfouri, University of Sheffield, Dept. of Materials, Mappin Street, Sheffield S1 SJD

D. McLean, 28 St. James Road, Hampton Hill, Middlesex TW12 1DQ

U.S.A.

A.S. Argon, Dept. of Mechanical Engineering, M.I.T., Cambridge, Massachusetts 02139

R. Asaro, Division of Engineering, Brown University, Providence, Rhode Island 02912

C. Herakovich, Virginia Polytechnic Institute and State University, Blacksburg, Virginia 24061

J.W. Hutchinson, Pierce Hall 316, Div. of Applied Sciences, Harvard University, 29 Oxford Street, Cambridge, Massachusetts 02138

E. Krempl, Dept. of Mechanical Engineering, Rensselaer Polytechnic Institute, Troy, New York 12181

T.G. Langdon, Dept. of Materials Science, University of Southern California, Los Angeles, California 90007

F.A. Leckie, Dept. of Mechanical Engineering, University of Illinois, Urbana, Illinois 61801

E.F. Masur, Division of Civil and Mechanical Engineering, National Science Foundation, Washington D.C. 20550

F. McClintock, Room I-304, M.I.T., Cambridge, Massachusetts 02139

A.K. Mukherjee, Division of Materials Science, Dept. of Mechanical Engineering, University of California, Davis, California 95616

T. Mura, The Technological Institute, Dept. of Civil Engineering, Northwestern University, Evanston, Illinois 60201

S. Nemat-Nasser, The Technological Institute, Dept. of Civil Engineering, Northwestern University, Evanston, Illinois 60201

E.T. Onat, Dept. of Engineering, Yale University, Becton Center, P.O. Box 2157, New Haven, Connecticut 06520

D.M. Parks, Dept. of Mechanical Engineering, M.I.T., Cambridge, Massachusetts 02139

J.R. Rice, Brown University, Div. of Engineering, Providence, Rhode Island 02912

J. Weertman, Materials Research Center, Northwestern University, Evanston, Illinois 60201

M.P. Wnuk, South Dakota State University, Brookings, South Dakota 57006

Yugoslavia

S. Sedmak, Faculty of Technology and Metallurgy, University of Belgrad, Karnegijeva 4, 11000 Belgrad